A Handbook of Food Packaging

A HANDBOOK OF FOOD PACKAGING

Second edition

Frank A. Paine
B.Sc., C.Chem., F.R.S.C., F.I.F.S.T., F.Inst.Pkg., F.Inst.D.
Secretary General
International Association of Packaging Research Institutes
and
Adjunct Professor
School of Packaging
Michigan State University

and

Heather Y. Paine
B.Sc., M.Sc., F.I.F.S.T., M.I.P.R.
Consultant Food Scientist

Published under the authority of
The Institute of Packaging

BLACKIE ACADEMIC & PROFESSIONAL
An Imprint of Chapman & Hall
London · Glasgow · New York · Tokyo · Melbourne · Madras

Published by
Blackie Academic & Professional, an imprint of Chapman & Hall,
Wester Cleddens Road, Bishopbriggs, Glasgow G64 2NZ

Chapman & Hall, 2–6 Boundary Row, London SE1 8HN, UK

Blackie Academic & Professional, Wester Cleddens Road, Bishopbriggs, Glasgow G64 2NZ, UK

Van Nostrand Reinhold Inc., 115 Fifth Avenue, New York NY 10003, USA

Chapman & Hall Japan, Thomson Publishing Japan, Hirakawacho Nemoto Building, 6F, 1-7-11 Hirakawa-cho, Chiyoda-ku, Tokyo 102, Japan

DA Book (Aust.) Pty Ltd., 648 Whitehorse Road, Mitcham 3132, Victoria, Australia

Chapman & Hall India, R. Seshadri, 32 Second Main Road, CIT East, Madras 600 035, India

First edition 1983

Second edition 1992

© 1983, 1992 Chapman & Hall

Typeset in 10/12 pt Times New Roman by Thomson Press (India) Ltd, New Delhi
Printed in Great Britain at the University Press, Cambridge

ISBN   0 216 93210 6   0 442 30862 0   (USA)

A catalogue record for this book is available from the British Library

Library of Congress Cataloguing-in-Publication data available

Preface

This is the second edition of a successful title first published in 1983 and now therefore a decade out of date. The authors consider the development of the right package for a particular food in a particular market, from the point of view of the food technologist, the packaging engineer and those concerned with marketing. While the original format has been retained, the contents have been thoroughly revised to take account of the considerable advances made in recent years in the techniques of food processing, packaging and distribution.

While efficient packaging is even more a necessity for every kind of food, whether fresh or processed, and is an essential link between the food producer and the consumer, the emphasis on its several functions has changed. Its basic function is to identify the product and ensure that it travels safely through the distribution system to the consumer. Packaging designed and constructed solely for this purpose adds little or nothing to the value of the product, merely preserving farm or processor freshness or preventing physical damage, and cost effectiveness is the sole criterion for success. If, however, the packaging facilitates the use of the product, is reusable or has an after-use, some extra value can be added to justify the extra cost and promote sales.

Many examples of packaging providing such extra value can be cited over the last decade. For example, those concerned with food production have produced new methods of processing and preservation which have had a great influence on filling and packaging methods. HTST pasteurisation, aseptic processing and reduced amounts of additives all place greater emphasis on getting the packaging right. Retailers are now required by consumers to meet the desire for a 'fresh' concept in food, and the better control of freezers and chill cabinets now demanded places more stress on the packaging process, particularly where 'sous vide' and modified atmosphere packaging are concerned. However, while every consumer wants a quality product delivered in a package adequate to maintain that quality for the necessary shelf-life, in almost every instance it is the product alone that the customer wants: the packaging, with a few exceptions, is an ancillary part of the transaction. Indeed, if the packaging appears to be over-elaborate, it may well create the impression that the cost of the product has been unnecessarily raised by the packaging, which will have a negative effect on sales. All packages must be easy and safe to handle, simple to open and use, and provide few problems in their disposal.

These considerations will influence factors such as the weight, size, shape, structure and the incorporation of convenience into the structure of the primary packaging and that used for transport and storage. The provision of convenience, whether of the primary packaging or the shipping container, is much wider now than in the early 1980s. Easy opening must now be tempered by seal integrity and the more recent risks of pilfering and malicious tampering.

Finally a far more sophisticated public must be given a cleaner, more modern approach to the graphics and promotional needs of the packaging than its 1980s counterpart.

This second edition, as its predecessor, is written for food technologists wishing to understand more fully those aspects of packaging technology that are relevant to the preservation, distribution and marketing of a particular food; packaging engineers wishing to know more about those aspects of science and technology that will influence the packaging process; and for students of food science and technology requiring an integrated approach to the subject.

Thanks are due to all those who in one way or another by argument, discussion and advice have contributed to the knowledge and understanding of the authors.

F.A.P.
H.Y.P.

Contents

1 Introduction to packaging   **1**

History   1
Definitions   3
The need for packaging   5
Designing successful packaging   8
  Product assessment   8
  The hazards of distribution   8
  Good package design and supermarket selling   10
  Marketing requirements   14
  Packaging and the self-service store   14
  The package and advertising   16
  The package and the price of the product   16
Packaging materials selection and machinery considerations   17
  Cost   17
The place of packaging within the marketing complex   19
  Materials utilization   20
  Machinery and line efficiency   22
  Movement in distribution   22
  Management   23
Properties and forms of packaging materials   23
Food process classification   26
References   32

2 Graphics and package design   **33**

Introduction   33
Management's role   33
Packaging and modern merchandising   34
Meeting customer and consumer needs   36
Beware the half truth and consider the alternative   37
Trends   39
Summary of consumer needs   39
The function of packaging graphics   40
The main printing processes   40
  Letterpress   40
  Flexography   44
  Lithography   45
  Gravure   47
  Silkscreen   48
  Ink-jet printing   49
  Hot die stamping and gold blocking   50
Factors affecting the choice of a printing process   50
References   52

3 Notes on packaging materials   **53**

Paper-based packing   53
  What is wood?   53
  Pulping   53

Beating   54
Paper testing   55
Types of paperboard   57
Plastics   59
    Thermosets   61
    Thermoplastics   61
    Polyesters   68
    Flexible packaging materials based on films and foils   68
    Materials tests on flexible packaging   70
Glass containers   78
    Properties of glass containers   78
    Sealing of glass containers   82
Metal packaging: the basic materials   83
    Cans and tin boxes   85
    Closures   92
Natural materials   94
    Round-stick joinery   94
    Straw   94
    Waxes and bitumen   95
    Stoneware   95
    Textiles   95
    Wicker baskets   96
References   96

4   Packaging machinery   97

Production and packaging line requirements   97
The property profile   99
Bottling   100
    Bottle feeding   103
    Bottle cleaning   103
    Filling liquids   104
    Filling dry goods (powders and granular material)   106
    Statistical recording—the legal requirements   110
    Capping of bottles and jars   114
    Labelling bottles and jars   116
    Case packing and sealing   116
    Shrink- or stretch-wrapping   117
    Palletizing   117
Canning operations   118
    Handling and storage of empty cans   118
    Cleaning empty cans   118
    Product preparation   119
    Filling   119
    Closing (seaming)   120
    Processing   121
    Cooling   124
    Handling and storage of filled cans   125
Wrapping operations   126
    Bags: manufacture, filling and closing   130
    Paper bags   131
    Film bags   131
    Open mesh bags   132
    Bag filling and closing equipment   132
    Bag-in-box packages   133
Cartoning   133
    Cartons for liquid products   134
    Cartons for solid products   135
    Cartoning systems   139

Form, fill and seal machines 141
Vertical form-fill-seal (f.f.s.) machines 142
Packaging machinery for vertical f.f.s. pillow packs 146
Horizontal f.f.s. machines 148
Packaging machinery for horizontal f.f.s. sachet packs 153
Thermoformed f.f.s. packs 153
Labelling 159
Purchasing, installation and operation of labelling machinery 162
Container handling 162
Standard case sealing, wrap-around case sealing and tray erection 162
Standard case sealing 162
Adhesive application systems 162
Wrap-around case sealing 163
Tray erection 164
Organization of packing lines 165
Further reading 166

5 Packaging for physical distribution 167
Introduction 167
Functions of a shipping container 168
Primary, secondary and tertiary levels of packaging 169
Fibreboard case performance 170
Board properties affecting stacking performance 175
Unitizing methods 181
Film wraps 182
Pallet stabilizing adhesives 184
Strapping materials and methods 184
References 186

6 Spoilage and deterioration indices 187
Biodeterioration 187
Effects of temperature on senescence 187
Microbial growth 188
Food spoilage and food poisoning 192
Moulds and yeasts 193
Preventing bacterial and mould growth 193
Insect infestation 194
Rodents and birds 194
Abiotic spoilage 195
The role of water in foods 195
Sorption isotherms 196
References 203

7 Fresh and chilled foods: meat, poultry, fish, dairy
products and eggs 205
Meat 205
Preparation of meat 205
Chilling and chilled storage 206
Cutting and boning 206
Deterioration of fresh and chilled meat 206
Chilled transport to the retail outlet 209
Wholesale packaging 209
Retailing 210
Vacuum packaging of meat 212
Modified atmosphere packaging 214
Poultry 216
Preparation and spoilage 216

Packaging   216
Modified atmosphere packaging of poultry   217
Fish and shellfish   217
Factory ships   217
Fish processors   217
Handling and transport   218
Chilling   218
Fish farming   219
Retailing   219
Deterioration of fresh and chilled fish   219
Prepackaging   221
Modified atmosphere packaging of fresh fish   221
Vacuum packaging of fresh fish   221
Milk   222
Quality and composition   222
Effect of temperature on bacterial growth   223
Pasteurization   223
Characteristic spoilage of pasteurized milk   224
Packaging   224
Returnable bottles   225
Other dairy products   226
Butter   226
Protection required   227
Dairy spreads   227
Eggs   227
Packaging requirements   227
Cream   228
Guidelines for packers, manufacturers, distributers and consumers   228
References   229

8   **Fresh fruits and vegetables (including herbs, spices and nuts)**   **231**
Fruits and vegetables   231
Variability   231
The growing process, respiration and ripening   231
Temperature   235
Composition of the atmosphere   237
Bacteriological conditions   241
Handling   241
Transport   241
Packaging   241
Prepacked fruit and vegetables   242
Modified atmosphere packaging   242
Prepared vegetables and salads   245
Fresh herbs and spices   245
Nuts and seeds   246
References   246

9   **Frozen foods**   **248**
Freezing   248
Commercial freezing methods   249
Storage and distribution   249
Protection needed by frozen foods   250
Types of package   251
Frozen meat and poultry   252
Freezer burn   253
The freezing process   253
Commercial freezing methods   254
Frozen poultry   254

Frozen fish   255
  Effect of freezing on fish   255
  Methods of freezing   256
  Storage   256
  Packaging frozen fish   256
  Frozen fish products   257
Frozen fruits and vegetables   257
Other frozen products   258
  Ice-cream   259
  Cook-freeze products   260
Guidelines for packers, distributors, retailers and consumers   261
Future trends   261
References   263

10 Heat-processed foods (including irradiated foods, etc.)   265

Heat processing   265
  Commercial sterilization   265
  Factors affecting resistance of microorganisms to high temperatures   265
  Factors affecting the rate of heat penetration   266
  Pasteurization   268
  Blanching   269
  General spoilage problems with canned foods   269
High-barrier plastics packaging   274
  Stepcan   275
  Letpak   275
Aseptic processing   276
  Aseptic canning   278
  Aseptic packaging using flexible materials   278
  Aseptic processing of milk and dairy products   280
  Aseptic processing of fruit and vegetable juices   282
  Problems with particulates   283
Sous vide   284
Packaging for microwavable foods   285
  Packaging materials   286
  Microwave active materials (susceptors and receptors)   287
Irradiation   288
  Methods of irradiation   289
  Applications   290
  Detection methods   292
Ultraviolet light   293
Ultrasonics   293
High pressure techniques   293
References   294

11 Dried and moisture sensitive foods   296

Reduction of available water   296
  Methods of drying   296
  Moisture levels of dried and moisture sensitive foods   298
  General spoilage considerations of dehydrated foods   299
  Oxygen scavengers   299
Active packaging systems   300
  Packaging requirements for different moisture levels   300
References   314

12 Other processed foods   315

Preservation by chemical means   315
  Classes of chemical preservative   315

Inorganic chemicals   315
Organic acids and their salts   315
Antioxidants   316
Antibiotics   316
Cured and smoked foods   317
Nitrite and nitrate curing   317
Colour and the curing process   317
Bacterial spoilage   319
Cured fish   320
Smoking   320
Fermented foods   321
Cheese   322
General requirements for cheese film wrappings   324
Yoghurt   325
Fermented meat and fish products   327
Vinegars, pickles, sauces and dressings   328
Other fermented products   330
Fats and oils   331
Spoilage mechanisms   331
Packaging requirements   332
References   334

13   Juices, soft drinks and alcoholic beverages   335

Fruit juices and beverages   335
Components and characteristics   336
Spoilage and its prevention   337
Packaging requirements   339
Beers and ales   342
Spoilage   342
Packaging materials   343
Cider   344
Wine   344
Packaging requirements   345
Distilled spirits   346
References   346

14   Developing packs for food   347

The product life cycle   347
Planning for change   350
Basic considerations for package development   350
Package structural development   351
Packaging coordination   353
Graphics   354
Packaging-line engineering   354
Cost of development   355
Developing a domestic food packaging for an export market   355
References   356

15   The economics of primary packaging   357

Cost comparisons   357
The economics of the glass primary package   359
Factors affecting cost   361
The economics of cans and canning   365
Economics of making tins   366
The ideal package shape for economy   366
Getting the most out of available material   367
The total pack concept   368

Costing tinplate packages for food worldwide 369
Food canning worldwide 370
Semi-scale production 371
The economics of cartons 374
Board selection: the economic considerations 376
Board selection for containment, compatibility and protection 378
Board selection for efficient running on a packaging line 379
Board selection for appearance and print quality 381
The economics of packaging with flexible materials 384
Economics of plastics moulded packs 386
Injection mouldings 388
Thermoforming 388
Compression mouldings 389
References 389

16 Using barrier materials efficiently 390
Transmission of gases and vapours through barrier materials 390
Theory 392
The variable factors associated with permeability measurements 395
Estimating the type of barrier required 402
Protection of a moisture-sensitive product 405
Packaging barriers and their relation to moisture changes in foods 407
The measurement of gas transmission rate 413
Measurement of water vapour transmission 416
Pests 417
Insect infestation 417
Rats and mice 421
The compatibility of foods with their packaging 421
Volatiles 422
Non-volatiles 423
References 425

17 Specification and quality control 426
What is quality? 427
Quality control 429
Process sequence control 429
Measurement, the assessment of quality 430
Measurement used for quality aspects 433
Sampling 435
Factors affecting quality in packaging 437
Package performance 437
Quality measurement and control of cartons 438
Quality control in a glass container factory 447
Quality checks on corrugated cases 452
Quality assurance 458
References 463
Further reading 463

18 Evaluation and testing of transport packages 464
Methods of evaluation 464
Journey hazards 464
Distribution systems 464
Obtaining data on journey hazards 465
The effect of environment on packages 466
Impacts 467

Crushing   469
The vibration hazard   470
Package test equipment   471
Mechanical testing   471
Climatic testing   472
Methods of using tests   473
Performance test schedules   474
References   476

Appendices   **477**

Appendix 1 European packaging legislation   477
Appendix 2 USA legislation   491

Index   **493**

1 Introduction to packaging

History

Man has many competitors for the food he produces. Animals, particularly rodents, insects and microorganisms (moulds, yeasts and bacteria), all cause wastage at various stages in the growth, harvesting, processing, storage, transport and sale of food. If microorganisms are permitted to flourish in food, they will make it unattractive, and it will waste by putrefaction, fermentation, or mould growth. These organisms, particularly bacteria, can affect food and render it poisonous to man, thereby causing sickness and even death. The provision of food which is good and safe to eat is therefore a duty of the food industry: the prevention of waste is essential to that industry and to a nation's economic well-being. Packaging plays a decisive role in achieving the objectives of safety and waste prevention.

The earliest packaging materials were probably leaves from larger plants which were used by early man to wrap meat from a kill. In early days when men lived by hunting and gathering, their nomadic existence necessitated some means of keeping food 'fresh' while they travelled. Animal skins were used for water and woven baskets were also employed.

It is perhaps inevitable that some of the landmarks in the fight to preserve food should concern the provisioning of armies. At the beginning of the nineteenth century Napoleon was finding it increasingly difficult to do this. The 'scorched earth' policy of his adversaries meant that it was not possible to live off the conquered territories, and the blockade by the British fleet deprived France of the sugar needed to preserve such produce as fruit. He needed packaged foods that could be taken along with his armies, and he offered a prize of 12 000 francs for a suitable method of preservation: this was won in 1810 by a Frenchman, Nicolas Appert, for the development of the 'canning' process, although he, at that time, used glass jars.

Later in the nineteenth century, the Great Trek West took place in America as the pioneers in their covered wagons spread out across the continent to establish their homesteads and plant their crops. Their survival until the first harvest was made easier by the dried and canned foods which they took with them. The American canning industry grew quickly to meet these demands and then to provision the rival forces in the American Civil War.

These examples of the early uses of packaged foods were concerned with survival, but they illustrate basic principles. Food must be available wherever there are people, and with modern population patterns this is seldom where it is grown. Food, in interesting variety, must be available all the year round, irrespective of the growing season. It must be presented in a way that is convenient to purchase and use, and in most instances this means that it must be packaged.

The choice of suitable packaging involves a number of considerations. For most food products there is an overriding objective: the package must provide the optimum protective properties to keep the product it encloses in good condition for its anticipated shelf life. Also to be considered are decisions which are subjective: the pack should be of the right shape and size and its graphics must attract the eye of the purchaser. The development and design of appropriate packaging has made it possible to offer the consumer a wide variety of food from which to choose, with complete confidence in its wholesomeness, whether it is seasonable or not.

In sophisticated societies, the food industry is the largest user of packaging at the consumer level. If we tried to draw a balance of the way in which the public interest has been served by this heavy involvement in packaging, it is evident that the advances in the hygiene and wholesomeness of food weigh heavily on the positive side.

Prevention of food waste is undoubtedly a vital objective for everyone, and can be applied at all stages between the grower and the home. In sophisticated societies, the food manufacturer producing packaged foods is concerned to reduce waste in pursuit of an efficient business. To achieve this he may advise on the husbandry of crops to ensure good yields, and provide rapid and effective transport of the crop to the processing and packaging plant to safeguard quality and quantity. In the well-publicized case of peas quick-frozen within two hours of harvesting, not only is loss of peas prevented, but also loss of nutrients, compared with some methods of so-called 'fresh' distribution. There is a further unseen benefit: all the pea hulls are left behind at the farm and the centres of population are kept free from much of the vegetable waste that would otherwise require disposal. Additionally, much of this waste can be used for animal food.

The packaging technique and choice of a pack with appropriate barrier properties is designed to prevent destruction of food by microbial or insect attack, depending upon its physical nature, and also to preserve the quality and nutritive value of many foods by the exclusion of oxygen and the control of moisture loss or gain. Packaging in 'portion' packs often helps the consumer to buy the quantity needed and no more. It is one of the ironies that these aids to the avoidance of food waste are more readily available in affluent societies than in the developing world, where considerable losses can occur: up to 25% (and sometimes more) of the food harvested can be lost through attack by pests (insects, rodents and microorganisms) often for lack of proper packaging or storage conditions.

The complex pattern of life in modern society requires many materials the availability of which is controlled by world supplies of raw materials, man-made restrictions on their availability, manufacturing capacity and consumer demand.

Definitions

To appreciate the place of packaging in the world economy, we must know what it is and how it functions. Packaging has been defined in several ways. Table 1.1 lists three of the more fundamental; (1) and (2) indicate that packaging contains and protects during transport and has an economic aspect. To ensure delivery, the package must at least provide information as to the address of the recipient, describe the product and perhaps explain how to handle the package and use the product. A little more thought and we recognize that packaging is part of the marketing process (Figure 1.1).

Marketing may be defined as the *identification, anticipation* and *satisfaction* of customer need profitably.

The basic function of food packaging is to identify the product and ensure that it travels safely through the distribution system to the consumer. Packaging designed and constructed solely for this purpose adds little or nothing to the value of the product; it merely preserves farm or processor freshness or prevents physical damage. Cost effectiveness is the sole criterion for success. If, however, the packaging facilitates the use of the product, is reusable or has an after use, some extra value can be added to justify extra cost and promote sales.

Packaging has also been described as a 'complex, dynamic, scientific, artistic and controversial segment of business'. Packaging is certainly dynamic and is constantly changing. New materials need new methods, new methods demand new machinery, new machinery results in better quality, and better quality opens up new markets which require changes in packaging. The cycle then starts again.

Thus, at its most fundamental, packaging *contains, protects and preserves*, and *informs*. At its most sophisticated, it provides two more functions—those of *selling* and *convenience*. In a world where the quality of products is high, in

Table 1.1 Definitions of packaging

(1)	A coordinated system of preparing goods for transport, distribution, storage, retailing and end-use
(2)	A means of ensuring safe delivery to the ultimate consumer in sound condition at minimum overall cost
(3)	A techno-economic function aimed at minimizing costs of delivery while maximizing sales (and hence profits)

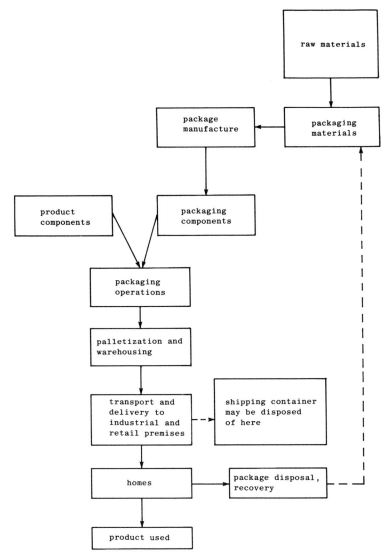

Figure 1.1 The marketing process and the packaging cycle.

many instances almost the only difference between competitive brands lies in the packaging, and only packaging influences the selling operations. We recall the last definition in Table 1.1: 'packaging is a techno-economic function aimed at minimizing costs of delivery while maximizing sales (and hence profits)'. At this level, the value of—or even the need for—the added functions is controversial, and as a result opinions vary as to whether packaging is a waste of material and energy, or is properly utilized for the conservation of goods and reduction of labour. There is no doubt, however, that in the post- fuel crisis world in which we now live, where materials are more expensive principally because of the extra cost of energy, we may well need to change our criteria of judgement. Containment, protection and information will always be essential in any packaging and these functions are basically conservational. How much we should spend on the 'selling' and 'convenience' functions and how far they are regarded as necessary, is a matter for discussion.

The need for packaging [1]

Efficient packaging is a necessity for every kind of food, whether it is fresh or processed. It is an essential link between the food producer and the consumer, and unless performed correctly the standing of the product suffers and customer goodwill is lost. All the skill, quality and reliability built into the product during development and production is wasted unless care is taken to see that the consumer gets it in prime condition. Proper design is the main way of providing 'a package which protects what it sells and sells what it protects'.

Thus, the packaging functions require specialized knowledge and skills, in addition to specific machinery and facilities, to produce a package which will provide most, if not all, of a number of basic requirements of which the following are the most important: containment; protection and preservation; communication; suitability for the packaging line, i.e. machinability; convenience in shape, size, and weight for handling and storage; adapted for the use of the product it contains; environmentally friendly in respect of manufacture, use and disposal. Moreover, these basics must be provided at one or all of the three levels of packaging usually employed, namely the primary pack, the secondary package or shipping container and the unit load.

These basic needs must now be examined in more detail together with some less important options.

- *Containment.* Obviously, the package must keep its contents secure between the end of the packaging line and the time when all the food has been eaten.
- *Protection and preservation.* The packaging must protect the food from both mechanical damage during handling and deterioration by the climate(s) through which the package will pass during distribution and storage in the home.
- *Communication.* All food packaging must communicate. Not only must the contents be identified and the legal requirements of labelling be met, but often the packaging is an important factor in promoting sales. Also, the unit load and the shipping container must inform the carrier about its destination, provide instructions about the handling and storage of the

food and inform the retailer about the method of opening the package and possibly even of the best way to display the product.

- *Machinability.* The majority of modern retail packages and many transport packages are today erected, filled, closed and collated on machinery operating at speeds of 1000 units or more per minute. They must therefore perform without too many stoppages or the process will be wasteful of material and uneconomic. Even when the numbers concerned are small and the items specialized, the need for a good performance in filling and closing operations is still important.

- *Convenience and use.* The most common impressions of convenience in retail packaging for foods are those of providing easy opening, dispensing and/or after use. Easy opening must be tempered by seal integrity. We must avoid the trap of producing an opening device which fails in transit, or of failing to provide sufficient control on the packaging line to ensure the device works 99% of the time. However, the provision of convenience is much wider than just these impressions. The shipping container as well as the primary packaging must provide convenience at all stages from the packaging line, through warehousing to distribution, as well as satisfying the needs of the user of the product.

We must also fit the packaging to the needs of the food and this involves answering questions such as:

- What age groups are we concerned with? Are they well informed, impulsive or irrational?
- What kind of packaging is used by the competition? Should we follow the general line or be different?
- What do distributors and retailers want from the packaging? Have they criticisms of our or the competition's packaging?
- Is the relationship between packaging cost and the selling price of the food correct or does it give a wrong impression of the position in the market?
- Is the possibility of pilfering, tampering or stealing such as to make an impact on the design of package?
- Is the possibility of an after use for the packaging worth considering as a sales incentive and will the packaging be generally considered as environmentally responsible?
- Do we want a strong brand identity?
- Is there a range of related products that might form a family resemblance in the packaging?
- How will the packages be set out on retail shelves, etc.?

Such considerations will obviously differ according to the food concerned and the customers who are expected to buy. They can only be answered by a well conducted survey of the market.

As industries grow the need for quality packaging undoubtedly increases. Hence there is likely to be more packaging in the world every year, rather than less. The pressure groups in affluent societies, particularly those concerned with the environment, frequently do not regard the package-making industries, or product manufacturers who use packages, as being sufficiently responsive to modern social needs. At least part of the reason for this is that packaging people everywhere have not made their aims as well known as they ought. The ordinary person in the street frequently understands the value of packaging far better than pressure groups would have us believe, and there is no doubt that the need for packaging in the coming years will be questioned critically and considerably in relation to costs. The challenge for packaging in the 1990s is to continue to serve people in the way that meets their needs and interests, and to make the facts about this known in a way that can be understood. To be effective, therefore, packaging must make the maximum contribution to the success of the marketing and distribution operations of which it forms a vital part, while at the same time be regarded as environmentally responsible.

In general, technical developments in the packaging of any food product arise from changes in four main areas:

(a) Availability of newer materials and improved constructions, e.g. improved flexible barriers through metallizing and co-extrusion and changes in thermoforming techniques, etc.
(b) Developments in food processing and/or packaging machinery such as aseptic processing and modified atmosphere packaging as well as faster and more accurate computer control of machines.
(c) Changes in methods of storage and distribution.
(d) Improvements in methods of management and control, such as the use of bar codes and just-in-time (JIT) deliveries.

Developments in marketing also influence packaging. If we define marketing as 'the identification, anticipation and satisfaction of customer need profitably' we realize the influence that customer lifestyles could have on the packaging of food. In many instances improved packaging can promote a marketing response to customer demands or even change lifestyles, and in the food area consumers have reacted strongly to such influences as:

(a) Malicious tampering, whether for blackmailing retailers or other reasons.
(b) Green issues such as organic farming, more acceptable methods of animal husbandry and the reuse and/or recycling of packaging before final disposal.
(c) Health lobbies (low fat and sugar diets; elimination of artificial colourings and reduction in preservatives, etc.).
(d) A desire to reduce meal preparation time to a minimum.

Designing successful packaging

In order to design successful packaging four sets of facts must be considered [2]:

● Product assessment
● The hazards of distribution
● Marketing requirements
● Packaging materials selection and machinery considerations.

Product assessment

All we need here are the answers to the question: How can the product be damaged or deteriorate? Some of these are obvious from a visual examination, some can be ascertained by simple measurements, whilst other information must be supplied by the designer and producer.

The more important facts required are:

(a) The nature of the product—the materials from which it is made and the manner in which these can deteriorate.
(b) Its size and shape.
(c) Its weight and density.
(d) Its weaknesses—which parts will break, bend, move about, become loose, scratched or abraded easily.
(e) Its strengths—which parts will withstand loads or pressures and which might be suitable for locating the product in the pack.
(f) The effect of moisture and temperature changes on the product, and whether it will absorb moisture or corrode.
(g) Compatibility—whether the product is likely to be affected by any of the possible packaging materials, which items can be packed together, with protection if necessary, and which items must not be packed together under any circumstances.
(h) Possibilities for dismantling complex products, how far stripping down may be carried out to reduce the package size to a minimum, and whether the required assembly, installation and use instructions will be such that the customer can handle them.

Table 1.2 summarizes these factors.

The hazards of distribution

Here we need the answer to the question of what happens to the package on its journey to the consumer. It is necessary to know the method of transport, the probable storage conditions, and the duration of both journey and storage. Important points to establish are:

(a) The type of transport—road, rail, sea or air.

Table 1.2 Product assessment

(1) **Physical state**	(2) **General nature**
Gas	Corrosive
Mobile liquid	Toxic
Viscous liquid	Volatile
Paste	Odorous
Liquid + solids	Perishable
Powder (free flowing?)	Sticky
Granules	Corrodible
Tablets	Fragile
Capsules	Abrasive
Solid block	Easily scratched

(3) **How can it be damaged?**	
By mechanical shock?	Fragility factor
By vibration?	Frequency range
By abrasion?	Surface finish
By crushing?	Safe load
By temperature changes?	Safe range
By moisture and relative humidity changes?	Critical values
By oxygen?	How?
By odours?	Which?
By light?	Fading
By spoilage?	Chemical changes
By incompatibility with materials?	
By rodents, or insects?	

(4) **How can the package be unsatisfactory?**

Admits dirt
Leaks
Not siftproof
Not compatible
    (a) transfers odours or flavours to product
    (b) causes corrosion of product
    (c) reacts chemically
    (d) loses strength in contact with product
Easily pilfered
Stains easily

(b) The degree of control over the transport—is it private or public transport?

(c) The form of transport—break-bulk, roll on-roll off (RORO), freight container, unitized load, postal, passenger train, etc.

(d) The mechanical conditions and duration of storage.

(e) The nature and intensity of the mechanical and climatic hazards in transport, storage, retailing and use.

(f) Whether handling aids are available for loading and off-loading at all points between maker and user.

(g) The importance of minimum cube (volume) in relation to transport costs.

Tables 1.3–1.5 summarize the possible hazards.

Table 1.3 Distribution hazards: mechanical hazards

Basic hazard	Typical circumstances
Impact	
(a) vertical	Package dropped to floor during loading and unloading on to or off nets, pallets, vehicle landing boards, etc.
	Package rolled over or tipped over to impact a face
	Fall from chutes or conveyors
	Result of throwing
(b) horizontal	Rail or road vehicle stopping and starting
	Swinging crane impacts wall, etc.
	Arrest by stop or other packs on chute or conveyor
	Arrest when cylindrical package stops rolling
	Result of throwing
(c) stationary package impacted by another	All above where circumstances cause the falling pack to impact another
Vibration	From handling equipment (in factory, depot and at transhipment points)
	Engine and transmission vibration from road vehicles
	Running gear—suspension vibration on rail
	Machinery vibration on ships
	Engine and aerodynamic vibration on aircraft
Compression	Static stacks in factory, warehouse and store
	Transient loads during transport in vehicles
	Compression due to method of handling, e.g. crane grabs, slings, nets, squeeze clamps, etc.
	Compression due to restraint
Racking or deformation	Uneven support due to poor floors, storage, etc.
	Uneven lifting due to bad slinging, localized suspension, etc.
Piercing, puncturing, tearing, snagging	Hooks, projections, misuse of handling equipment, or wrong method of handling

Good package design and supermarket selling

Since much packaging is designed for the domestic consumer, we must take note of current food distribution systems. Figure 1.2 shows the flow from primary food producer to final consumer [3] (the flow of food is virtually all one way, but some parts of the chain could permit the use of returnable packages). The retail scene in the sophisticated parts of the world has changed considerably over the last 30 years. Shops have become bigger and have combined into multiples which in their turn have developed the self-service principle through supermarkets to hypermarkets dealing not only with food but with non-grocery items as well. This has made the package a vital part of the sales armoury of the food packer.

Table 1.4 Distribution hazards: climatic hazards

Basic hazard	Typical circumstances
High temperature	Direct exposure to sunshine Proximity to boilers, heating systems, etc. Indirect exposure to sun in sheds, vehicles, etc., with poor insulation High ambient air temperature
Low temperature	Unheated storage in cold climates Transport in unheated aircraft holds Cold storage
Low pressure	Change in altitude, particularly in unpressurized aircraft holds—aircraft pressurization failure
Light	Direct sunshine UV exposure Artificial lighting

Liquid water
  (a) fresh
	
	Rain during transit, loading and unloading,
	   warehousing and storage
	Puddles and flooding
	Condensation and ship sweat, etc.

(b) polluted	Salt sea spray—deck cargo, lightering surf boats, etc. Salt water puddles on docks, etc. Bilge water and sea water in holds Industrially polluted puddles and spray, e.g. at chemical works
Dust	Exposure to wind-driven particles of sand, dust, grit, etc.
Water vapour	Humidity of the atmosphere, both natural and artificial

In recent years, independent shops have accounted for only a small share of the total sales of groceries, while multiples and cooperatives of one kind or another dominate grocery sales. This gives the packaging buying departments of the multiples and cooperatives considerable purchasing power, which in turn puts pressure on food packers to get their marketing (and hence their packaging) right.

Essentially, supermarkets offer the consumer a wide choice at low prices, and it is as well to remember that since the early 1950s progressively more consumers have availed themselves of these twin benefits. Fast turnover and bulk buying can support a wider product range, while lower prices reflect the substantial reduction in the wage bill brought about by self-service techniques. Considerable economies are effected in goods handling by enabling manufacturers and other suppliers to make one-drop economic deliveries directly into these larger retail units, and further savings in labour overheads are achieved by prepackaging at the food processor's factory.

Food packaging in particular has evolved to meet supermarket requirements. In addition to protecting against contamination, modern food packag-

Table 1.5   Distribution hazards: other hazards

Basic hazard	Typical circumstances
Biological	

(a) microorganisms
    fungi
    moulds
    bacteria
	Are ubiquitous and adapt themselves to varied conditions. Require moisture and generally will not grow at relative humidities of less than 70%. Will grow over a wide range of temperatures

(b) beetles
    moths
    flies
    ants
    termites
	In general high temperatures are more favourable for development than low ones and, below 15°C, development is unlikely. A relative humidity (r.h.) of 70% is very favourable for most insects but some will develop at below 50% r.h. Infestation usually starts from eggs laid on packaging materials, penetration then being made by the small newly hatched insects. Migration from adjacent packs or from natural habitat (particularly in tropical localities) may occur

(c) mites	As for insects, but they are less tolerant of dry conditions (few survive and develop slowly at about 60% r.h.) and they develop over a lower temperature range
(d) rodents (rats, mice)	May be present in warehouses, transit sheds, storage areas, holds, etc. Will attack most materials to keep in condition, and softer materials for making nests (or for food)
Contamination by other goods	
(a) by materials of adjacent packs	Obliteration of marking, printing etc., by rusty metalwork—strapping, wire bands. Effects of damp packing materials, especially hessian, on non-water resistant materials, adhesives and metal parts
(b) by leaking contents of adjacent packs	Damage to containers of liquids, powders and granulated substances may result in leakage of the contents. The effect of the resultant contamination on adjacent packs can range from the spoiling of external appearance to complete disintegration of a pack and its contents, depending on the nature of the contaminant, the packing materials and the contents of the pack contaminated
(c) radioactivity	

ing offers protection against breakage, against tampering with the contents, and even, where needed, against the adverse effects of light. Moreover, the surface design of the supermarket package is so conceived as to carry a wealth of product information to the potential purchaser. Contents, weight, price, instructions for preparation, suggestions for use, and other relevant details are all dovetailed into a pattern of direct communication.

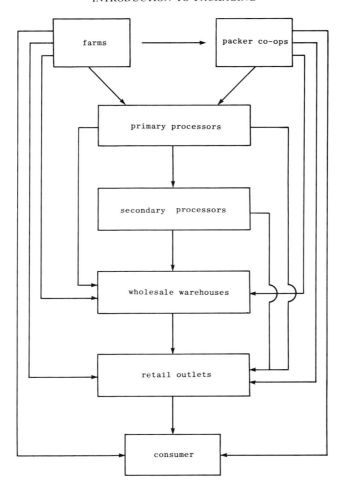

Figure 1.2   Distribution systems for foodstuffs.

Supermarket packaging is cost-effective in the sense that the package is pivotal to the supermarketing philosophy of wide choice at low cost. By eliminating much of the expensive labour element that bedevilled the distribution chain for centuries, the prepackaged self-service supermarket food product attracts less total expense and therefore helps to keep prices down.

Above all, the function of the modern food package is to hold down supermarketing costs, and to offer the consumer a tangible price benefit. Government statistics suggest that this object has been achieved, but in an inflationary climate the margin between cost and price is a slim one.

Improved packaging has made a considerable contribution towards efficiency and improvement in volume sales per employee in food outlets.

Table 1.6   The top ten retailers in Europe (turnover and sales per employee)

Organization	Country	Year	Turnover (£m)	Sales per employee (£1000)
CarreFour	France	1987	5.74	142
Dee Corp	UK	1987	4.84	64
J. Sainsbury	UK	1987/1988	4.80	88
Marks & Spencer	UK	1987/1988	4.58	69
Migros	Swiss	1987	4.55	74
Tesco	UK	1987/1988	4.12	82
Karstadt	Germany	1987	4.06	61
Promodes	France	1987	3.54	116
Casino	France	1987	3.46	86
Coop AG	Germany	1986	3.43	83

Table 1.6 gives some figures for the top ten retailers in Europe. Note the variability in the sales per employee in the different organizations.

Marketing requirements

Here we need answers to such questions as: What is the product for? What is a convenient quantity? Where is it to be sold? Why does it need packaging?

A more detailed checklist is given in Table 1.7. In addition, specific considerations may require to be taken into account, and the more important ones are as follows.

The package and the image.   The package projects the image, not only of the product which it contains, but often also of the company which sells the brand. The pack must therefore enhance and maintain the projection for which it has been designed, both to the customer at the point of sale and to the retailer.

It is essential that the image that the package projects is completely integrated with the advertising for the food product itself. If the package design does not fit in with the image that consumers have of the food itself, the resulting confusion will not promote sales.

Packaging and the self-service store

Because labour costs account for a high proportion of the potential profit to be made by both the manufacturer and the retailer, the tremendous growth of self-service throws responsibility on the packaging designer. The designer must ensure that a pack at the point of sale has the type of appeal which will make customers pick it up and put it into their shopping baskets. In other words the pack must stimulate impulse purchasing. It must be easy to handle and attractive to see and hold (e.g. squeeze packs, plastic outers that are finger-contoured) and it has to stack easily.

Packs must also be easily recognized end-on, since the profit per linear foot of shelf often means that packs are stacked this way towards the customer. The

Table 1.7 Marketing considerations

(1) The product	(4) The package	(5) Convenience and use
What is the competition? (a) Packages used? (b) Quantities sold? (c) Price bracket? What are the selling points of (a) Competitive products (b) Our product	(a) *Primary package* Size, shape, weight Standard, gift, seasonal Bag, envelope, pouch, sachet Rigid or folding box, card pack (blister, skin, etc.) Metal container Glass container	(a) *Primary package* Inspection before purchase? Easy opening? Reclosure? Measured dose? Dispensing aid? For storage out of sight? or for use where attractiveness important?
(2) The retail service Self-service? Department store? Mail order? Doorstep? How do competitors retail?	Collapsible tube (metal or plastics) Plastics container (blown, injection or thermo-form) Moulded pulp container Composite tube	Easy grip? Special function? e.g. spray, cook-in-pack, squeeze use, etc. Disposal?
(3) The customer Age Sex Income group Social level Location (home, local, regional, national or export)	(b) *Transport package* Size, weight, shape, number of units Wooden case or crate Fibreboard case or drum Sack (paper, textile, plastic) Metal drum Glass carboy Plastics drum Bale	(b) *Transport package* Weight? Shape? Display conversion? Disposal? Returnable? After use? Pallet movement? Containerized transport? Hand holds? Fork truck? Slinging?

introduction of 'sell pack' techniques typified by the use of cage pallets may also require recognition in design.

The package and advertising

Colour predominates in advertising on TV, cinemas, posters and many of the press media, so the package must be made of materials that take and hold colours. The marketing function for any product embraces advertising and in many retail fields, particularly food, the packaging is more widely seen and read than any other form of advertisement. Since it is often the only means of recognition at the point of sale, the public must constantly be made aware of the package as well as the product.

The package design must take the limitations of the advertising medium into account. A full colour representation of an appetizing dish which is effective on the shelves of the supermarket may not integrate well with a black and white picture in a newspaper.

Advertising is consequently a specialist job and for this reason many specialist advertising agencies exist to supply the food industry with designs that provide the 'right' message. Since there are laws controlling advertising, there is a Code of Practice in the UK which says that all adverts must:

- be legal, decent, honest and truthful
- be responsible to the consumer and society
- follow business principles of 'fair competition'.

These guidelines should be followed in all labelling and descriptions of food offered for sale.

The package and the price of the product

We have to understand that the cost of packaging bears only a general relationship to the cost of the end product. A manufacturer will charge what he believes—through the use of market research—that the customer is prepared to pay for his goods. The profit mark-up can therefore be disproportionately high, and this may be because the cost of the pack is only a relatively small percentage of the ex-factory price. The packaging manufacturer will always be under pressure to produce at the lowest possible price, because this enables his clients, the food processors, to make marketing adjustments to their product, such as price cuts or bonus offers to retailers.

Increasing pressure is exerted in affluent societies today by a large number of consumer groups on behalf of the consumer. As a result, the public is putting products and their packs under considerable scrutiny. Packs which, for example, deliberately seem to make the contents of an inner container appear to be bigger than could be assessed visually should be avoided. Research indicates that products which are packed in expensive and bigger-than-necessary outers fail more frequently than more conventionally packed brand lines.

Packaging materials selection and machinery considerations

Among the factors playing an important part in selecting a packaging system for a specific product are:

- Production methods
- Display requirements
- Economic considerations
- Marketing needs
- Product characteristics
- Properties of the packaging materials.

All these must be taken into account irrespective of the market involved. In considering *production methods* it must be decided whether the packaging materials can be handled on existing machinery or sometimes even whether an existing machine will do the entire job.

There is also the question of how efficient the packaging line will be if a particular size and/or shape of package is chosen. The degree of skill required by the packaging line operators may also be an important factor. For example, if an existing package is redesigned, will the present operators be able to handle it with or without retraining, or will new personnel be needed?

Although *display requirements* are largely the responsibility of marketing and advertising, certain technical factors must be taken care of before final selection, e.g. the relation of the packaging material to the printing process must be examined.

The size and shape of the package is also important. Insistence on a large face panel on a carton might make the display stand out but could complicate the simpler types of dispensing methods for the consumer. Compromise is often needed in this area.

In general, in selecting a primary package for a particular food four basic questions must be answered:

(1) What must the package achieve?
(2) What types of package are available that can do this?
(3) What are the pros and cons of each of the potential packages in the context of the achievement required?
(4) What costs will result for each possibility in relation to the other elements of the distribution system?

Cost

It is important to note that any packaging assessment must include a definition of *optimum* quality standards. The question of cost must not compromise these standards. Quality assurance and control are dealt with in detail in chapter 17.

The overall packaging cost is made up of a number of different factors.

- Packaging material cost (delivered to factory)
- Storage and handling cost of the empty packages or materials
- Filling cost (including quality control and handling of filled packages)
- Storage cost of the filled packages
- Transport costs of delivering filled packages
- Insurance costs involved in transport
- Losses due to breakage or other spoilage of the product (including loss of goodwill)
- Effect of the package on sales figures.

In practice, decisions often have to be made on specific areas. Typically, a cost comparison between two systems (see Table 1.8) would focus on the following:

(a) *Packaging material (or container) prices.* To be strictly comparable, quotations should be obtained at the same date (or as close to each other as possible), for the same quantity, with identical terms of sale. Primary, secondary and tertiary packaging must all be included.

(b) *Machinery cost.* The capital cost will be depreciated over a number of years, in accordance with the company's policy.

Table 1.8   Cost elements in packaging

Development costs
  Definition of requirements
  Initial search for concepts
  Design, both for function and graphics
  Model and sample production
  Market testing of prototype
  Preparing buying specifications
  Preparing quality inspection methods and introducing a quality assurance service
  Overcoming 'teething troubles'

One-time costs
  In addition to the development costs this will include the principal tooling costs for production such as dies, moulds, packaging line equipment, modifications, etc.

Materials costs
  Primary, secondary and tertiary packaging

Packaging machinery and process costs other than one-time costs
  This will include purchase or hire of packaging machinery, costs of writing equipment off in an appropriate period, and labour costs of all kinds

Distribution costs, including insurance.

(c) *Machine efficiency.* This should be output achieved as a percentage of the theoretical maximum output (e.g. if a machine can run at 1000/h but gives an actual output of 800 in 1h, then its efficiency is 80%). Losses are due to many causes, including jams, minor delays, change-over time, operator fatigue and the need to make adjustments.

(d) *Machine speed.* As purchased, this should always be in excess of initial requirements, to allow for growth later.

(e) *Line efficiency.* If two machines, each running at 85% efficiency, are linked in sequence the line efficiency cannot be greater than 72.25% (85% of 85%), unless we arrange to hold a buffer stock between operations, so that if the first machine stops, production can continue. (If the second machine stops, the buffer stock will increase.)

(f) *Single or multipath packaging line.* In a sequence of machine operations, it is desirable for each machine to run slightly faster than the one before it. The required output may be achieved by a single fast machine, or by two or more slower machines in parallel. These alternatives will usually result in different costs, because of differences in capital, labour requirements and efficiency. Remember that if the entire production comes from one machine, a major breakdown stops production completely, whereas a breakdown on one machine in a group of three permits two-thirds of the operation to continue.

(g) *Labour cost.* Manning requirements and rates of pay, including supervision and maintenance, must be considered.

(h) *Inflation.* It is sensible to try to make some allowance for the effects of inflation. It is, however, certain that a machine bought today will cost less than a similar one bought next year. A slight labour saving this year is likely to become a larger labour saving (in money terms) next year because of increased wages. However, if a material saving is made by a pack change this year, the saving may increase or may turn into a loss next year, depending upon the relative price movements of materials, and these cannot be reliably predicted.

These guidelines on the knowledge needed to produce cost-effective packaging underline the great importance that the packaging of a product can have on the success of the business producing it (see also chapter 15).

In our discussions we have assumed that each business has only one product, a situation which is rarely encountered. Most businesses have several products, and to stay up to date it is vital that the packaging of each should be revised at suitable predetermined intervals.

The place of packaging within the marketing complex

We have already seen that packaging plays a vital part in distribution, and is the major factor in ensuring that the quality that is obtained at the end of the production line is maintained until the product reaches the hands of the ultimate customer. To repeat an earlier definition, packaging in product distribution is aimed at maximizing sales (and repeat sales, and so profits), while minimizing the total overall cost of distribution. It can be regarded, therefore, as a benefit to be optimized rather than merely a cost to be minimized. Packaging must be considered in relation to four major factors in industry: materials utilization, machinery and line efficiency, movement in distribution, and management and manpower.

Materials utilization

It has been estimated [4] that about 150 million tonnes of several materials valued at something like 120 billion pounds sterling are used to package all the products consumed by the world's population of 5 billion. Table 1.9 gives the approximate world consumption in 1985 and an estimated figure for the year 2000. The increases in materials usage will be much lower in advanced industrialized countries than in those at present in the process of development and Table 1.10 illustrates this on a per capita basis for four typical stages of development. The differences between the various types of country are

Table 1.9 World consumption of packaging materials (millions tonnes) [4]

Packaging material	1985	2000
Wood	20	21.5
Metal	15	17.5
Glass	35	48
Paper and board	64	107
Plastics	16	53
Total	150	247

Table 1.10 Per capita consumption of packaging materials according to income in the year 2000 (kg/capita per year) [4]

Packaging material	Type of country			
	Advanced industrial	Less advanced industrial	Developing	Underdeveloped
Wood	10	5	3	2
Metal	8	4	3	1
Glass	45	10	5	2
Paper and board	70	45	10	5
Plastics	30	20	6	3
Total	163	84	27	13

considerable and the typical stages are defined as:

(1) An advanced industrial country has a population of 500 million with a per capita income of 10 000 US dollars in 1987.
(2) A less advanced industrial country has a population of 740 million and a per capita income of 5000 US dollars.
(3) A developing country has a population of 2800 million and a per capita income of only 1000 US dollars.
(4) An underdeveloped country has a population of 2164 million and a per capita income of 100 US dollars.

For comparison the average packaging consumption per capita for the world's population in 1987 was 30 kg and that estimated for 2000 is 40 kg. Wood and metal are expected to fall from 4 kg and 3 kg to 3.3 kg and 2.8 kg respectively while the other media rise; glass from 7 to 7.8 kg; paper and board from 12.8 to 17.3 kg; plastics from 3.2 to 8.6 kg.

Thus, we see that, overall, paper and board packaging are the most used materials (approx. 40%) with glass and plastics competing for second place. There are differences between countries at the same stage of development, however, and these are illustrated for some EC countries in Table 1.11.

It should be remembered that in all these tables the figures are in terms of weight of material used and that in terms of numbers of containers or area of wrapping the proportions would be very different. For example a 0.5 litre glass bottle even of the single trip type would weigh 15–20 times a 1.5 litre PET bottle.

Of course the reasons for the variability of material consumption has a lot to do with the resources of the particular country or area. Well-wooded areas use timber- and paper-based indigenous materials while the Middle East areas have few trees and paper-based packaging must be largely imported.

Table 1.11 Per capita consumption of major packaging materials in some EC countries (kg/capita per year) [4]

	Country					
Packaging material	Germany*	France	UK	Netherlands	Denmark	Spain
Metal						
tinplate	7.8	10.4	11.7	17.2	19.5	7.7
aluminium	1.6	0.9	1.0	0.5	1.0	1.0
Glass	45.3	54.0	31.0	41.0	20.0	26.5
Paper and board						
paper and paper bags	5	4	4.2	3.5	11	2.7
carton board	6.6	6.8	9	6	15	6.4
corrugated board	33	32	25	30.6	41	23
Plastics	20	15.8	13	15	18	10

*Figures given relate to what was West Germany.

Table 1.12 Per capita consumption of packaging materials in the USA, Japan and Europe (kg/capita per year) [4]

Packaging material	USA	Japan	Europe	World
Wood	21.4	13.8	11.4	4.0
Metal	15.4	15.2	10.2	3.0
Glass	59.8	18.2	37.5	7.0
Paper and board	104.7	84	44	12.8
Plastics	23.4	20.1	15.1	3.2
Total	230	154	118	30

The final table (Table 1.12) in this section shows that per capita consumption of packaging is greater in the USA than in Japan and Europe and all three exceed the rest of the world by between 3 and 7 times. The scope for improvement is far greater therefore in the developing countries.

Machinery and line efficiency [1]

The ability to run packaging materials efficiently on a packaging line is very much related to the smooth operation of all the mechanisms involved. The packaging material must feed properly into the forming section of the machine. It must then enclose the product adequately and seal it efficiently. It must then be collated and placed into its transport container, and finally into a unit load or into a freight container for despatch, without the production of large quantities of waste. Jam-ups on packaging lines and lack of control not only produce waste packaging material, but at best they require the product which has been inefficiently packed to be packed again, and at worst they may have damaged products to such a degree that they can no longer be sold.

Quality assurance on production and packaging lines is therefore vital and the 1990s will undoubtedly see better control systems produced for packaging lines. This has already started in the field of automatic weighing by means of minicomputers, and microprocessor-controlled systems will be the subject of much development.

Movement in distribution

Once packed, the product may be stored in a warehouse, but ultimately the package leaves the factory for the outside world to meet the mechanical hazards of transport and the climatic difficulties of storage before it reaches a supermarket, an engineering workshop, a factory or a shop somewhere in the world on the way to its ultimate user. The study of the movement of goods and the hazards which exist is therefore a very important part of the packaging scene.

Early studies were concerned with developing tests which reproduce either the hazards of transport, or the effects of those hazards on the pack or the product, enabling an evaluation of packages for particular systems of distribution to be made. They also include the examination of specific products for the protection they need against mechanical and climatic hazards, largely to answer one of three basic questions:

(1) How can a package which does not behave satisfactorily in protecting its contents be improved without a corresponding increase in expenditure?
(2) How can a package which is performing very satisfactorily be reduced in cost without reducing its performance characteristics?
(3) How can a new product which has never been packed before for this market be packaged economically?

The answers to these questions are of considerable significance to every producer of goods, and to all nations so far as export is concerned. The International Standards Organization (ISO) has now produced standards for all the main tests and for test schedules.

Management

Materials, machinery and movement are three vital aspects of the packaging complex, and are welded together by an appreciation in the management area of the importance of packaging in ensuring quality to the customer. All the care and skill exercised in making a particular product can be entirely wasted if the packaging is incorrect in some way—if it fails to protect the product en route, if it fails to sell the product in the marketplace, and if it costs too much to produce.

Management must ensure that the design brief for any product incorporates a requirement to design the packaging as well, and provides a statement of the markets in which the product is to be sold. The designer of the equipment, particularly if it is a mechanically or electrically actuated machine, can then assess how much strength must be built into the product to withstand the hazards of transport and how much the packaging should be relied on to ensure safe delivery. Very frequently a more expensive, better-designed product can lead to lower cost packaging which, overall, will sell the better product at a lower cost to the ultimate user.

Properties and forms of packaging materials

The continued success of any packaging material is dependent in the end on its properties and performance, in relation to cost. Table 1.13 indicates the properties of the main packaging media and their possible advantages and disadvantages. While paper and board are undoubtedly the best source of stiffness and printability at lowest cost, plastics have considerable advantages in many other areas. Glass, of course, has the specific advantage of considerable inertness towards foods, while metal provides the greatest strength. Table 1.14 attempts to show the way these materials can be made into packaging forms. Note that, having both the flexible and the rigid form,

Table 1.13 Comparison of the principal packaging media [6]

Wood
+ Withstands tremendous pressure and keeps its shape
+ Offers a cheap way of shipping oddly shaped heavy equipment
+ Enhances food and beverages in gift boxes
+ Offers good impact strength
− Can only be used for solids
− Is more expensive than other raw materials
− Is inappropriate for high speed packaging

Metal
+ Offers the best barrier properties after glass
+ Can be microwaved when coated or sandwiched between vinyl
+ Provides the tensile strength needed to operate aerosol sprays
+ Can be re-used as a container and offers 'collectible' image
− Limits re-usability (sardines, nuts, cat food)
− Can affect taste of food or beverages

Glass
+ Offers tremendous barrier properties
+ Reinforces consumer security
+ Conveys the 'feel' of crystal, and creates a good impression
+ Provides the impermeability necessary for certain medical uses
− Weighs more than any other packaging material
− Can break in filling, shipping, palletizing, storage or use

Paper
+ Billboards the product
+ Makes aseptic paperboard packaging possible when laminated with plastic
+ Is microwaveable
+ Can contain a variety of geometric shapes
− Degrades quickly
− Offers few barrier properties

Plastic
Injection moulding
+ Gives the widest possible variety of crisp shapes
+ Allows for greater detail
+ Conveys a quality impression and is reusable
+ Can hold spring and check valves, making pump dispensers possible
− Needs high tolerances to function correctly
− Requires larger expenditures in tooling

Thermoforming
+ Offers cost benefits over glass and injection moulding
+ Requires less development time than other methods
+ Allows many packages to be immediately filled at any temperature
− Can make a given item seem inexpensive
− Generates waste in the production process
− Is generally less attractive than injection moulding

Table 1.13 (*Continued*)

Blow moulding
+ Can often be manufactured quickly, with a small amount of material
+ Offers greater versatility in shapes than glass or paperboard
+ Can hold a wide variety of liquids
− Does not allow corners to be crisp

Flexible packaging
+ Offers the advantages of plastic, metal and paper
+ Maximizes barrier properties
+ Increases shelf life over certain other plastics
+ Can be shipped in light weight sheet form
− Involves high start-up manufacturing costs

Table 1.14 Possible forms of packaging in various materials

Form	Material				
	Paperboard	Metal	Plastics	Glass	Wood
(Rect)angular packaging forms					
Boxes	×	×	×	−	×
Trays	×	×	×	−	×
Cartons	×	−	×	−	−
Cases	×	×	×	−	×
Crates	×	×	×	−	×
Pallets	×	×	×	−	×
Cylindrical and non-angular packaging forms					
Bottles	−	−	×	×	−
Jars	−	−	×	×	−
Tubs	×	−	×	−	−
Canisters and cans	×	×	×	−	−
Tubes	×	−	×	× (ampoules)	−
Drums	×	×	× ⎫		×
Casks	−	−	× ⎬ carboys?		×
Kegs	−	×	× ⎭		×

Flexible packaging

	Papers (plain, coated and impregnated)	Plastics films (plain and coated)	Metal foils (plain and coated)	Paper/ plastics/ foil laminates	Textiles
Labels	×	×	×	×	×
Wrappers	×	×	×	×	×
Bags	×	×	−	×	×
Envelopes	×	×	−	×	−
Sachets	×	×	−	×	−
Pouches	×	×	−	×	−
Collapsible tubes	−	×	×	×	−
Sacks	×	×	−	×	×
Bales	−	×	−	×	×

plastics can cover the whole range—no other medium can do this. Plastics also of course have the great advantage that specific properties may be added and combined by using multicomponent structures.

Rigid, angular types of packaging are mostly produced in paperboard, metal, plastics and wood, while the majority of cylindrical packs are of glass or plastics. Glass and timber do not appear in the flexible packaging field, but the combinations of the various flexible materials are considerable. We can see from this that plastics materials are well placed to provide almost every conceivable type of packaging, and in fact much of the packaging development over the past 25 years has been due to the influence of plastics on packaging of all kinds. Metal, glass, paper and board, as well as other materials, are often used in conjunction with plastics. Indeed, there are few packages today that do not contain more than one material in their make-up, but without the plastics component they would be far less effective. In the future the distinctions between the materials will be even further reduced.

Food process classification

Although many attempts at a classification for food processes have been made, until recently none has been completely successful. If such a classification

Table 1.15 A three-level classification of food process technologies

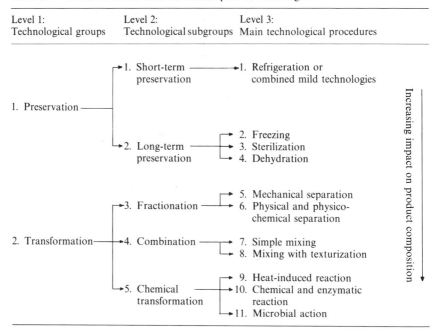

Level 1: Technological groups	Level 2: Technological subgroups	Level 3: Main technological procedures
1. Preservation	1. Short-term preservation	1. Refrigeration or combined mild technologies
	2. Long-term preservation	2. Freezing
		3. Sterilization
		4. Dehydration
2. Transformation	3. Fractionation	5. Mechanical separation
		6. Physical and physico-chemical separation
	4. Combination	7. Simple mixing
		8. Mixing with texturization
	5. Chemical transformation	9. Heat-induced reaction
		10. Chemical and enzymatic reaction
		11. Microbial action

Increasing impact on product composition

existed it could be most useful to packaging and food technologists alike. A most useful paper has recently appeared which may fill the gap. One of the major problems has been the difficulty of uniquely placing a specific food in a particular class. One may, for example, find in the same list products classified according to the raw material used (e.g. dairy products, cereal-based food, fruit derivatives, etc.) and at the same time classified according to the processing technology used (e.g. frozen foods, fermented foods, baked goods, etc). This is confusing; should ice-cream be a frozen food or a milk-based food? Peri's paper [5] outlines a consistent classification with two characteristics:

● No ambiguities—only one correct classification for each item
● A hierarchical ranking of items.

The scheme, which classifies any food product according to both its raw material and its processing, is outlined in Table 1.15 which shows three levels. The first level distinguishes preservation methods (which inhibit or reduce the

Table 1.16 Classification of unit operations

Part I: Fundamental unit operations

Technological procedure (see Table 1.15)	Unit operations
1. Refrigeration, combined mild technologies	Contact, convective or evaporative cooling; modified atmosphere conditioning; pasteurization; irradiation
2. Freezing	Freezing by contact or with cold air or by cold liquids; cryogenic freezing
3. Sterilization	Sterilization in cans; aseptic processes; UHT treatments by steam injection, steam infusion, friction heating, microwave heating; microfiltration; ultrafiltration
4. Dehydration	Freeze-drying; air-drying; foam-mat drying; spray drying; vacuum drying; drum drying
5. Mechanical fractionation	Sedimentation; centrifugation; filtration; ultrafiltration; pressing; sieving; sifting; wet rendering; thickening
6. Physical and physico-chemical fractionation	Liquid–liquid and liquid–solid extraction; sulpercritical fluid extraction; crystallization; winterization; distillation (steam, vacuum, molecular); stripping; flocculation; pervaporation; coagulation; churning; gel-filtration; dialysis
7. Simple mixing	Mixing; blending; kneading; homogenizing
8. Mixing with texturization	Emulsion and foam formation; homogenizing
9. Chemical transformation by heat treatment	Roasting; toasting; cooking; frying, grilling; baking; extrusion-cooking
10. Chemical transformation by chemical or enzymatic reactions	Hydrolysis; oil neutralization; hydrogenation, transesterification; various hydrolases, isomerase reactions
11. Chemical transformation by microbial action	Liquid and solid-state fermentation; alcoholic and acid fermentation; immobilized cell fermentation; cheese and sausage ripening

Continued

Table 1.16   (*Continued*)

Part II: Complementary unit operations

Purpose	Unit operations
1. Size reduction	Milling; grinding; crushing; rolling; homogenizing; cutting; spraying; impact breaking
2. Forming	Forming; moulding; extruding; spinning; agglomeration; briquetting; sintering; tableting; pelleting; granulating
3. Thermal conditioning	Heating; cooling; tempering
4. Physical state conditioning	Melting; thawing; prilling; dissolving
5. Water content conditioning	Evaporation; freeze-concentration; reverse osmosis; hydration; diluting; wetting
6. Separation of minor components and impurities	Peeling; coring; destoning; brushing; washing; rinsing; cleaning; hulling; husking; deboning; adsorption; desorption; degassing; filtration; microfiltration; sieving; sedimentation; cyclones; sifting; flotation; flocculation; clarification; centrifugation; ion-exchange
7. Addition of minor components	Carbonication; coating; dipping
8. Selection	Grading; sorting (by shape, colour, etc.)
9. Conditioning	All operations connected with conditioning and packaging; e.g. bottling; filling; sealing; dosing; corking; labelling
10. Transfer	Transfer of liquids, powders, particulates, solids; pumping; stirring

Table 1.17   A three-level raw material classification

Level 1: Raw-material groups	Level 2: Raw-material subgroups	Level 3: Raw-material species, breeds, varieties
1. Animal	1. Meat	Beef, pig, horse, sheep, chicken, rabbit, game, etc.
	2. Milk	Milk from various species
	3. Seafoods	Shellfish, various fish species, etc.
	4. Other	Honey, eggs, etc.
2. Plant	5. Cereals	Wheat varieties, maize varieties, barley varieties, rice varieties, other species and varieties
	6. Legumes	Various species and varieties
	7. Oil-bearing plants	Oilseed species and varieties, olive varieties
	8. Fruits	Various species and varieties
	9. Vegetables	Various species and varieties
	10. Sugar-bearing plants	Sugarbeet, cane
	11. Tropical plants	Coffee, cocoa, etc.
	12. Officinal plants, spices	Various species and varieties
	13. Microorganisms	Various genera, species and strains
3. Mineral	14. Water	
	15. Salts	

Table 1.18 A synoptic table of food processes and products

		Preservation				Fractionation		Transformation				
			Long-term preservation					Combination	Mixing with texturization reaction	Chemical transformation		
Raw material	Shelf-stable natural products	Short-term preservation	Freezing	Sterilization	Dehydration	Mechanical separation	Physical separation	Simple mixing		Heat-induced reaction	Chemical and enzymic react.	Microbial action
Animal												
Meat		Fresh meat	Frozen meat	Canned meat products	Smoked, dried meats, ham, freeze dried, salt-cured meats	Tallow, lard, mechanically deboned meats	Meat extracts, gelatin		Meat spreads restructured meats	Convenience foods, bologna, cooked sausages	Meat hydrolysates	Dried or semi-dried sausages
Milk		Raw and pasteurized milk		Sterilized, UHT, condensed milk	Whole and skim-milk powder, whey powder	Cream, butteroil	Butter, casein, whey protein, lactose, milk protein co-precipitates	Infant milk preparations, flavoured milk drinks	Ice-cream, sherbets, processed cheese		Lactose hydrolysates, casein hydrolysates	Cheese, cultured dairy products
Seafoods		Fresh fish and shellfish	Frozen fish and crustaceans	Canned fish and shellfish	Smoked, salt-cured dried, freeze-dried fish	Fish oil, fish meal, caviar	Fish oil, fish protein concentrates		Fish spreads	Convenience foods fish sticks	Fish protein hydrolysates	Fish sauces
Other (honey, eggs)	Honey	Fresh and pasteurized eggs, pasteurized honey			Dried, freeze-dried eggs							
Plant												
Cereals Wheat						Wheat flours and by-products, bran, germ	Gluten, starch		Dried and fresh pasta products	Chemically or air-leavened bakery products, cooked-extruded products		Bakery products, yeast or naturally leavened

Continued

Table 1.18 (*Continued*)

Raw material	Shelf-stable natural products	Short-term preservation	Freezing	Sterilization	Dehydration	Mechanical separation	Physical separation	Simple mixing	Mixing with texturization	Heat-induced reaction	Chemical and enzymic react.	Microbial action
				Preservation		Fractionation				Transformation — Chemical transformation		
Maize				Canned sweet corn		Starches, corn syrup, corn germ	Corn oil, meal, zein, corn gluten	Blend and fortified corn products		Popcorn, cooked-extruded products, breakfast cereals, caramel	Dextrines, dextrose, high fructose corn syrups, sorbitol	Lactic acid, mannitol, methyl glucoside
Barley							Malt extract, soluble toasted barley extract			Toasted barley	Malt	Beer, whisky, fermented cereal products, alcohol
Rice and others						White rice, flours		Flour blends, compound feed		Rice puddings, parboiled rice, puffed, flaked rice		
Legumes	Legume seeds		Frozen fresh legume seeds	Canned legumes		Protein concentrates	Guar gum, arabic gum, locust bean gum					
Oilseeds	Oilseeds						Defatted flours, oil, protein concentrates and isolates, lecithin, salad dressing, shortenings	Mixed oils	Margarine, mayonnaise, dressings, coffee creamers	Toasted, salted peanuts and other oilseeds, puffed seeds	Fatty acids, glycerol, hydrogenated oils, shortenings	
Olive				Canned olives	Dried olives	Virgin olive oil	Olive oil					Table olives
Nuts										toasted salted nuts		
Fruits	Fresh fruits		Frozen fruits	Canned fruits and purées	Dried, sun-dried, freeze-dried, dehydro-frozen fruits	Fruit juices, essential oils, aqueous essences, concentrated juices, nectars	Essential oils, pectin, colouring matters, aroma fractions and concentrates	Fruit beverages, soft drinks, liqueurs	Fruit puddings, fruit ice	Jams, jellies marmalades		Fruit wines, ciders, vinegars, distilled spirits, alcohol

Source												
Vegetables		Fresh vegetables	Frozen vegetables	Canned vegetables and purées	Dried, freeze-dried dehydro-frozen vegetables	Tomato juices and concentrates	Leaf protein		Ketchup	Potato chips, french fries, convenience foods		Pickled vegetables, sauerkraut
Sugar-bearing plants							Sugar, molasses	Soft drinks, liqueurs, aromatized syrups		Caramel, confections, candies, nougats, creams, fudge	Inverted sugar	Rum, alcohol
Tropical plants					Dried cocoa and coffee beans, leaves	Cocoa powder, cocoa butter	Instant coffee, tea powders, decaffeined coffee		Chocolate, chocolate spreads	Roasted coffee		
Officinal plants, spices	Aromatic seeds	Fresh spices and aromatic herbs			Whole and ground dried aromatic spices, seasonings		Essential oils, oleoresins, water soluble extracts emulsions					
Micro-organisms			Frozen microbial strains		Dried yeast, algae, freeze-dried starters		Yeast extracts, agar-agar, single-cell protein					
Mineral		Mineral water, rocksalt					Marine salt	Dietetic salts, substitutes				

deterioration of the food) from transformation technologies (which modify the raw materials to produce foods with new or improved functional, nutritional and/or sensory characteristics).

At the second level the preservation methods are divided into short- and long-term while the transformation processes are distinguished under three subheads: *fractionation, combination* and *chemical transformation technologies.* At the third level eleven further technological procedures are delineated according to increasing impact on the composition of the food product; four related to preservation and seven to transformation procedures. These criteria provide a hierarchical order of technologies into which any food product can be classified.

This combines with a classification of fundamental and complementary unit operations (Table 1.16, parts I and II) and a three-level classification of raw materials (Table 1.17) to complete the system. The fundamental unit operations (Table 1.15) give more detail on the methods of preservation such as sterilization, freezing and chemical transformation by heat, enzymes or microorganisms, and the complementary unit operations (Table 1.16) deal with operations such as size reduction, peeling, coring and grading. Together, these allow a synoptic table to be constructed (Table 1.18).

This classification can be used to examine the consistency of the technological choices of a food company and a guide to product innovation.

It will also prove useful in pinpointing the areas where protection by the packaging or by avoiding deteriorating influences in distribution (e.g. by changing the transport mode or the turn round time) can improve the shelf life.

References

1. F.A. Paine, *Packaging Design and Performance*, Pira International (1990).
2. F.A. Paine (ed.), *Fundamentals of Packaging*, revised edition, Institute of Packaging (1981).
3. G.G. Bird and K.J. Parker (eds), *Food and Health*, Elsevier Applied Science, London (1980).
4. M. Pariat, J.P. Pothet and M. Veaux, *J. Packaging Technol. Sci.* II(4) (1989), 183–192.
5. C. Peri, A synoptic table of food processes and products, *Ital. J. Food Sci.* II(2) (1990), 69–78.

2 Graphics and package design

Introduction

When we defined marketing in chapter 1 we did not specify:

- Who identified the customer needs?
- Who satisfied them?
- Who anticipated them?
- Who fixed the profit?

In the case of the producer of food it is the marketing department who must identify the needs of the customer. Anticipation, satisfaction and profit lie in the provinces of R&D, production and finance respectively while motivation is provided by the customer.

Management's role

The critical role of management in any business must be to guide the firm to a better future [1]. This can only be done by identifying the factors which have the greatest influence on the company's competitiveness and developing a strategy which will transform the vision, imagination and skill of the employees into commercial results. The firm's corporate resources must be deployed in the best way and any additional needs to assist development introduced. Management must select the markets or market segments in which success is likely and devise the tactics by which it can be achieved.

The balance in packaging between preserving the food and presenting it attractively must be struck, while compatibility problems with the possible packaging materials and forms must be avoided, commensurate with the food's hygienic needs. The means by which the company objectives may be achieved fall under three headings, marketing, production and finance, each of which needs its own strategy. Thus the overall marketing strategy will be concerned with:

- Advertising
- Promotion
- Public relations
- Sales
- Packaging.

Hence packaging, and in the first instance this means packaging design, both structural and graphic, is one of the most important tools of the marketing mix. It contributes to the positioning of the product in the market [2], its presentation and even, on many occasions, to the use of the product after purchase.

Package design is therefore one of the critical elements in merchandising today. This has not always been so. Fifty years ago, sales of mass consumption products depended far more on their inherent quality than they do today, and the differences in quality between different brands of the same kind of product were much greater. Consumers could taste, see or feel the difference, and they would pay more for a high quality product than for one of low quality.

Today, in the sophisticated areas of the world, the situation is different. A common technology has emerged and differences in quality between many mass consumption products are no longer apparent to the average consumer. The battle for consumer preference now hinges not just on quality and price, but on other elements of 'the marketing mix', such as advertising, promotion and packaging. When quality and price become equal they cancel themselves out, and the decisive factors determining the sale are often these other elements. With the rise of self-service merchandising, a product on the supermarket shelf must sell itself. The advertising messages that were absorbed before entering the store will often strongly influence choice; but at the moment of truth, when the consumer reaches for a particular product, it is the packaging that often tips the scale.

Thus, packaging's role is far greater than providing protection or making the product easier to use. It is also more than an attention-getting device. A good package can express and create the impression and ideas about a product (the 'image') that the manufacturer wishes to implant in the consumer's mind, to make it seem different and better than competing products.

Packaging and modern merchandising

A good way to gain an understanding of the role of packaging in modern merchandising is to look back at the development of retailing and see how packaging's role has changed in response to the changes in shops and in the market situation.

Until the nineteenth century there was little or no prepackaging of consumer goods. Food and other household and personal items were delivered in bulk, to a small shop or a stand in the market. Each batch of goods would be directly scrutinized by the consumers. If consumers liked what they saw (or had no choice), they bought it. If the decision was influenced by factors other than the visible quality of the product, they were probably factors such as the personality of the merchant and the location of the store.

Then, in the last years of the nineteenth century, prepackaged foods, soaps, household articles and other items began to appear on retailers' shelves. At the same time, the concept of brand names was developing, as manufacturers saw that the only way to win consumers' confidence in their products would be by building up a positive personality by advertising their virtues, convincing consumers that their products were as good as those of the old-fashioned products sold in bulk. The recognition of the importance of brand identity was reflected in the design of the early consumer product packages. They were heavily 'brand-related'. They placed the main emphasis on the name of the producing company, and often on the man who owned it, using his name or even his picture. The number and variety of packaged products then proliferated, and by 1920 it was not unusual in the industrialized markets to find shops with about half their goods packaged.

Manufacturers also found that they could sell only so much of a particular product. To expand sales, they began introducing several varieties of the same basic product—different flavours of chocolate bars, for example. Manufacturers also diversified into new lines. The result: promotion could no longer focus almost exclusively on the brand; it had to talk about the different varieties sold under the brand name. The new need was reflected in package designs. Packaging became 'product-related' as opposed to brand-related. The development of product-related packaging was helped by advances in printing. When colour lithography was used more extensively in packaging about 1930, coloured pictures of the product appeared, increasing the emphasis on what the consumer could expect to find inside the package.

The third major phase in the history of packaging was a direct result of the development of the self-service shop and the supermarket after the 1939–1945 war. Now the packages were on their own. The salesman who could talk to the consumer about the products had disappeared; the package had to sell the products. Customer selection may be partly predetermined by advertising before the purchaser enters the store or it may be on impulse and the price will also play a big part, but even so visual attraction will always be an important consideration.

One result of this was that packages had to become more informative. This, and the development of labels with actual colour photographs on them, reinforced the product illustration trend that was already under way. Before long the modern supermarket emerged—vast emporiums with row after row of products, where as many as 4000 different items vied for the consumer's attention. To win it, package designers began using bright colours and bold designs. Visual impact became vital. This concern was intensified by the introduction of the so-called 'private brands'—the chain stores' own brands. Because these were generally not advertised outside the stores, the package became especially important as a device for making the consumer aware of the brand.

The modern concept of packaging is based on the understanding that what really interests the consumer is not the brand, or the ingredients, or even the product itself. Essentially, the consumer is interested in the benefits he thinks he will get from using the product. Thus, a package of baby food might be more interesting to the potential purchaser—a mother who wants her child to be healthy—if it carried a picture of a baby bursting with good health, rather than a picture of the grain and milk from which the food was made, or a bowl of the prepared food. Finding out what benefits consumers are looking for, and then supplying them, has become a central element in marketing strategy for mass-consumption products. Package design has become a major vehicle for implementing that strategy.

While the history of packaging as a communication medium shows that it has progressed from concentration on brand, then on product, and finally on the consumer, all three of these elements are still components of modern package design, or at least of the thinking that goes into that design. In the simplest terms, the three elements express *who sells what to whom* [3]. Much of the change that has taken place in package design is in the sophistication with which these components are elaborated and used.

If we examine several modern packages, all successful in the market, we may well find that one focuses attention on the brand, another on a consumer benefit, and another on the product. The relative weight of these elements in any particular package is—or should be—the result of a careful analysis of the market situation and a precise definition of marketing goals. Even though the creation of any good marketing strategy must be consumer-related, the actual package that is designed as a result of that strategy may not (overtly) show a consumer benefit at all. The 'benefit' idea may have been implanted in the consumer's mind through other media, such as television advertising, so the marketing men may decide that the package should highlight the brand. Any package should be created and analysed in the context of the total marketing mix. If it is not, a package that looks good may not be any use at all as a salesman for the product.

In recent years the concept of packaging being 'convenient' has become somewhat less dominant in the food field and 'fresh, healthy' food has become more important. While convenience, easy-open packaging and easy-to-prepare foods are still in demand, the requirement for the fresh and healthy concept has led to growth in microwaveable modified atmosphere packaging (MAP), sous vide and ready meals.

Meeting customer and consumer needs

First we must note that the final customer (the consumer) and the user are not always the same. A product manufacturer uses packaging to contain the product that he wants to deliver to a customer who uses the product.

Since packaging is an activity practised by almost every industry and is universally used to deliver every type of product to the final user or customer it might be worthwhile trying to classify the 'users' of packaging [2].

There are at least four kinds of 'user' of packaging:

(1) Makers and packers of products
(2) Distributors, carriers and warehousemen
(3) Retailers of all kinds of product
(4) Consumers.

Obviously the detailed cUStomER requirements will be very dependent on the product but there are certain points which must be considered for every product and end-user. Every end-user wants a quality product delivered in a package adequate to maintain that quality for the necessary life of the product, at a price which is economic and reasonable in relation to any possible alternatives. In almost every instance it is the product alone that the customer wants; the packaging, with the notable exceptions of products like cosmetics and gifts, is an ancillary part of the transaction. In addition, while customers will recognize that the packaging must be adequate, if it appears to be over elaborate it may well create the impression that the cost of the product has been unnecessarily raised by the packaging and this will have a negative effect on sales. All packages must be easy and safe to handle, simple to open and use, and provide few problems in their disposal.

These considerations will influence factors such as the weight, size and shape of both the primary packaging and that used for transport and storage.

However, many of the factors influencing consumer requirements will vary according to their age, sex, income and social level. The location of their home will also influence their ideas as to the appropriateness of a package for any product. They will have different colour preferences and associations: may want different sizes and quantities and sometimes want special considerations such as tamper evidence, child resistance or anti-pilfering devices built in.

Beware the half truth and consider the alternative

A recent report which claimed to have polled more than a thousand shoppers about the level of service they received in high street shopping centres made the following comments:

When asked '*Are you satisfied with the service you are getting in shops?*', over a quarter of shoppers said they were dissatisfied; 35% complained that they had to wait too long at check-out desks and 29% felt that they had great difficulty in getting help from sales assistants.

This all sounds very negative but it could be put another way using the same data:

When asked '*Are you satisfied with the service in shops?*' almost three-quarters of shoppers said they were satisfied; 65% felt that the time to pay at check-out was reasonable and over 70% appreciated the help they got from sales assistants.

That sounds almost encouraging. Probably the truth lies somewhere in between. It is the old story of whether the bucket is half-full or half-empty; or whether you are a pessimist or an optimist. However, the facts indicate that improvement is needed in any event, in two quite different ways. First, more efficient check-out arrangements, currently being tackled by bar code recognition and billing coupled with more efficient staff training, and second by providing better and more information about products both to assistants in the shop and on the package.

The latter may need to be improved by better labelling of items like foods, etc. since it is almost impossible in self-service supermarkets for the staff to cope completely. The changes that are occurring in our retailing practices need to note these facts in spite of the somewhat biased attitudes of certain reports.

Consumers not only have expectations about their purchases but they also have expressed their views on what they require of packaging at governmental level. The Packaging Group of the UK Waste Management Council set out a UK Code for the packaging of consumer goods in the mid-1970s (see Table 2.1).

The italics in the preamble have been added to bring the document more up to date. These principles could be made into more precise requirements by a second level code in which the methods by which packaging could be evaluated for compliance were set out. This has never been done and is the

Table 2.1   UK code for the packaging of consumer goods

The prime function of packaging is to enable consumers to receive products (*particularly foods*) in good condition at the lowest reasonable price. Any manufacturer, distributor or retailer concerned with the design or use of packaging has a responsibility to ensure that there is a regular review of packaging having regard to the economics of the total manufacturing/distribution chain and to consideration of reuse (*recycle*) and disposal. Marketing and commercial considerations should be reconciled as far as possible with economy in the use of materials and energy, and the environment.

1. Packaging must comply with all legal requirements
2. In containing a product the package must be designed to use materials as economically as practicable, while at the same time having regard to protection, preservation and the presentation of the product
3. Packaging must adequately protect the contents under the normal foreseeable conditions of distribution and retailing and in the home
4. The package must be constructed of materials that have no adverse effect on the contents
5. The package must not contain any unnecessary void volume nor mislead as to the amount, character or nature of the product it contains
6. The package should be convenient for the consumer to handle and use. Opening (and reclosure where required) should either be obvious or indicated, convenient and appropriate for the particular product and its use
7. All relevant information about the product should be presented concisely and clearly on the package
8. The package should be designed with due regard to its possible effect on the environment, its ultimate disposal, and to possible reuse and/or recycling where possible.

Table 2.2 Examples of lightweighting [4]

1. Cans for beans in tomato sauce: in 1970 cans weighed 68.9 g
   in 1980 "   "   58.4 g
   in 1990 "   "   56.6 g
2. Glass bottles for milk (in the UK): pre-1939 a 1 pint bottle weighed 538 g
   in 1950 "   "   "   "   397 g
   in 1960 "   "   "   "   340 g
   in 1990 "   "   "   "   245 g
   A movement into other lighter, less costly materials, such as coated paperboard and plastics films, has taken place.
3. Cans for beer: in 1950 a standard beer can weighed 91 g and was made entirely from tinplate, in 1985 using aluminium with tinplate ends the weight was 20 g and in 1990 it was 17 g.
4. Even the recent 1.5 litre PET bottle has been reduced in weight from 66 g to 42 g by using a design which eliminates the (originally necessary) base cup.

reason why the eight statements remain as rather trite expressions of intention.

Clause 1 is legislative in nature, clauses 5, 6 and 7 are concerned with convenience and information for consumers, clauses 2, 3 and 4 outline the protective and preservative character of packaging, and clause 8 the possible environmental effects.

The protective aspects of packaging are specifically designed to reduce and prevent waste by preserving perishables and preventing damage. For example, in countries where there are good packaging systems, food losses between the sources of supply (farms, plantations, lakes, seas, etc.) and the consumer average about $2\frac{1}{2}\%$. In areas such as the less developed countries where packaging techniques, for whatever reason, are poor, losses of 25–40% of the total harvest are not uncommon.

Clause 2 of the Code makes a statement that has dominated the package making industries since the late nineteenth century: lightweighting should be implemented to reduce material usage and costs. Packaging, with the possible exceptions of the gift and cosmetics markets, is not something the product maker wants to sell, nor something that the customer wants to buy; they both need it only for the utilitarian purposes of protection in transit, convenience and information. Consequently all sides work for an optimum cost to achieve the objectives required by the particular market. Thus, lightweighting, first within the initial packaging material and later by switching to an alternative medium, is the normal method of progress (see Table 2.2).

Trends

Another important way of anticipating customer needs is by studying trends in particular fields. It is necessary to be able to recognize a genuine trend if we are going to use it as a possible means of motivation. A trend can be defined in this sense as 'a positive movement in a particular direction with long-term potential'; if there is only a short-term potential it is not a trend, merely a novelty and will not last. Typical trends developing over the last half-century were (are still?):

- Growth of self-service
- Growth of the DIY movement on the light engineering trade
- Development of the bar code check-out system
- Changing methods of selling retail products
- Increase in the number of married women working.

The ability to recognize these trends and the developments stemming from them has made great changes in methods of retailing and the packaging needed.

Summary of consumer needs

Perhaps at this point we might record what consumers appear to need in food product terms and try to relate these to packaging. From a general point of view consumers want:

1. Food and drinks that are safe, nutritious, do not make them ill and are not damaged in distribution
2. Food that, when kept in the larder, refrigerator or freezer as appropriate, does not 'go bad'
3. 'Fresh' food to be available all year round
4. Convenience foods that require little or no preparation other than heating
5. Packaging which is just right for the job and does not mislead.

Obviously price will also be of interest but the warning given by Theo Melis of Unilever [5] should be kept in mind: 'The bitterness of poor quality remains long after the sweetness of low price has been forgotten.'

The function of packaging graphics [6]

Packaging graphic design provides appeal (for presentation and display) and also information related to directions for use (warnings, shelf life, batch number, destination, storage conditions in the warehouse, transit and handling). Printing is the principal means of fulfilling these functions; materials may be preprinted, or printed or over-printed during the product packaging operation. A check list for pack graphic design is given in Table 2.3.

Where the essential function is one of graphic appeal, effective design depends on a number of factors, of which the main ones are:

- A knowledge of the product
- A knowledge of the packaging materials to be employed (functional and

Table 2.3 Pack graphic design: a check list (courtesy Siebert/Head Ltd., London) [6].

Design objectives	New design
	Redesign/updating:
	design continuity/analysis of existing elements
	Range integration/extension
Consumer requirements	Size/size impression
	Appropriateness of pack for product
	Importance of package appearance in use
	Preferred/associated colours
	Usage instructions
Retail display requirements	Shelf sizes
	Estimated display life
	Position of display:
	shelf
	counter
	dispenser
	check-out
	window
	Position in relation to purchaser:
	eye level
	above/below
	Display by brand group/product group
	Panels most often seen:
	front/back/top/bottom/sides
Wholesale display requirements	Need for coding/simplification on outer packaging
(including cash and carry)	Problems of cut cases
	Degree of branding necessary
Production requirements	Number of colour/varnish
	Size of print area
	Method of reproduction
	Materials
	Quantity
	Size run
	Cost limitations
Special requirements	Additional future copy
	Foreign/multilanguage copy
	Price flashes
	Promotional flashes
	Special requirements for back, sides, top or bottom
	Need to tie in with advertising/point of sale
	Inclusion of bar code area
Legal requirements	Cautionary information to be included
	Legislation to be observed:
	appropriate designation style and size
	weight statement and form
	ingredients copy, size and order
	maker's name and address
	illustrations
	special claims
	hazard warning requirements
	inclusion of 'e' symbol
Restraints	Materials
	Inks
	Pack forms

structural detail, such as dimensions and tolerances of the item, and area to which decoration may be applied)
- A knowledge of the methods of distribution and selling
- A knowledge of the printing methods which could be employed
- The quantities required per order and per year.

The main printing processes

A basic knowledge of the main printing processes is essential in packaging, since there is little packaging completely devoid of printing, and such knowledge of the characteristics of the various processes helps considerably in understanding their effective application in package design.

In all types of printing, ink has to be applied selectively to certain areas of the paperboard, film, foil, tinplate, or other packaging material. This ink is first applied to the printing unit, which may be a cylinder, a plate, or a collection of individual printing elements, and is then transferred to the material by direct contact. There are several methods employed to effect this selective application of ink.

Letterpress

In this method of printing, ink is applied to the raised printing agent (Figure 2.1)—type, line, solids or halftone plates—contained in a forme. The ink is then transferred to the paper, board or other substrates by impressing the inked forme against it. Various machines are employed which run at a range of press speeds. Platen machines (Figure 2.2) for example, in which the forme and substrate are impressed together flat, with overall contact, may be of the hand-fed type running at about 800 impressions per hour. Auto-platens, with automatic feeding, are capable of 3000 impressions per hour.

In the modern flatbed machine (Figure 2.3) the bed reciprocates, and on the forward stroke the printing elements, either individual type or blocks, pass under a series of inking rollers before being carried under an impression cylinder around which the paper to be printed is wrapped. The paper is squeezed between cylinder and 'forme', and the ink is transferred from the printing areas to the paper. On the return stroke of the flatbed, the cylinder, which rotates continuously, is raised to clear it. These print at speeds from

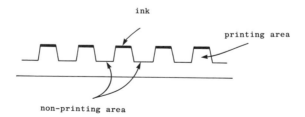

Figure 2.1   Relief process.

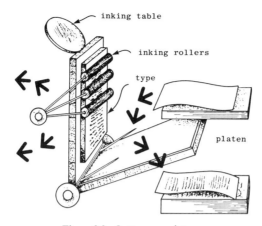

Figure 2.2 Letterpress platen.

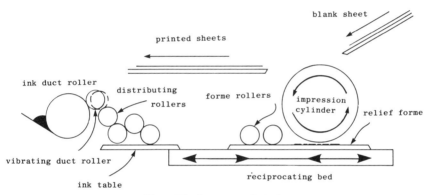

Figure 2.3 Letterpress flatbed.

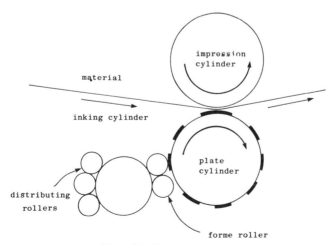

Figure 2.4 Letterpress rotary.

about 1200 impressions per hour (if they are on large machines) to about 4500 impressions per hour on smaller high-speed two-revolution machines.

Rotary sheet-fed machines (Figure 2.4) are capable of even higher speeds of up to 6000 impressions per hour. The forme on these presses is a curved plate and the substrate is printing at the nip between printing plate and impression cylinder. Web-fed rotary presses have been in use mainly in the newspaper field, but also for reel-fed packaging work, particularly on carton board. Offset letterpress printing is also used, where the ink is transferred from a curved letterpress relief plate to a rubber blanket on an offset cylinder, and thence to the paper at the nip between blanket and impression cylinder. Most of the machines used for this purpose are offset litho presses with sufficient rebate on the plate cylinder to take a thin relief plate. The process is particularly suitable for the printing of line and solid work on carton board, and in label printing.

Original letterpress plates may be copper, zinc, magnesium or plastics, such as photo-polymers (Dycril, Nyloprint). Duplicate plates may be electrotypes, which are sometimes chromium-faced, and stereotypes, often nickel-faced or rubber or plastics, such as PVC. The fact that only the printing areas apply pressure to the paper in the printing operation results in a very sharp outline to the print and slightly more ink at the edges, which gives the print a degree of crispness. Where these characteristics are important, e.g. in advertising, letterpress is dominant, but it is also very widely used in carton, wrapper, and label printing.

Flexography

The old name for flexography was aniline printing, which derived from the use of inks consisting of alcohol solutions of aniline dyestuffs. Flexography nowadays generally means printing material on the reel using flexible plates and evaporation drying inks.

The process is simple (Figure 2.5) and akin to web rotary letterpress printing

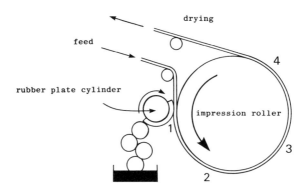

Figure 2.5 Flexographic process. 1, 2, 3, 4: colour stations.

in that the flexible plate has raised image areas and is linked by a roller rotating in the duct containing a thin ink which is transferred to the web of the substrate at the nip between printing and impression cylinders. It is dried by heated air as the web passes under a hood. There is some penetration of the ink into the stock if this is absorbent. The amount of impression should be minimal, for, as in letterpress printing, squash occurs at the edge of the image. The image is usually line and solid, although coarse halftone screens have been printed satisfactorily. The process has been improved so that better solids and finer detail can be printed by use of an engraved 'Anilox' roller, which may also have a doctor blade, to transfer ink from duct to printing plate. Flexography is used for printing paper (particularly kraft bags) and for film and foil.

Sheet-fed flexography is employed in printing flat lined corrugated cases at machine speeds of 12 000 impressions per hour, using kick feeders. In this type of work moisture-set or water-reducible inks may be used, although recently interest has been shown in the use of quick-setting oleoresinous inks to achieve higher gloss and rub resistance.

Lithography

As the name implies, stone was originally employed for this process, the printing image being drawn directly thereon, but litho printing is now done almost entirely from metal plates.

The inventor of the process, Alois Senefelder, discovered towards the end of the eighteenth century that a certain type of porous stone would pick up a greasy ink when dry, but if the surface of the stone was damp it would no longer accept the ink (Figure 2.6). If, therefore, the image to be printed was drawn on the dry stone with a greasy crayon and then the stone was damped, an inking roller passed over the surface would deposit ink on the crayoned areas (the printing areas) but not on the damped non-printing areas. The image could then be transferred to paper and the printing areas recharged with ink for the next print.

Early litho machines operated on the flatbed principle and were as slow as letterpress flatbeds. The stones reciprocated under damping rollers, inking rollers, and an impression cylinder. The offset principle, almost universal

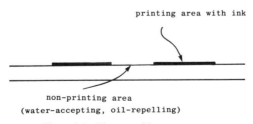

Figure 2.6 Planographic process.

today, was devised by the tinplate printers, who found that their stones were being damaged by the rough edges of the tinplate sheets and it was difficult to get even contact between the stone and the tinplate. They overcame this trouble by transferring the ink image on the stone on to a rubber blanket wrapped round a cylinder, and then retransferring this image to a tinplate sheet by passing the sheet between the blanket cylinder and an impression cylinder. Although offset, this was still flatbed and therefore slow.

It was not until the discovery that certain metal surfaces could be treated to be either water-repellent (hydrophobic) or to wet readily (hydrophilic) to produce a continuous film of water which is oil-repellent, that the flatbed principle could be replaced by faster rotary offset (Figure 2.7), using a flexible metal plate on one cylinder, a rubber blanket on the second cylinder, and an impression cylinder with a suitable resilient surface.

Suitable metals for litho plates include zinc and aluminium but a more logical, though expensive, surface comprises one metal (usually copper) for the printing areas and a different one (chromium or stainless steel) for the non-printing areas. In this way one can select metals which have inherent preferences for grease (printing areas) and water (non-printing areas) instead of taking an 'indecisive' metal such as zinc and modifying its properties as required.

It should be noted that 'offset' does not necessarily mean litho, or vice versa, but in practice nearly all litho printing is done offset, and there are few letterpress offset machines in operation. Offset printing gives a thinner printed ink film than direct letterpress (or litho) and needs inks of higher tinctorial strength to give the required effect, but the introduction of the resilient blanket

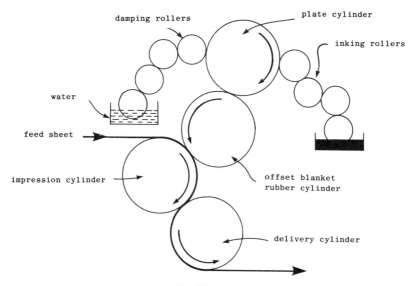

Figure 2.7 Offset lithography.

between plate and paper enables a good print to be obtained on a comparatively rough surface.

Lithography lacks the flexibility of letterpress, since the printing surface is prepared as one plate by chemical and photographic processes, and it depends for uniform results on the correct balance between ink and water being maintained by the printer. To enable the non-printing areas to retain sufficient water to repel the ink on the inking rollers, the plate surface is grained and this tends to reduce the sharpness of the image. Nevertheless, there is little to choose between the processes as far as the finished result is concerned and many jobs are done equally well by either.

Sheet-feed litho presses run at speeds of 3000–7000 impressions per hour. This printing process is widely used for the printing of cartons and labels in packaging, as well as for the printing of books, brochures, magazines, etc. The offset principle allows fairly rough stocks to be printed with fine halftone illustrations. On non-porous materials, such as foil, litho inks dry entirely by oxidation/polymerization; in tin-plate printing, they dry by passage of the printed tin-plate through an oven.

Gravure

This process came into prominence with the increasing use of films and foil, especially for packaging. Gravure combines the use of a volatile ink, suitable for non-porous surfaces, with the ability to reproduce accurate tone values, which flexography cannot yet do regularly (Figure 2.8). It began as a means of printing wallpapers in the late eighteenth century and has always been rotary, either from a cylinder whose surface is recessed to produce the printing areas, or a thin copper plate suitably etched and then wrapped around a cylinder.

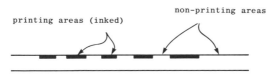

Figure 2.8 Intaglio.

The entire surface is flooded with ink, either by dipping into a reservoir or spraying, and then as it rotates against a flexible blade called a 'doctor', the peripheral surface (non-printing area) is wiped clean, leaving ink only in the recessed printing areas. The material to be printed is passed between this doctored cylinder surface and a resilient impression cylinder, when most of the ink in the recesses is lifted out and transferred to the paper (Figure 2.9).

Gravure cylinders are expensive to make, but can produce a large number of prints at high speed, and are therefore used on longer runs. The rapid drying of the ink enables subsequent operations such as waxing, varnishing, laminating, or cutting and creasing, to be performed as an 'in-line' operation on a composite machine, and the completed work can be re-reeled immediately

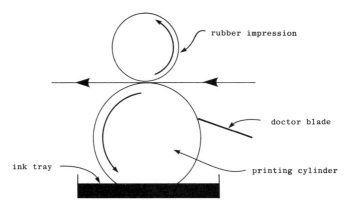

Figure 2.9   Photogravure.

afterwards. Gravure has many advantages in that the process allows in-line varnishing and other finishing operations, such as waxing, coating, lamination, cutting and creasing to be done on reels.

Sheet-fed gravure machines are used in packaging work, particularly where dense colours and metallic inks are being printed on carton work, in which case some of the printing may be done by litho, and other special effects (such as gold printing) may be done by gravure instead of the usual bronzing process.

Silkscreen

Screen printing is a stencil process (Figure 2.10) of increasing importance, in which fluid ink is pushed through the stencil, which is supported on a nylon or wire screen, using a rubber squeegee (Figure 2.11). It has evolved from poster printing on machine-glazed paper, to printing on containers of various shapes and composition, from steel drums to polythene bottles for domestic usage and for much point-of-sale display work.

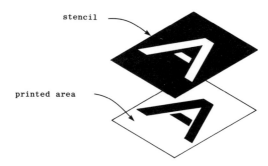

Figure 2.10   Stencilling.

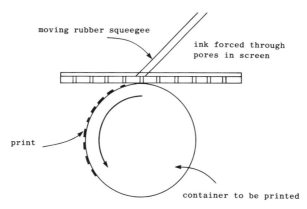

Figure 2.11 Screen process.

Ink-jet printing [7]

Ink-jet printing is a versatile non-contact technique which has grown dramatically over the last decade or so. Because there is no contact between the print head and the substrate and a good selection of inks is available, quality results can be achieved on practically any material e.g. glass, metal, plastics, paper and board, shrink-wraps and even on food products themselves.

In continuous-ink-jet (CIJ) printing, as the ink is forced out of a nozzle, it breaks up into a stream of electrically charged droplets which are deflected by an electric field as they pass through the print head (see Figure 2.12). The amount of deflection and the movement of the material being printed produces a dot matrix pattern from which characters and numbers, etc. can be made. CIJ printing is very fast and can print batch numbers, best before and date codes at speeds enabling it to keep up with the fastest bottling and canning lines.

Valve-jet printing is simpler. Here the characters are formed from a row of nozzles, one for each drop in a column of the dot matrix. The ink drops are released only when the print head receives a signal, the timing, sequence and duration of which are electronically controlled. This method can handle the printing of large characters up to 75 mm high at speeds up to 100 m per minute. Typically it is used for coding shipping cases, trays and shrink-wrapped packs.

Hot die stamping and gold blocking

The print dies used in this process are also raised, as the process is allied to relief, although there is no ink receptive area. In this process (Figure 2.13), a carrier (usually a plastic film like polyester) is coated with a foil or pigment, which under *pressure* and *heat* can be transferred to another surface. This process has certain advantages (no drying or surface preparation, quick for

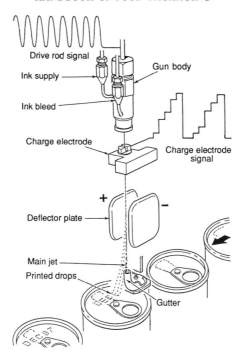

Figure 2.12 The principle of ink-jet printing.

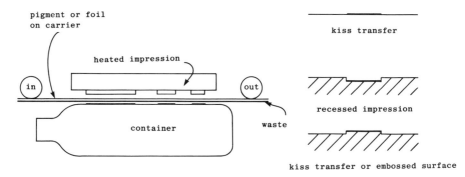

Figure 2.13 Hot die stamping and gold blocking.

colour changes, no cleaning down). It can print on raised level or recessed surfaces. The plates may be silicone rubber or metal. Soft plates provide for a kiss transfer, whereas metal plates can be used to give a distinct impression on to which the pigment or foil is deposited. The metallic foil consists of a carrier, release coating, lacquer, a metallized layer deposited by vacuum and a hot melt type adhesive specially for the material being decorated.

The carrier may be glassine, cellulose film, or polyester film. The choice of carrier relates to speed and transfer temperature. Polyester (e.g. Melinex) is the most common base. The lacquer components protect the print surface and hot melt gives adhesion, the heat being transferred via the printing die.

Factors affecting the choice of a printing process

In the 1950s, the types of printing suitable for the various processes were clearly defined, so that there was little doubt about which process should be used to give the best results on any particular job. For example, letterpress was used for bold striking designs for advertising, whereas litho was the natural choice for water-colour reproductions and pastel shades. Improvements to the processes, however, have so extended the scope of each, that there is a considerable degree of overlap and the design no longer determines which process will be used.

Other important factors influencing the choice are:

1. Material to be printed—metal, glass, paper, board or plastic
2. Quality of print required
3. Quantity to be printed
4. Form in which printed material is required
5. Special requirements such as potential toxicity, rub resistance and freedom from odour.

If the material to be printed has a smooth non-porous surface, the choice of process usually lies between flexography or gravure, since these employ evaporation-drying. Precipitation-drying is not feasible because the solvent cannot soak away after it has precipitated the resin and pigment, and oxidation-drying inks would be very prone to set off between the smooth surfaces during the lengthy drying period.

If high quality print is required on these surfaces, gravure would be preferred to flexography although the latter can now be very good. If the surfaces are smooth but porous, high quality print would necessitate letterpress, litho or gravure, but if the surfaces are rough and porous, the offset principle embodied in litho machines would give a better print than direct impression.

The length of run is an important factor influencing the choice of process used. For very short runs indeed (which is rarely the case with package printing) silk screen is the cheapest, and applicable to a variety of surfaces. Short runs, especially those with a preponderance of type matter, favour flexography, to avoid the cost of producing litho plates or gravure cylinders.

For medium-length runs litho, flexography or letterpress would be suitable, but, as the run increases, litho is preferred owing to the higher running speed of the rotary litho machine. Gravure merits consideration at this stage.

Beyond this, flexography and gravure become increasingly economic due to high running speeds.

If the customer requires the printed material in reel form, then evaporation-drying is needed. The pressure built up in a reel of printed work will almost certainly produce set-off if slower (oxidation) drying inks are used as in litho printing. Moisture-set inks applied by letterpress are a possibility here if conditions met subsequently in the packing machines do not preclude their use because of their lower degree of rub resistance.

The size of the design which is to be printed may also influence the choice of process. The production of letterpress original plates is expensive, but duplicate plates, stereotype, electrotype or plastic are comparatively cheap. The cost of a litho plate will depend on whether it consists of a single large design or a multiplicity of identical small designs.

Where special requirements such as rub resistance or light-fastness are concerned, the properties of the ink film laid down by the various processes must be considered. Offset litho prints a thinner layer than direct letterpress or gravure, and this thinner layer should be more rub-resistant. On the other hand, a litho ink requires to be stronger tinctorially because of its thinness, and to print a strong colour the proportion of varnish may not be high enough to ensure rub resistance.

There are many stages involved in the production of a pack, from the initial assessment of its functional requirements, construction, and surface design, through the printing, cutting to shape, forming and making-up of the completed object; but the visual appeal of the pack depends to a great extent on the surface decoration, and printing is an important part of pack production. However, the printing operation cannot be considered as a separate operation and an end in itself; it must be integrated with other operations and suited to functional requirements and production conditions.

References

1. E.R. Corner, private communication.
2. F.A. Paine, *Package Design and Performance*, Pira (1990).
3. D. Judd, B. Alders and T. Melis, *The Silent Salesman*, Octagram Books, Singapore (1989).
4. Data from members of INCPEN, London.
5. T. Melis, World of exteriors, in *Food Packaging Technology International No. 1*, F.A. Paine (ed.), Cornhill Publications, London (1989), pp. 58–67.
6. F.A. Paine (ed.), *Fundamentals of Packaging*, revised edition, Institute of Packaging (1981), p. 99.
7. L. English, *Ink-Jet Printing*, Domino (UK) Ltd., Bar Hill, Cambridge CB3 8IU, UK.

3 Notes on packaging materials

Paper-based packaging

What is wood?

The outer and inner layers of bark in a tree conceal a layer which contains the plant's food, and it is by means of this layer that the tree itself grows. The main woody part of the tree consists of bundles of cellulose fibres running vertically up the trunk, held together by a material called *lignin*.

The woody part of the tree consists of about 50% cellulose fibres, 30% lignin, 16% carbohydrates and some 4% of other materials such as proteins, resins and fats. It is principally the cellulose which is eventually made into paper. This is composed of individual fibres which are finer than human hairs and are a few millimetres in length at the most. These fibres are about 100 times as long as they are thick. Lignin is the chemically complex substance which holds the fibres together. It is useful, perhaps, to think of it as the glue holding the tree in one piece.

Pulping

To be useful to the papermaker, the raw material must be reduced to a fibrous state. This operation is called pulping, and there are two basic methods: *mechanical* and *chemical* pulping. In both processes, the bark is stripped from logs cut to a suitable length at the appropriate stage in their growth. Logs for mechanical pulping may be used directly in 1.2 m (4ft) lengths or, alternatively, they may be chipped, i.e. converted into pieces of uniform size about 15–20 mm long.

Two methods of mechanical pulping are employed. In one, the logs are pressed against the surface of a large revolving grindstone, kept wet by a stream of water which also removes the fibres. In the other system, the wood chips are passed between the two plates of a disc refiner with specially treated surfaces, very close together and rotating at high speed. In this way the wood chips are reduced to individual fibres, but the water-soluble impurities only are removed, and most of the lignin still remains; many fibre bundles and some damaged fibres are also left in the pulp. Much grinder and disc-refined wood pulp is used for newsprint, although substantial quantities are employed as a mixture with chemical pulp for making certain kinds of board. Mechanical pulp is normally made from softwood, typically spruce.

Chemical pulping starts from chips, and removes all materials other than the cellulose fibres by chemical action and solution. The chemicals convert the lignin to a soluble form which is removed by washing. This produces cellulose fibres of a higher purity than those produced by the mechanical processes— they are generally much less damaged and the fibre bundles are fewer. Several different chemical pulping processes are used, and the quality of the pulp depends upon the process, as well as the kind of wood. For packaging purposes, three chemical processes are of major importance. These are the 'kraft process' which retains most strength in the fibres, the so-called 'sulphite' process which is less strong, and the 'semi-chemical' process.

The kraft process.   In the kraft or sulphate process, wood chips are digested in a solution of caustic soda and sodium sulphate for some hours, which dissolves out the lignin, leaving the cellulose fibres to be washed. The name comes from the Swedish word 'kraft', meaning strength. In its early days, kraft paper was always associated first with a brown colour and second with long fibres and what was called a 'wild look through'. Because the fibres were long, they did not form a uniform sheet and, when held up to the light, the paper had an uneven density.

Sulphite pulps.   The sulphite process uses sulphur dioxide and calcium bisulphite, which are mixed with the chips in aqueous solution and heated to about 140°C. Once again the lignin is dissolved out, leaving the fibres, and after digestion the mass is washed with water and then bleached with another chemical, such as calcium hypochlorite, before pressing into pulp sheets. This gives a very pure cellulose fibre, although the resulting pulp is not as strong as that from the kraft process.

Semi-chemical processing.   In the semi-chemical process the wood chips, which are usually from beech or birch trees, are treated partly by chemicals and partly mechanically to reduce them to fibres, hence the name semi-chemical. This semi-chemical pulp is often used for the manufacture of the fluting medium for corrugated board.

Beating

Once the pulp has been produced, it may have to be bleached to make it white, or coloured, or treated in other ways. One of the most important processes in the pre-preparation of fibres for paper making is the so-called beating process. The object here is to rub and brush the individual fibres, and cause them to split down their length in such a way as to produce a mass of thin fibrils which will enable them to hold together in the matted paper more strongly. This process is called *fibrillation*. The greater the degree of fibrillation we can induce, the higher the strength of the paper can be. Different pulps respond

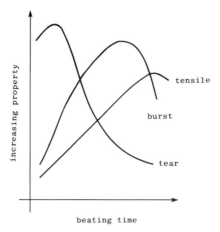

Figure 3.1   Effect of beating on strength properties.

differently to this treatment. Softwood fibres will fibrillate to a greater extent than hardwood fibres, and hence softwoods are potentially able to produce stronger papers. Beating also cuts the fibres and this is undesirable, for although some cutting may be necessary for good running on the machine, the strength of the paper is reduced. Furthermore, beating breaks up any fibre clumps and refines them into individual cellulose fibres. The art of beating packaging pulps is to maintain a high proportion of fibrillation and a low proportion of cutting in the pulp, to give the desired properties in one finished paper. Figure 3.1 gives an idea of the effect of beating on paper strength characteristics in respect of burst, tensile and tear strength.

Paper testing

Paper is still one of the most widely used packaging materials, and is found both as a wrapping material and in converted form, such as bags and sacks. The main types of packaging papers are listed in Table 3.1. In either case the paper may be modified by treatment or by coating in order to tailor the properties. The main functional requirements may be modified by the circumstances of use and the effects of conversion. For example, the sharp creases made during the manufacture of self-opening satchel (SOS) bags produce weak spots, and so two plies of thinner paper or other material, other than a single sheet, are used to reduce this effect.

It is essential to identify the properties which are important in the use of the material and the effect of variations in those properties. Variability can cause difficulties in packaging and handling operations, and poor performance in use. When these essential properties have been identified and acceptable limits of variability determined, they may be incorporated into buying specifications.

Table 3.1 Main packaging papers

Basic material	How made?	Weight range		Tensile strength		Properties and uses
		lb/1000 ft²	kg/1000 m²	lb/in width	kg/m	
Kraft papers	From sulphate pulp on softwoods (e.g. spruce)	14–60	70–300	MD[a] 14–65 CD[a] 7–30	250–1150 125–535	Heavy-duty paper, bleached, natural or coloured; may be wet-strengthened or made water-repellent. Used for bags, multiwall sacks and liners for corrugated board. Bleached varieties for food packaging where strength required
Sulphite papers	Usually bleached and generally made from mixture of softwood and hardwood	7–60	35–300	Very variable		Clean bright paper of excellent printing nature used for smaller bags, pouches, envelopes, waxed papers, labels and for foil laminating, etc.
Greaseproof papers	From heavily beaten pulp	14–30	70–150	MD 10–25 CD 5–12	180–450 90–215	Grease-resistant for baked goods, industrial parts protected by greases, and fatty foods
Glassine	Similar to greaseproof but super-calendered	8–30	40–150	MD 8–30 CD 5–16	140–535 90–285	Oil and grease-resistant, odour barrier for lining bags, boxes, etc., for soaps, bandages and greasy goods
Vegetable parchment	Treatment of unsized paper with concentrated sulphuric acid	12–75	60–370	12–80	215–1450	Non-toxic, high wet strength, grease and oil-resistant for wet and greasy food, e.g. butter, fats, fish, meat, etc.
Tissue	Lightweight paper from most pulps	4–10	20–50	Low strength		Lightweight, soft wrapping for silverware, jewellery, flowers, hosiery, etc.

[a]MD, machine direction; CD, cross direction.

Routine testing of incoming materials will detect any widespread deviation from specification and smaller deviations on a long-term basis, but in-batch variations can only be detected with certainty if large-scale sampling is employed. This is usually uneconomic. Reel materials also present difficulties in sampling, since these can only be taken from the outer layers of the reel. Where possible, and certainly if there is any question of litigation, the sampling procedure given in BS 3430 should be adhered to. All paper testing should be carried out in standardized conditions, and samples should be allowed to reach equilibrium prior to testing (this normally takes 24 h). The conditions normally used are 23°C, 50% r.h.

A list of the most frequently used tests is given in Table 3.2. In all instances the standard method is outlined and a reference given. It is possible to carry out simple tests to get an indication of the order of some properties, but results obtained from these should only be used as a guide, and the full test procedure used to confirm the results.

Types of paperboard

Terminology. The line of demarcation between paper and paperboard is rather loose. ISO define paper with a grammage above 250 g/m^2 as paperboard, or simply as board. However, in some parts of the world the dividing line is placed at 300 g/m^2. There are many anomalies: e.g. blotting *papers* are often heavier than 300 g/m^2 and fluting medium and some liner *boards* used for corrugated are less than 250 g/m^2.

Structure and general properties. Typical paperboards consist mainly of cellulose fibres of which the most common source is mechanical and chemical pulps derived chiefly from wood. Material recovered from used papers and boards (secondary fibres) are also widely used in the cheaper grades of material. Straw and esparto grass are also used for some boards. The structure of typical boards can vary from a single homogeneous layer or two to eight plies, each of which may or may not have the same fibre composition depending on the strength, stiffness, optical, bending or other properties required for the finished board. Typical structures for white lined folding box boards for cartons are shown in Figures 3.2, 3.3 and 3.4. The top ply in a typical board is of bleached chemical wood pulp which will give the necessary surface strength and printability. An underliner of a lower grade (essentially white pulp) will prevent show-through of the grey/brown colour of the middle plies of secondary fibre. The back may be similar to the middles but where extra strength or printability is required, a better grade of chemical/ mechanical pulp mixed in the appropriate ratio, should be used. It is very important that all the plies are well bonded and this is dependent on achieving the right degree of mechanical entanglement and hydrogen bonding in the boardmaking process. Table 3.3 gives some details and uses of specific types of paperboard.

Table 3.2   Selected test methods

	Test, etc.	Material	Reference[a]
Pre-test procedures	Sampling	Paper	BS 3430
		Board	
	Conditioning	Paper	BS 3431
		Board	ISO/R 187
	See also:		
	Laboratory humidity ovens		
	Injection type		BS 3718
	Non-injection type		BS 3898
	Laboratories, controlled atmosphere		BS 4194
Tests for mass and density	Grammage	Paper	BS 3432
		Board	FEFCO 2 and 10
	Mass	Wax on paper	BS 4685
Tests for strength properties	Adhesive strength	Adhesive and adherends	BS 847
	Burst strength	Paper and board	BS 3137
		Board	FEFCO 4
		Paper (wet)	BS 2922
	Folding endurance	Paper	BS 4419
	Heat-seal strength	Plastics/laminates	PIFA 2/74
	Ply-bond strength	Paper	TAPPI
			RC 364
	Puncture strength	Paper, board	BS 4816
	Stiffness	Paper	BS 3748
	Ring stiffness	Paper	ASTM D-1164
	Tear strength		
	Initial	Paper	ASTM D 827
	Internal	Paper	BS 4468
	Tensile strength and stretch	Paper	BS 4415
Test for surface absorption and permeability properties	Abrasion (see rub-proofness)		
	Absorption of water by	Paper, board	BS 2644
	Absorption of wax by	Paper	ASTM D 668
	Friction coefficient = 1/slip	Paper	TAPPI T 503
	Moisture content	Paper	BS 3433
	Permeability	Paper to air	BPBIF p. 13
	(see also water		BS 2925
	vapour)	Sheet materials	ISO 2556
	transmission	to gas	BS 2782-514A
	Resistance to grease	Paper	ASTM D 722
	Resistance to oil	Paper	BPBIF RMT 1
	Roughness	Paper	BS 4420
	Rubproofness	Paper	BS 3110
	Water vapour transmission	Sheet materials	BS 3177

Table 3.2 (*Continued*)

	Test, etc.	Material	Reference[a]
Analytical tests	Acidity/alkalinity (pH)	Paper	BS 2924
	Ash	Paper	BS 3641
	Chlorides	Paper	BS 2924
	Contraries copper, iron	Paper	BS 1820
	Sulphates, reducible	Paper	BS 1820
Dimensional tests	Curl	Paper	BPBIF RM 4.2
	Dimensional stability	Paper	BPBIF RM 53
	Thickness	Paper	BS 3983
		Board	BS 4817
Optical tests	Brightness Whiteness Reflectance Opacity Gloss	Paper	BS 4432
	Light fastness	Paper	BS 4321
Miscellaneous materials tests	Crease quality	Board	BS 4818
	Odour and taint	Packing materials	BS 3755
	Resistance to blocking	Paper	ASTM D-918

[a]The methods referred to are taken from:

ISO/R   International Standards Recommendations;

BS   British Standards. In some instances the number of the standard is followed by a further reference, e.g. BS 2782.509 is Method 509 in BS 2782;

ASTM   American Society of Testing and Materials Standards;

TAPPI   Technical Association for the Pulp and Paper Industry (USA) Standard. Where TAPPI Standards have been developed by ASTM, the latter reference has been used;

FEFCO   Fédération Européenne des Fabricants de Carton Ondule Test Methods;

BPBIF   Technical Section of the British Paper and Board Industries Federation Test Methods;

PIFA   Packaging and Industrial Films Association Standards;

BPBMA   British Paper and Board Manufacturers' Association.

Copies of all British Standards specifications may be obtained from The British Standards Institute, 2 Park Street, London WIA 2BS.

Table 3.3 Types of paperboard

| Board type | Liner | Construction | | | Appearance | Uses |
		Underliner	Middle plies	Back		
Unlined chip	All plies of repulped waste chemical/mechanical pulp mixtures				Grey flecked. Waste from households, used containers, offices, newspapers, etc.	Rigid and folding boxes for soaps, detergents, hardware, electrical goods, boots, shoes, also showcards, stationery, tubes and bookbinding
Brown lined chip	Dyed waste pulp	Repulped waste chemical/mechanical mixture			Brown/grey body, sometimes known as British strawboard	Some rigid box use in hardware, electrical, shoe trades; also showcards and bookbinding
Kraft lined chip	Unbleached chemical pulp	Repulped waste chemical/mechanical mixture			Brown/grey body, more uniform surface than BLC	As above with emphasis on show card use for all markets
Cream lined chip	Semi-bleached chemical + mechanical pulp	Repulped waste chemical/mechanical mixture			Cream/grey body	Used for folding box use in food, pharmaceutical and clothing industries
No. 2 white lined chip	Bleached chemical pulp	Repulped waste chemical/mechanical mixture			Off-white liner with grey of body showing through	One of the most popular grades of board for economical folding cartons; used in cartons for

Type				Appearance	Uses
No. 1 white lined chip	Bleached chemical pulp	Semi-bleached chemical + mechanical	Repulped waste	White liner/cream underliner/grey body	breakfast cereals, detergents, clothing, including shirts and hosiery, toilet tissues, some foodstuffs, hardwares, toys and games
White lined Manila (triplex)	Bleached chemical pulp	Semi-bleached chemical + mechanical	Repulped waste · Semi-bleached chemical + mechanical	White/cream/grey/cream Back may be all mechanical	Some cheese cartons and other foods, pharmaceuticals
Duplex	Bleached chemical pulp	Semi-bleached chemical + mechanical or all mechanical	Increased chemical over middle possible	White/cream body known as white lined folding boxboard	Folding cartons for cigarettes, frozen foods and other foods, pharmaceuticals, cosmetics, toiletries, biscuits, cakes, etc.
Double white lined	Bleached chemical pulp	Semi-bleached chemical + mechanical	Bleached chemical	White/cream/white smooth on both sides	Coated for high finish for greater printability and superior quality
Solid white		All plies of fully bleached chemical pulp		White	Frozen and speciality foods, especially liquids, cosmetics, pharmaceuticals
Kraft liner	90% + unbleached kraft			Brown both sides	⎫ Mainly for corrugated boards
Test liner	90% unbleached or bleached kraft	Repulped waste from corrugated boards		White or brown top side, brown-grey under side	⎭

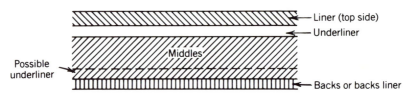

Figure 3.2 Structure of a typical multi-ply paperboard.

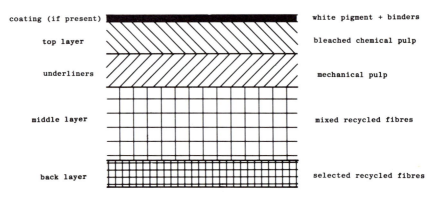

Figure 3.3 Ply construction of white lined chipboard.

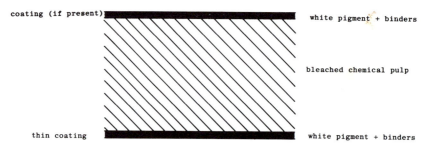

Figure 3.4 Construction of fully bleached solid board.

Plastics

The use of plastics in packaging has increased markedly over the last few decades, particularly for foods and drinks. In 1990 the consumption of plastics worldwide was estimated at 85 Mt. By 2000 AD it is expected to rise to 100 Mt. The reasons for this are:

(a) Lower costs than other materials
(b) Lower energy content
(c) Wide range of properties

(d) More scope in forming and shape
(e) Light weight coupled with strength
(f) Easier disposal after use.

More than 30 different plastics are used in packaging but the most common ones are polyolefins, polyvinyls and polyesters. They may be divided into *thermosetting* and *thermoplastic* resins.

Thermosets

There are only three thermosets used to any extent in packaging. Phenol formaldehyde and urea formaldehyde are used mainly for bottle closures, while glass-reinforced polyesters are used for large containers.

Phenol formaldehyde resins are dark brown or black in colour and rather brittle. Various fillings such as chopped fabric or wood are added to improve the impact strength or to reduce costs. *Urea formaldehyde resins* can be obtained in white or pastel colours, but are more expensive; cellulose is used as the filler because of its whiteness.

Both materials have good chemical resistance and are widely used for closures. They are insoluble in organic solvents, are attacked by strong acids and alkalis but are resistant to weak acids and alkalis. Urea formaldehyde resins are particularly favoured for closures by the cosmetics industry because of the wide colour range available and their resistance to oils and solvents. Phenol formaldehyde is widely used for pharmaceutical closures because it is more resistant to water.

Glass fibre-reinforced polyesters have high strength to weight ratios and good resistance to outdoor weathering. In general, they have good chemical resistance, and resistance to solvents. They are resistant to most organic and inorganic acids, except strong oxidizing acids, and weak alkalis. They are, however, hydrolysed by strong alkalis. They have been extensively used for storage tanks and for large transit containers.

Thermoplastics

Low density polyethylene (LDPE) accounts for the biggest proportion of the plastics used in packaging. One of the reasons for its widespread use is its versatility. It can be extruded into film, blown into bottles, injection moulded into closures and dispensers of all sorts, extruded as a coating on paper, aluminium foil or cellulose film, and made into large tanks and other containers by rotational casting.

Low density polyethylene is relatively inert chemically and almost insoluble in all solvents at room temperature. Some softening and swelling can occur, with hydrocarbons and chlorinated hydrocarbons. Permeability is low for water vapour but many organic vapours and essential oils pass rapidly through low density polyethylene. Its permeability to oxygen is fairly high so where oxidation is likely to be a problem, low density polyethylene is not suitable.

Linear low density polyethylene (LLDPE) is generally stronger and tougher than LDPE but has similar properties.

High density polyethylene (HDPE) has a higher softening point than low density polyethylene and is harder. Its barrier properties are also superior to those of low density polyethylene. For an equal wall thickness, high density polyethylene gives bottles a greater rigidity than low density polyethylene. Alternatively, equivalent rigidity can be obtained at lower wall thickness and often at lower cost in spite of the fact that high density polyethylene is the more costly on a weight basis.

Polypropylene (PP) is similar chemically to low density polyethylene and high density polyethylene. It is harder than either, however, and has a less waxy feel. It can be injection moulded, blow moulded and extruded into film and sheet. The sheet can be thermoformed to give thin-walled trays of excellent stiffness. Polypropylene has excellent grease resistance and is also more resistant to solvents than low density polyethylene. Toluene and xylene, however, will cause swelling. Polypropylene is not subject to stress cracking and it differs in this respect from both the polyethylenes. Its softening point is higher than both polyethylenes, but it is still easily able to withstand steam sterilization.

One outstanding property of polypropylene is its resistance to fatigue when flexed. This means that integral hinges can be formed in an injection-moulded article, which will stand up to a very large number of flexings. Laboratory tests have, in fact, proved that even a million flexings will not cause failure of a well-designed hinge. In packaging, the obvious way of utilizing this property is to mould a box and lid in one shot. The absence of subsequent assembly processes can often lead to substantial cost savings.

Although polypropylene is a rigid polymer, it is more resilient than polystyrene. The resiliency of polypropylene has been found useful in the design of screw-on closures, for it permits the moulding of slight undercuts into which can be snapped decorative inserts, ideal for the closures of cosmetic containers. In a less resilient material, such undercuts would prevent the moulding being removed from the mould without damaging the undercut. Another consequence of polypropylene's resilience is the possibility of designing a linerless closure. A thin sectioned diaphragm or fin is moulded on to the inner surface of the closure in such a position that it bears down on the upper surface of the bottle neck. A rigid material would not 'give' enough to take up inequalities in the glass surface and so would not form a good seal. On the other hand, a more flexible material such as low density polyethylene would 'give', but would not press back strongly enough to form a seal. Linerless closures not only reduce direct costs, by eliminating the wad, but they can also reduce labour and inventory costs. Polypropylene has a low impact strength at low temperatures, but copolymers are available in which low temperature impact strength is improved. The improvement is sufficient to enable polypropylene copolymers to be used for the injection moulding of beer crates and soft-drinks crates.

Ionomers describes a family of polymers in which there are ionic forces between the polymer chains, as well as the usual covalent bonds between the atoms in each chain. Although these interchain forces are strong, they are not sufficient to hold the molecules together when the polymer is heated, and ionomers are still thermoplastics and not thermosets.

The first commercial ionomer was 'Surlyn' A, a polymer of ethylene. It is similar in many of its properties to polyethylene, but because of the ionic interchain forces it has a high melt strength and therefore excellent drawing characteristics. Ionomers can be used in extrusion coating, and very thin coatings can be obtained. Skin packaging is another field where the high melt strength of 'Surlyn' A is useful. Chemically it is resistant to weak and strong alkalis, but suffers attack by acids. Hydrocarbons cause swelling but it resists attack by ketones and alcohols.

The presence of ionic forces between chains also modifies the crystalline structure of the material and 'Surlyn' A is more transparent than low density polyethylene. The polar nature of the material also means that printing is easier.

TPX is also a polyolefin and belongs to the same family as polyethylene and polypropylene. Like them it is resistant to acids and alkalis as well as to many solvents. It is softened by hydrocarbons and is subject to stress cracking in the same way as polyethylene. One difference between TPX and polyethylene and polypropylene is its clarity, which is almost as good as perspex. The softening points of the polyolefins range from low density polyethylene which is below the boiling point of water, through high density polyethylene and polypropylene, to TPX as the highest. Its specific gravity is the lowest of the four materials, being only 0.83.

The impact strength of TPX is better than that of polystyrene but below that of polypropylene. Permeability of TPX to gases and water vapour is higher than that of either polyethylene or polypropylene. It can be injection moulded, blow moulded and extruded into sheet. Thermoforming of sheet is difficult, however, because TPX has a narrow melting point range.

Polyvinyl chloride (PVC) has good gas barrier properties and is a moderate barrier to moisture vapour. It can be blown into bottles, although this is a relatively recent development because PVC has a softening point only a little below the temperature at which it begins to degrade. Chemically, it is resistant to weak or strong acids and alkalis. It is soluble in esters and ketones and is attacked by aromatic hydrocarbons. PVC has excellent oil and grease resistance and this led to the first wide-scale use of PVC bottles for salad oil in France.

Polyvinylidene chloride (PVdC) copolymer is a copolymer of vinylidene chloride with vinyl chloride used either for film or as a coating. Its outstanding property is its low permeability to water vapour and gases. As a shrinkable film, polyvinylidene chloride has been used for wrapping poultry, hams and similar items and for the in-store wrapping of cheese. It can be heat-sealed but it is not particularly heat stable when heated for any length of time over about 60°C.

Polyvinyl alcohol (PVA) is unusual for a plastic in being soluble in water. It is utilized, therefore, in the manufacture of film sachets used to give controlled dosage in water. The sachet plus contents is simply added to the required amount of water, the sachet dissolves and the contents are released. This is particularly valuable where the contents are toxic or where there are other reasons for not touching them by hand.

Ethylene vinyl acetate copolymer (EVA) is a polymer with the flexibility of PVC, but this flexibility is inherent and no plasticizers are necessary. It has a greater resilience than PVC and a greater flexibility than low density polyethylene. This makes it particularly suitable for snap-on caps. Permeability to water vapour and to gases is higher than for low density polyethylene and the solvent resistance is lower. Stress cracking resistance, however, is good.

EVA has a high impact strength down to quite low temperatures. EVA film has a greater tendency to blocking than low density polyethylene, so that a higher percentage of anti-blocking additives is necessary. It can be heat-sealed or high-frequency welded, but for the latter greater power is needed than with PVC.

Polystyrene (PS) is a colourless, transparent thermoplastic, hard and with a fairly high tensile strength. It softens at about 90–95°C and is intrinsically brittle. It is resistant to strong acids and alkalis and is insoluble in aliphatic hydrocarbons and the lower alcohols, but is soluble in esters, aromatic hydrocarbons, higher alcohols, ketones and chlorinated hydrocarbons. It is a poor barrier to moisture vapour.

The brittleness of polystyrene has already been mentioned. However, toughened or high-impact grades are available. The higher impact strength is achieved by blending synthetic rubbers, usually styrene-butadiene or polybutadiene, with the polystyrene, either chemically or mechanically. These high impact strength grades no longer possess the clarity of basic polystyrene, but the chemical properties are almost unchanged.

Acrylonitrile-butadiene-styrene (ABS) is similar in many respects to toughened polystyrene. There is some overlapping in impact strength between the very high impact grades of toughened polystyrene and the lower end of the ABS spectrum. ABS is more expensive than polystyrene. It is used for trays and tote boxes, especially for those of large area where rigidity and a minimum of warping are required.

Polycarbonate has a high impact strength, a high softening point, and the merit of clarity. It is resistant to weak acids and alkalis, but is slowly attacked by strong ones. Polycarbonate is soluble in aromatic and chlorinated hydrocarbons but is insoluble in the paraffins.

Cellulose acetate (CAc) is sensitive to moisture and is not dimensionally stable. Its tensile strength is about the same as that of polystyrene, but its impact strength is slightly better. Mechanical properties generally, however, are altered by moisture pick-up. It is only slightly affected by weak acids and alkalis, but strong ones cause decomposition. It is soluble in ketones, esters and alcohols. Like polystyrene, cellulose acetate has a high clarity and is a poor barrier to moisture vapour. It is used extensively in the manufacture of windows for cartons, because it is easily stuck to the board, giving good seals even at high speeds.

Acrylic multipolymer (XT polymer) has been suggested in the USA as a bottle blowing material suitable for food and pharmaceuticals. Impact strength is moderate and dependent on bottle shape and manufacturing conditions. Oil and grease resistance is high, as is resistance to acids, alkalis, detergents and aliphatic hydrocarbons. Resistance to aromatic and chlorinated hydrocarbons is poor, however, and products containing high concentrations of alcohol should also be avoided. Gas and odour permeabilities are low, but water vapour permeability is higher than that of polyethylene or PVC. Bottles made from XT polymer have good contact clarity when filled but are somewhat hazy when empty.

Lopac is the trade name for a material made by Monsanto Chemical Co. Ltd., and is another copolymer with an acrylic base. The main monomer is methacrylonitrile, with small percentages of styrene and methylstyrene. It is still in the development stage as a possible material for packaging carbonated soft drinks. It is a hard, rather brittle polymer but has excellent barrier properties, good resistance to creep, and is very clear.

Barex is the trade name for a material manufactured by Vistron Division of Standard Oil of Ohio. It consists mainly of acrylonitrile copolymerized with methylacrylate, together with a small percentage of a butadiene/acrylonitrile rubber. This is also a clear polymer with good creep and barrier properties and good impact strength. Again it was developed as a bottle blowing material for carbonated drinks.

Nylons (polyamides) were first prepared by the condensation of di-acids with di-amines and were characterized by a number derived from the number of carbon atoms in the parent compounds. Thus nylon 6,6 is the condensation product of adipic acid and hexamethylene diamine, both of which have six carbon atoms in the molecule. Later, methods were developed for the manufacture of nylons by the condensation of certain amino acids. Nylons prepared by this route are characterized by a single number derived from the number of carbon atoms in the parent amino acid.

In general, nylons are tough materials with high tensile strength and good resistance to abrasion. They also have high softening points and can withstand steam sterilization (up to about 140°C) and dry heat to even higher temperatures. They retain their flexibility at low temperatures so that they have a wide temperature range of use. Nylons are slightly hygroscopic and their mechanical properties are altered somewhat by water absorption. The effect is not permanent and the properties are recovered on drying.

Nylons have fairly high moisture vapour permeabilities but are very good gas barriers, and nylon films are thus used in laminates for vacuum packaging. Nylons are also good barriers to odour. Chemically, nylons are resistant to weak acids but are attacked by concentrated mineral acids. They are resistant to alkalis, even at high concentrations, and are particularly resistant to organic solvents, oils and greases.

The transparency of nylon films is excellent, especially when biaxially oriented, but their gloss is only fair although this, too, is improved by orientation.

Polyesters

The *polyethylene terephthalates* (PET) are undoubtedly the most important of these materials. They can be used in film form for boil-in-the-bag and other applications, but must be orientated to develop the full tensile strength. They are not easily heat-sealable and are therefore often laminated to polyethylene film for bag-making purposes. Since PET was first used in the late 1970s to produce a clear lightweight shatter-resistant beverage bottle, it has probably grown faster than any other plastic for this use.

There are two kinds of PET extrusion and thermoforming products, an amorphous variety (APET) and the partially crystalline polyethylene tere-phthalate (CPET). APET is a clear transparent sheet while CPET is opaque. The former is used for bottles etc. and the latter is used in trays for microwave use. There is a wide range of materials under this heading and most manufac-turers have their own particular varieties.

Polytetrafluorethylene (PTFE) is smooth and waxy to the touch, has a very low coefficient of friction and excellent non-stick properties. It is a very tough plastic with a wide temperature-of-use range ($-100°$ to $+200°C$). It is extremely inert chemically, being resistant to almost all chemicals.

Polytrifluorochloroethylene (PTFCE). A copolymer of this material is used in film form with the commercial name of Aclar. It has the lowest water vapour permeability of any polymer film and is also a good barrier to gases. It retains its flexibility down to temperatures of around $-195°C$ and has a softening point between $185°C$ and $205°C$ according to grade and crystallinity.

Polyvinylfluoride (PVF) has excellent resistance to solvents, acids and alkalis and can even be boiled in strong acids and alkalis without losing its strength. It is unaffected by boiling in carbon tetrachloride, acetone, benzene and MEK for 2 h and is impermeable to oils and greases. It is strong, flexible and is extremely resistant to failure by flexing. The water vapour permeability of PVF film is low, as is its permeability to gases and to most organic vapours.

Laminates with PVC can be vacuum-formed, and PVF film has been used in packaging, giving close conformity to the shape of the product.

Table 3.4 gives some properties of typical plastics used in packaging.

Flexible packaging materials based on films and foils

Flexible packaging materials provide an alternative solution to the distri-bution of many types of goods for which crush protection is not important.

Table 3.4 Some properties of plastic materials commonly used in packaging

Plastics material	Density (kg/m³)	Water absorption (24 h) (%)	Water vapour transmission rate (38°C, 90% r.h.) (g/25 μm per m² d)	Oxygen transmission rate (23/25°C, 50% r.h.) (cm³/25 μm/m² d. atoms)	Printability	Transparency	Resistance to sunlight (outdoors)
Acrylonitrile butadiene styrene	1010–1100	0.2–0.45	—	780–1100	Excellent	Poor	Poor
Acrylics (polymethyl) methacrylate	1100–1200	0.1–0.4	—	3000	Excellent	Excellent	Excellent
Cellulose acetate	1220–1340	1.7–7.0	155–630	1800–2400	Excellent	Excellent	Excellent
Cellulose acetate butyrate	1150–1220	0.9–2.2	470–630	9400–16000	Excellent	Good	Good
Polyamides	1010–1190	0.3–2.8	63–340	40–1400	Good	Fair–good	Fair–good
Polycarbonate	1200	0.15	172	4500	Excellent	Excellent	Good
LDPE/LLDPE	900–930	0.01	16–24	7100–7800	Good	Poor–fair	Fair–good
HDPE	945–965	0.01	4.7	2100–2900	Good	Poor	Poor–fair
Polypropylene (homopolymer)	900–910	0.01–0.03	11	2400–3800	Poor	Fair	Poor
Polypropylene (copolymer)	890–910	0.03	—	—	Good	Fair–good	Poor–fair
Polyvinyl chloride (unplasticized)	1350–1600	0.04–0.4	14–80	80–300	Excellent	Good	Excellent
Polyvinyl chloride	1160–1400	0.15–0.75	80–500	80–9000	Excellent	Fair–good	Fair–good
Polystyrene (unmodified)	1040–1070	0.01–0.03	110–160	3900–5500	Excellent	Excellent	Fair–good
Polystyrene (toughened)	1030–1070	0.05–0.07	120	2700	Excellent	Poor	Fair–good
Styrene acrylonitrile	1060–1080	0.15–0.25	—	—	Excellent	Excellent	Fair
Polyethylene terephthalate	1340–1390	0.1–0.2	16–20	47–94	Good	Excellent	Excellent
Polyethylene vinyl alcohol copolymer	1120–1210	Very hygroscopic	24–120	0.2–1.6 (0% r.h.) 13–23 (100% r.h.)	Good	Good	Good
Polyacrylonitrile copolymer	1150	0.28	60–80	12	Good	Excellent	
Polyvinylidene chloride copolymers	1640–1740	0.1	0.3–3	0.5–9	Good	Good	Poor
Phenol formaldehyde	1240–2000	0.03–1.2	—	—	Fair	Poor	Fair
Melamine formaldehyde	1470–1520	0.1–0.8	—	—	Good	Poor	Good
Urea formaldehyde	1470–1520	0.4–0.8	—	—	Good	Poor	Good

Table 3.5 Flexible packaging materials: possible components

Plastic films	Non-plastic webs	Coatings and adhesives
Polyethylenes	Papers	Cellulose esters
Polypropylenes	Paper-like webs of mixed	Cellulose ethers
Polyvinylchlorides	cellulose and plastics	Rubber hydrochloride
Polyvinylidenechlorides	Papers made of plastics	Chlorinated rubbers
Polyvinyl acetates	Bonded fibre fabrics	Chlorinated polyolefins
Polyvinyl alcohols	Cloths and scrims	Natural and synthetic rubbers
Polyesters	Spun bonded fabrics	Natural and synthetic waxes
Polycarbonates	Regenerated cellulose films	Natural and synthetic
Polyurethanes	Aluminium and steel foils	bitumens and asphalts
Polystyrenes		Natural and synthetic resins
Polyallomers		Adhesives of all types
Phenoxies		Prime, key, bond or sub coats
Ethylene vinyl acetate		Latex bound mineral coatings
copolymers		Deposited metal layers
Ethylene ethyl acrylate		
copolymers		
Fluoro and chloro-fluoro		
hydrocarbon polymers		
Ionomeric copolymers		
Vinyl copolymers		
Block and graft copolymers		

They function primarily in retaining the goods, separating them from their environment whilst identifying and displaying them to advantage. They are used as containers for liquids, pastes, granules and solids, e.g. strip packs, sachets, bags and sacks; wrappers or liners for packs of other materials, e.g. twist wraps, shrink wraps, stretch wraps, parcels, box liners; and labels and closures, e.g. diaphragm lids on tins, bottles, plastics containers. Flexible packaging materials are used essentially for high-volume machine-packed goods where the forming and closing of the containers is part of the operation of the filling line. The paper sack, the squeeze bottle and the plastics collapsible tube are also examples of the use of flexible materials in packaging. Flexible packaging materials are produced as webs from one or more of a great number of possible starting materials [1] (Tables 3.5, 3.6). The method of manufacture of the base materials, additives, and the way the components are assembled, all affect the final properties.

Materials tests on flexible packaging

Factors influencing the selection of a test method. The choice of a test method depends on (1) the *objectives* (is the test required for comparison or checking of materials?); (2) the *relevance to performance* (test properties should be correlated to actual performance in use); and (3) the *precision* required. The more precise and statistically valid the results required, the more complex the test equipment and its calibration, the better qualified the test staff and

Table 3.6 Flexible packaging materials: assembly of components (adapted from [2] by courtesy of the author)

Possible forms in which components layers enter the combining operation	Webs, films (continuous or in sheet form)
	Solvent solutions
	Water solutions
	Emulsions
	Dispersions
	Plastisols
	Hot melts
	Resin pellets or powders
	Vapours
	Fibres
	Encapsulated products
Possible methods of assembly of component layers	
(1) Coating	Roller
	Trailing blade
	Air knife
	Rod
	Extrusion
	Coextrusion
	Curtain
	Dipping
	Brush
	Electrostatic movement
	Magnetic jump
	High vacuum deposition
	Powder fusion
	Spray
	In-situ polymerization
	Polymerization from gas phase
(2) Methods of joining	Hot melt setting
	Adhesive setting
	Heat welding or bonding
	Coextrusion
	Mechanical interpenetration

supervisory staff must be; the greater the number of replicate tests, the stricter the sampling procedure, and hence the more expensive the tests. It therefore follows that, whether we are comparing material or checking materials against a specification, we must decide on the order of difference that will matter. On this depends the precision with which the test must be done and hence the cost of the test. Where the order of difference is large, the 'test' may be no more than a simple visual inspection. To find a small difference, repeated tests using accurate equipment and statistical techniques may be required.

Availability of standard methods. When comparing materials precisely, the method used for testing each must be the same, and for this reason standard methods are used. Any divergences from standard procedures should be recorded. When including tests in a specification, the method should be one for

Table 3.7 Selected test references [3]

Test, etc.	Material	Reference[a]
Pre-test procedures		
Sampling	Paper, board	BS 3430
Conditioning	Paper	BS 3431
	Board	ISO R-187
See also:		
Laboratory humidity ovens		
Injection type BS 3718		
Non-injection type BS 3898		
Laboratories, controlled		
atmosphere BA 4194		
Conditioning	Packages	BS 4686 Pt. 2
Tests for mass and density		
Grammage	Paper	BS 3432
	Board	FEFCO 2 & 10
Mass	Wax on paper	BS 4685
Density	Plastics	BS 2782-509
	Cellular materials	BS 4370
Test for strength properties		
Adhesive strength	Adhesive and adherends	BS 847
	Adhesive tape	BS 3887
Burst strength	Paper and board	BS 3137
	Board	FEFCO 4
	Paper (wet)	BS 2922
Column crush strength	Corrugated board	FEFCO 8
Crush strength	Plastics	BS 2782
Flat crush strength	Corrugated board	BS 4687
Folding endurance	Paper	BS 4419
Impact strength	Plastics	BS 2782-306
Ply bond strength	Paper	TAPPI RC 364
Pressure	Aerosols	BS 3916
	Steel drums	BS 1702
Puncture strength	Paper, board	BS 4816
Shear strength	Adhesive	BS 647
	Plastics sheet	BS 2782-305
Stiffness	Paper	BS 3748
Ring stiffness	Paper fluting media	AST D-1164
Tear strength		
Initial	Paper	ASTM D827
Internal	Paper	BS 4468
	Plastics	BS 2782-308
Tensile strength and stretch	Aluminium foil	BS 3313
	Paper	BS 4415
	Plastics film	ISOR-1184
	Plastics	BS 2782-301
	Twine	BS 2570
Tests for surface absorption and permeability properties		
Abrasion (see rubproofness)		
Absorption of water by:	Desiccants	BS 2540
		BS 2541
	Paper, board	BS 2644
	Plastics	BS 2782-502
Absorption of wax by:	Paper	ASTM D688

Table 3.7 *(Continued)*

Test, etc.	Material	Reference[a]
Friction coefficient = 1/Slip	Plastics	BS 2782-311 A
	Paper	TAPPI T503
Moisture content	Paper	BS 3433
Permeability (see also water vapour	Paper to air	BPBIF p. 13
transmission)		BS 2925
	Sheet materials to gas	BS 2782-514A
		ASTM D1434
Resistance to grease	Paper	ASTM D722
Resistance to oil	Paper	BPBIF RMT1
Roughness	Paper	BS 4420
Rubproofness	Paper	BS 3110
Water vapour transmission	Sheet materials	BS 3177
	Sachets	BS 2782-513 A-P
	Packages	ASTM D985
		ASTM E96
Analytical tests		
Acidity/alkalinity (pH)	Adhesives	BS 647
		BS 844
	Desiccants	BS 2540
		BS 2541
	Paper	BS 2924
	Woodwool	BS 2548
Ash	Paper	BS 3641
Chlorides	Paper	BS 2924
Contraries (as arsenic, benzoic acid,	Paper	
borates, casein, copper		BS 1820
formaldehyde, lead, etc) in	Plastics	BS 2782-Pt.
Styrene monomer	Polystyrene	BS 2782-403
Sulphates	Paper	BS 2924
Sulphur, reducible	Paper	BS 1820
Dimensional tests		
Curl	Paper	BPBMA p. 46
		BPBIF RMT 2
Dimensional stability	Paper	BPBIF p. 53
	Plastics	BS 2782-106
Thickness	Paper	BS 3983
	Board	BS 4817
	Plastics	BS 2782-512
	Foil	BS 3313 Pt. 2
Optical tests		
Brightness, whiteness, reflectance,	Paper	BS 4432
opacity		
Lightfastness	Paper	BS 4321
Miscellaneous materials tests		
Crease quality	Board	BS 4818
Odour and taint	Packing materials	BS 3755
Resistance to blocking	Paper	ASTM D-918
	Plastics	BS 2782-301A & D
Shrink ratio	Plastics film	PFMS 4/68

Continued

Table 3.7 (*Continued*)

Test, etc.	Material	Reference[a]
Container, package, etc., tests		
Compression strength	Containers	FEFCO 50
		ISO R2872
Impact, horizontal	Packages	ISO R 2244
		BS 4826 Pt. 5
Impact, vertical	Packages	BS 4826 Pt. 4
Stacking	Packages	BS 4826 Pt. 3
Vibration	Packages	ISO R2247*
		BS 4826 Pt. 6
Water vapour permeability	Packages	ASTM D985
		D1251
General testing	Pallets	ASTM D1185
Low pressure	Packages	ISO R2873
Water spray	Packages	ISO R2875
Rolling	Packages	ISO R2876
Instrumented drop	Packages	PIRA method
Instrumented vibration	Packages	PIRA method
Cushioning materials tests		
Creep	Cushioning material	BS 4433 Pt. 3 in draft
Dynamic characteristic	Flexible cellular materials	BS 4433 Pt. 9
		ASTM 1596
	Loose fill material	PIRA method
Biological tests		
Mould growth	Cushioning material	BS 1133 Sec. 12
		App. G
	Packaging	PIRA method

[a]ISO   International Standards Organisation.
BS   British Standards (in some instances the number of the standard is followed by a further reference, e.g., BS 2782-509 is Method 509 in BS 2782.)
ASTM   American Society for Testing and Materials Standards.
TAPPI   Technical Association for the Pulp and Paper Industry (USA) Standards (where TAPPI Standards have been developed by ASTM, the latter reference has been used).
FEFCO   Federation Européenne des Fabricants de Carton Ondulé Test Methods.
BPBIF   Technical Section of the British Paper and Board Industries Federation Test Methods.
PFMS   Plastic Film Manufacturers' Association.

which both contracting parties have the necessary equipment, use the same procedure and express the results in the same way. Some selected tests are given in Table 3.7 [3]; sources for test methods are given in Table 3.8.

Influence of the nature of the material on test procedures.   In many materials, there is a relationship between *temperature and moisture content*, and other properties. Tests on such materials must be preceded by conditioning, and performed under standard conditions.

Paper, wood and other materials will show different results, depending on the *grain direction*. (This is the dimension parallel to the direction of the material when travelling through the machine which made it. It is often called

Table 3.8   Useful sources of test methods

1. *International (ISO) Standards*
   Usually re-issued as National Standards

2. *National Standards*
   (a) *British Standards* describing methods, e.g. BS 3755, Assessment of Odour in Packaging Materials
   (b) *British Standards* specifying materials, containers, etc. may contain test methods not readily available elsewhere, e.g. BS 1820 Vegetable Parchment, Appendix F, has a test for determination of arsenic
   (c) *ASTM standards* Part 15: Paper, packaging, cellulose etc.

3. *Trade associations' publications*
   (a) *Test methods*, e.g. BPBMA series of proposed procedures (PPs)
   TAPPI standards
   FEFCO series of test methods
   (b) *Voluntary standards for materials, etc.* e.g. Packaging Films Manufacturers' Association series of standards contain sections giving detailed test methods.

the machine direction (MD) and is distinguished from the cross direction (CD).) It will generally be necessary to test in both directions.

The more *variable* the material, the greater the number of random samples that must be taken, using correct sampling procedures.

Properties that may need to be considered.   These are summarized in Table 3.9. The tests carried out on flexible packaging materials must be chosen to examine those properties which are relevant to the application. For example, when reels of film are used on a form-fill-seal machine the coefficient of friction and the stiffness must be suitable [4]. Friction must not be too low or the web will wander too high, and web breakage may occur. Stiffness is important in forming a neat-looking pouch. Bags must be easy to open, and if the filled bags are to be stacked, the slip, which is the reciprocal of coefficient of friction, should not be too low or the stack will not be stable.

For shrink-wrapping film, slip and shrink characteristics (the percentage shrink in each direction), shrink tension and thickness are important. For over-wrapping, stiffness is important as the cut-off length of film must be fed round the object to be wrapped without allowing the edge of the film to drop.

The *coefficient of friction* (Table 3.10) [5], which is an important property for a variety of applications, is determined from the force required to pull a block (sledge) covered with film at a fixed speed on a flat surface covered with the same film. The difference between UK and USA standard methods is in the size of the sledge and the speed at which it moves. *Tensile strength* (Table 3.11), together with elastic modulus, percentage elongation at break and yield strength (the load at which the film continues to stretch without an increase in pulling force) are measured on a tensile tester by pulling a rectangular strip of film 15 mm wide until it breaks.

Table 3.9 Properties of flexible packaging materials

Protective properties	Production properties	Appearance properties
Water vapour permeability	Strength properties	Printability
Gas permeability	Resistance to blocking	Haze and gloss
Thermal conductivity	Freedom from static electricity	Resistance to abrasion
Compatibility with product	Forming and/or creasing	Resistance to ageing
Freedom from taint, odour,	properties	Resistance to fading
toxicity	Flammability	Attraction of dust by
Strength properties	Thickness variation	static charges
Puncture strength	Heat shrinkability	
Tear strength	Dimensional stability	
Stiffness	Air permeability	
Tensile strength and	Sealing temperature, pressure	
stretch	and dwell time	
Impact strength	Adhesion properties	

Table 3.10 Coefficients of friction (BS 2782, Part 3/1965, method 311H): results from a PIRA study

Material	Thickness 0.001 in	mm	Coefficients of friction Static	Dynamic
Polythene normal slip	2.5	0.064	0.13	0.12
Polythene low slip	2.5	0.064	0.13	0.12
Polythene high slip	2.5	0.064	0.73	0.90
Polypropylene C	0.7	0.018	0.32	0.34
Polypropylene O	0.5	0.013	0.76	0.94
PVC	1.0	0.025	0.23	0.27
PVdC	1.2	0.030	0.79	0.52
Nylon 11	1.2	0.030	0.63	0.86
Polyester	2.0	0.051	0.28	0.34
Oriented polystyrene	1.7	0.043	2.1	1.47
Paper	3.0	0.076	0.31	0.3

Table 3.11 Tensile strength (25 mm wide strips, 150 mm long): results from a PIRA study

Material	Thickness 0.001 in	mm	Breaking load (kg)	Elongation (%)
Aluminium foil/paper	1.5	0.038	4.3	2.7
Paper, waxed	2.0	0.051	8.1	4.6
Cellulose triacetate	1.0	0.025	5.0	23.0
Unplasticized PVC	0.8	0.020	5.7	34.0
Cellulose MSAT	0.9	0.023	6.5	37.0
Polypropylene	0.75	0.019	6.6	105.0
Saran	2.2	0.056	11.9	113.0
Polyester	2.0	0.051	23.2	167.0
Polythene	1.5	0.038	2.1	444.0
Nylon	1.1	0.028	5.5	534.0

Table 3.12 Dart impact strength (BS 2782, Part 3/1961, method 306B): tables adapted from reference 6

Materials	Thickness (mm)	Impact strength (gf)
Polyethylene	0.07	130
Polyethylene	0.04	65
Polypropylene	0.07	64
Polypropylene	0.5	320
Polypropylene (oriented)	0.03	750
PT cellulose	0.03	430
Cellulose laminate	0.1	633
Paper	*Basis wt* (g/m^2)	
Sulphite/sulphate	75	105
Sulphite	75	75
Beleny sulphite/sulphate	70	80
Drevity sulphite	75	58

Impact strength is measured by the dart impact method (Table 3.12) [6]. Films are variable in this property and the test determines the weight of dart required to break 50% of the samples. A hemispherical-headed dart is dropped from a fixed height on to an area of film held in a frame and weights are added to the dart until it breaks through the film. Each drop requires a fresh area of film approximately 10 in square and ten drops are made with at least six different dart weights, so this test requires a large area of film.

Stiffness or *flexural rigidity* (Table 3.13) of films cannot be measured by the methods used for papers as the forces involved are too small. For films, stiffness is measured by the force required to push a strip of film 10×200 mm through a slot 5 mm wide with a rounded knife blade 2 mm wide. The film is slowly forced into a semicircle round the indentor, at which point it drops through the slot. The maximum force required is recorded [5]. The test does not measure a basic property of the material, but is dependent on the nature of the layers of which the film or laminate is made up, its coatings, any printing, and its thickness.

The *thickness* of films may be determined either by measuring 10 thicknesses with a micrometer, or, where the density of the film is known, by weighing a known area and calculating. This is the best method for polyethylene shrink film, where the thickness may vary by $\pm 10\%$ in places and a large number of micrometer readings would be required for a good mean. The *shrink ratio* of such film is determined by measuring the changes in dimensions of a square of film before and after immersion in an oil bath at a controlled temperature.

The measurements for *haze and gloss* require specialized optical instruments, and do not normally require more than visual inspection by the user. The same applies to print. *Tests for abrasion* may be carried out using the Pira

Table 3.13 Flexural rigidity, Handle-O-meter method: results from a Pira study L.C. Chong [5]

Flexural rigidity (N/m)	Materials	Thickness	
		0.001 in	mm
Less than 1	Orientated polypropylene	0.5	0.013
	PVC plasticized	0.8	0.020
	Nylon 6	1.5	0.038
	Polyethylene	1.0	0.025
1–5	MSAT 300	–	–
	MXXT/S 400	–	–
	Coated polypropylene	0.8	0.020
	Orientated polystyrene	1.0	0.025
	Aluminium foil	1.0	0.025
	LD polyethylene	2.0	0.051
	Cellulose/PE laminate	2.0	0.051
5–15	MSAT 600	–	–
	Foil/polythene	2.0	0.051
	MSAT 300/MSAT 300	–	–
	Cellulose/PE laminate	3.0	0.076
	Polypropylene/PE laminate	3.5	0.089
	Nylon/PE laminate	4.0	0.100
10–20	Polyester	2.0	0.051
	LD polythene	5.0	0.127
	Paper/PE laminate	3.0	0.076

rub tester. *Tests for compatibility* with the product for taint and odour are carried out using taste panels and by gas chromatography [6].

These are the basic tests for testing flexible packaging materials. The properties of the material which are important can only be decided from its application, the type of machinery on which it will be filled, the product, transit and display conditions. These properties must then be laid down in the specification and the appropriate tests selected.

Glass containers [1]

Properties of glass containers

Glass as a packaging material has the advantages of chemical inertness, clarity, rigidity, resistance to internal pressure, heat resistance and low cost. Its disadvantages are its fragility and heavy weight. Let us now consider each of these in more detail.

Chemical inertness. Chemical inertness is relative, but we are justified in regarding glass as being unaffected by, and having no effect on, most products likely to be packed. The only liquid which reacts rapidly with glass at room temperature is hydrofluoric acid, and for organic liquids such as oils and solvents there is no detectable reaction.

Water and aqueous solutions do react with glass but at an extremely low rate at room temperature. The reaction is a displacement of some hydrogen in the water by an equivalent amount of sodium, thus giving sodium hydroxide and imparting a very slightly alkaline reaction to the water. For normal purposes, the amount of alkalinity is negligible, even over long periods of time. The reaction is speeded up at elevated temperatures, and repeated high temperature sterilization will extract appreciable amounts of sodium from the glass.

For products which are extremely alkali-sensitive (such as certain drugs or transfusion fluids) specially treated bottles are available. One treatment, known as 'sulphating', consists of filling the bottles with sulphur dioxide at a temperature of 500°C. The acid gas reacts quickly with the sodium in the surface layers of the glass to give sodium sulphate which is later washed away with water.

Glass is also a complete barrier to water vapour and to gases. However, there still remains the possibility of loss or pick-up of gases or vapours via the bottle closure.

Clarity. The clarity of glass has been an important factor in sales, especially in the self-service environment. While not every product benefits from visibility, a great number do, especially in the realm of food and drink. If protection from light is required for some reason, then coloured glasses are available. The particular colour required will depend on which part of the spectrum needs to be excluded. If exact information on this point is lacking, then a fairly safe course of action is the use of an amber glass, since normal amber glass, only 2 mm thick, excludes practically all light of wavelengths less than 450 μm.

For cosmetic creams, a more pleasing effect is given by opal glass. This is not opaque, merely translucent, but in thick sections it does prevent visibility of the contents.

Rigidity. For certain products, the rigidity of glass is a disadvantage: there are certain powders, and products such as washing-up liquids, where a squeezable container is required so that the container can act as a convenient dispenser.

In many other instances, however, the rigidity of glass is an advantage. It makes the container easy to handle on the filling line, and it retains its shape during all phases of marketing. The rigidity of glass also means that outer containers can be less rigid, because the glass containers themselves can take the load of other packages stacked on top of them. Also, a jar does not change volume appreciably under stress. This is particularly important under conditions of vacuum filling which could cause trouble with less rigid containers.

Resistance to internal pressure.   This is a particularly important property for the packaging of products such as carbonated beverages (beer, soft drinks, etc.) and aerosols. Glass has been found to be a perfectly satisfactory material for such products, although there are examples of bottles which have exploded in use. The usual reason for such failure is a flaw of some sort in the bottle. Such flaws may be submicroscopic in nature (and undetectable by normal commercial means) or may be surface flaws due to abrasion. Another reason for failure is bad distribution of the glass in the bottle; the main thing to avoid is abrupt changes in cross-section and sharply radiused corners. This brings us to the effect of bottle design on bursting strength. The shape of bottle best able to resist internal pressure is the sphere, but as this is impractical for use as a packaging container, the next best is the cylinder and this is, of course, widely used.

Heat resistance.   Resistance to high temperatures is an important property in many fields of packaging. Since glass will withstand temperatures up to about 500°C, its high temperature resistance is obviously adequate for any packaging use. The main situations in packaging where high temperature resistance is necessary are (1) hot filling; (2) cooking or sterilization in the container, and (3) sterilization of empty containers by steam or dry heat.

Hot filling may be necessary because the product is viscous at ordinary temperatures (e.g. peanut butter), or may be carried out in order to maintain sterility (e.g. jams, which are filled hot in order to prevent mould growth—the surface of the jam must subsequently be kept free from recontamination, usually by vacuum filling). Temperatures are not usually above 100°C and may often be appreciably less. Some other materials such as plastics can sometimes give trouble, not because they melt but because there is the risk of distortion, whereas glass is normally perfectly satisfactory for this type of use.

Cooking or sterilization in package is typified by the use of glass containers for the preservation of fruit and vegetables, or for the bottling of beer, where the beer has to be pasteurized in the bottle. Glass has been used for these applications for a very long time, with satisfactory results.

Sterilization of empty containers may be carried out by steam or boiling water, or by dry heat, in a hot air oven. It is especially important for multi-trip food or beverage containers which must obviously be both clean and sterile before re-filling.

Since glass will withstand temperatures of up to about 500°C, it would seem that there ought to be no problems in the use of glass under high temperature conditions. There is, however, another factor to be considered, in addition to the absolute value of the temperature reached in use. The phenomenon known as thermal shock, as its name implies, refers to the effect of sudden temperature changes. The effect of a sudden change is minor if both surfaces of the glass are heated or cooled simultaneously. The trouble arises when one surface only is rapidly chilled or heated. Let us consider what happens to a glass bottle at uniform temperature, when one surface is suddenly chilled. The cooled surface will immediately try to contract, but is prevented from doing so freely by the warmer glass to which it is, of course, still firmly attached. The cold surface, therefore, is in a state of tension, while the warm surface is in a state of compression.

The wall thickness of the glass container plays an important part in determining the magnitude of the stresses induced in the container because the thicker the wall, the longer the time taken for heat to travel through the glass and the greater the differential temperature between the inner and outer surfaces. Bottle shape also has an important effect on resistance to thermal shock, the stresses being highest near the join between the base and the side wall. Thermal shock resistance is improved by avoiding too abrupt a join at this point. Instead, there should be a gentle curvature or insweep.

Because shape and wall thickness both have an effect on thermal shock resistance, it is difficult to give exact figures, but the following give some idea of orders of magnitude of sudden temperature drops which can be accommodated by various capacity bottles, without breakage.

Winchesters, quart flagons, etc.	25–35°C
Medium size multi-trip bottles, such as milk bottles, beer bottles, etc.	30–40°C
Medium size single-trip bottles (with thinner walls), e.g. wine and beer bottles	45–60°C
Medium size lightweight food jars, such as pickle, sauce and jam jars	50–70°C
Small round bottles or vials for pharmaceuticals, etc.	60–80°C

These temperature ranges relate to drops in temperature. It has been found in practice that sudden temperature *increases* of one surface produce lower stresses than sudden temperature drops, so that the quoted figures relate to the worst cases likely to be encountered. For most practical purposes, the figures above are quite adequate and the behaviour of glass containers in use bears out this assumption.

Cost. The raw materials for glass are not expensive, are indigenous and the manufacturing process is not labour-intensive. However, capital costs and fuel costs are high. The deciding factor is the mass production techniques where production is carried out 24 h per day until the furnace is worn out, thus keeping overheads to a minimum.

Fragility. This is one of the two main disadvantages of glass. The impact strength of glass is an important factor throughout the life of the container but particularly during filling (especially on high-speed machines), during distribution, and when in the hands of the consumer.

During filling, it is not necessarily the bottle which breaks that is the most important, although there could be loss of an expensive product and hold-ups on the machine. A more subtle risk is chipping of the glass, due to a number of high speed impacts, none of which is severe enough to cause outright failure. Such chips of glass may fall inside the container and are then very difficult to detect. In the case of a product such as a cosmetics cream, there is a risk that the consumer will receive a wound which could lead to a claim for damages. The situation with a baby cream would be even worse. The presence of glass chips in a food product can also lead to severe consequences. Much can be done, of course, to minimize shocks on the filling line. Conveyors and guide rails can be cushioned or lubricated and modern designs of plant also concentrate on eliminating the movement of bottles as far as possible.

As far as distribution is concerned, the bottles should be cushioned in some way if heavy knocks are anticipated. On the whole, experience shows that glass containers have adequate impact strength provided that they are well designed and protected. A point to be noted is that when bottles filled with liquid are dropped on to a rigid surface there will be, in addition to the normal impact stress in the glass, an effect due to the momentum of the liquid. At the moment of impact this will be converted into an instantaneous internal pressure, which can be high. Cushioning around the bottle will help to prevent the sudden destruction of momentum and so reduce the peak internal pressure. Nevertheless, it is important to design any bottle containing liquids to withstand fairly high internal pressures.

Heavy weight.   The weight of glass bottles, their second disadvantage, has been consistently reduced during this century, particularly during the post-war period. Earlier bottles were heavy because it was difficult to control the distribution of glass, hence a large amount of glass was used for each bottle in order to ensure sufficient strength at every point. Even so, the lightest one-pint milk bottle yet produced still weighs about 290 g (10 oz) to hold 20 oz of milk. This should be compared with a plastics milk bottle which weighs about 15 g (0.5 oz) for the same capacity. In some ways the weight of glass bottles is a positive point. They are more easily handled on fast packaging lines, for example, while some products (such as cosmetics) seem to have more consumer appeal when packed in a heavier material.

Sealing of glass containers

Although glass is a complete barrier to moisture vapour, gases and odours, the product can still deteriorate if the sealing of the bottle is at fault. The main features of a good bottle closure are as follows:

(1) It should prevent loss of the contents or any constituent of the contents.
(2) It should prevent penetration of any substance from outside the container.

(3) The closure material should not react in any way with the contents of the container.

(4) It should be easy for the consumer to reach the contents. This may be by removal of the closure, but could be by piercing a hole in it (as with some closures where a thin diaphragm is placed in the centre of a cap).

(5) It may have to make a good re-seal, as with the closure on a jar of instant coffee.

(6) It may have to be pilfer-proof, i.e. it must be obvious, visually, if the closure has been removed and replaced.

(7) It should harmonize with the container. A well-designed closure can add considerably to the sales appeal of the pack.

There is a wide range of closures available for the sealing of glass containers, but they can be divided into three broad groups.

(a) *Normal seals.* These are closures whose main function is to give a good seal when internal and external pressures are approximately equal. They will usually be able to withstand reasonably small changes in pressure, such as might be caused by changes in ambient temperature.

(b) *Pressure seals.* Pressure seals are closures which are designed to withstand high internal pressures, such as occur in carbonated beverages.

(c) *Vacuum seals.* Vacuum seals have to give an airtight seal where the pressures inside the containers are appreciably lower than those outside. The seal is usually maintained by the higher external pressures so that if there is a loss of vacuum inside, the closure may leak.

Metal packaging: the basic materials [7]

The main metals used in packaging are mild steel sheet, tinplate, terne plate, galvanized mild steel sheet, stainless steel, aluminium alloys and aluminium.

Mild steel plate (usually cold-reduced strip-mill steel) is the principal metal-drum-making material. Terne plate (a mild steel coated with a tin/lead alloy) and galvanized steel, with an electrolytic deposit of zinc, are also employed. For special contents, aluminium alloy sheets, commercially pure aluminium sheet, and stainless steel (normally an 18–8 nickel chrome steel) are used for drums and kegs.

Tinplate is the principal material for metal boxes and cans. It is mild steel (i.e. low-carbon steel) coated on both sides with tin. The base steel plate or strip is manufactured by rolling hot steel ingots down to a strip with a thickness of 1.8 mm (0.07 in). The strip is then pickled continuously in a bath of hot dilute sulphuric acid to a finished gauge of 0.15–0.50 mm (0.006–0.020 in). The sheet is finally annealed and temper-rolled to impart the required hardness and surface finish. Since the early 1960s, economies have been produced by reducing the thickness of the steel base. These economies depend on the fact that further cold reduction of the sheet produces a material having a greater intrinsic stiffness; hence a thinner sheet can be used for some applications. The plate is then known as 2CR (double cold reduced) or DCR (double reduced).

The tin coating was originally applied by running the steel plate through a bath of molten tin (hot dipping process). Although this method is still employed, the majority is now made by a continuous electroplating process (giving electrolytic tinplate). Hot dipping gives a comparatively thick coating, with a lower limit of about $22\,g/m^2$ ($11\,g/m^2$ on each side). This figure corresponds to the old figure of 1 lb per basis box (1 basis box = 31 360 square inches of tinplate, i.e. 62 720 in^2 total surface area). The electrolytic method is more flexible than hot dipping and produces coatings down to $5.6\,g/m^2$ ($2.8\,g/m^2$ on each side). In addition, it is possible with the electrolytic process to produce differentially coated tinplate, i.e. tinplate with a different coating weight on either side. Such material is used when the protective requirements for the outside and inside of a can differ. The lowest feasible tin coating can then be used for each surface, leading to overall reduction in tin coating costs. Manufacture of electrolytic tinplate is a highly technical operation, requiring extensive initial capital investment. In spite of this, it has almost entirely ousted hot-dipped tinplate for can manufacture, because of the savings in tin coating costs.

Hot-dipped tinplate possesses a naturally bright finish, whereas electrolytic tinning produces a rather dull coating. Electrolytic tinplate can, however, be brightened by heating, momentarily, either in a bath of hot oil or by electrical induction. This process is known as flow brightening. Flow brightening not only improves appearance but also the resistance of the plate to corrosion. The dull matt finish of the as-plated sheet is due to a porous finish, and the protective properties of the coating are much improved by melting the tin to give a more coherent finish. After flow brightening, electrolytic tin coatings are treated to remove any tin oxide formed during the process. They are then treated in chromic acid, dichromate or chromate/phosphate solutions to stabilize the finish. This is because tin oxide is often affected by subsequent storage or baking of the tinplate.

The high-speed equipment now used for can making necessitates the use of a film of lubricant on the tinplate surface. Cotton seed oil, or synthetic oils such as di-octyl sebacate or di-butyl sebacate are normally used. The oil film is extremely thin—of the order of a few tenths of a millionth of a millimetre—and the lubricant must be compatible with any subsequent lacquer coating used.

The final step in tin economy is the elimination of tin entirely, so producing *tin-free steel* (TFS). This material is produced by electrolytically coating mild steel plate with a chromium/chromium oxide film by a process developed in Japan. The chromium/chromium oxide layer is even thinner than the thinnest tin coatings normally used, and must be lacquered before it can be used in the manufacture of containers. It is, however, satisfactory for protecting the steel from rusting during transit and storage prior to can manufacture.

The cost of TFS is lower than that of tinplate, but is increased by the necessity for lacquer coating. If plain tinplate can be used for packaging a particular product, therefore, it will normally be cheaper to use. On the other hand, if the tinplate has to be lacquered, then lacquered TFS will normally be cheaper.

Blackplate is the name given to the base plate of mild steel before tinning. It can be used for container manufacture, using the techniques of welding or cementing for the production of the side seam. Both its surfaces have to be protected with some sort of coating, otherwise rusting easily occurs.

Aluminium and its alloys have long been used for the manufacture of rigid containers, although not to the same extent as tinplate. Aluminium is also used in the form of foil (both alone and laminated to other materials), and for collapsible tubes. With aluminium, seamless bodies can be produced by impact extrusion, particularly if the depth is larger relative to the diameter. To achieve adequate stiffness, an alloy (with about 1% of manganese usually, although there are other alloying materials) is necessary. Aluminium cannot be soldered by any suitable technique but it does not rust tinplate. Although light and pleasant to handle, its resistance to chemical attack is limited. Its use for any purpose must be carefully considered in relation to the product.

Cans and tin boxes [8]

Tinplate is still the most common raw material for tins and cans. This is mild steel sheet with a very thin layer of tin on each of its surfaces. Blackplate is not often used, but it is most easily described as tinplate without the tin, i.e. mild steel sheet, available in the same range of thickness as tinplate. Tin-free steel is like tinplate, but the tin has been replaced by other corrosion-resistant metals such as chromium. Aluminium is being used in increasing quantities.

Tinplate and aluminium allow designers scope to produce intriguing effects through the use of surface finishes which can be produced with lacquers and varnishes. An example is the well-known Golden Syrup can where the tinplate, lacquered with a transparent gold lacquer, shines through to create a golden glow.

Some cans for food can be produced with an internal liner, often in white, to enhance the appearance of the food.

Combining the metal can with other materials such as plastic or foils can create an entirely new image. Cans for powdered milk, for example, have an aluminium foil tagger sealed on the closure ring conferring pilfer resistance and easy opening as well as an impression of hygiene. Plastic lids fitted over normal end seams and used for re-closure after opening are both decorative and functional.

Until about 1970 the soldered three-piece can was standard. Since then much development has taken place. The construction of the three-piece can, however, has not changed fundamentally in over 150 years. A flat rectangular sheet of tinplate was formed into a cylinder and the resulting join was soldered to form a side seam and circular discs of tinplate were then mechanically secured to flanges made at the ends of the cylinder by a rolled seam, known as a double seam. One end was fitted by the can maker and the other by the packer after filling.

Advances have been made in techniques but not in principle. Since about 1985, solder has been eliminated almost everywhere by electrical resistance welding and laser welding is a prospect for the future. Automation has increased the speed of manufacture and better, more secure mechanical joints have been developed. In addition to the improved three-piece can, two-piece cans have been developed and today there are three major types of open top food and beverage can in use:

(1) The three-piece welded can (Figure 3.5)
(2) The two-piece drawn and redrawn (DRD) can (Figure 3.6)
(3) The two-piece drawn and wall-ironed (DWI) can (Figure 3.7).

Easy opening ends, first for beverages and then with full aperture for foods, have been developed and are now made out of steel as well as aluminium which was used for the first developments. The steel easy-open ends are becoming more popular now that the initial problems have been largely overcome.

Making metal boxes is all about producing good joints in sheet metal and there are three basic types of body construction: the built-up body, the hooked or seamed corner body and the seamless or drawn body construction.

The *built-up body* (Figure 3.8) must have at least two components, the side wall and the base. It may have a third or even a fourth component, but it must have at least two. The side wall is referred to as the body and the base is called the bottom or end. A two-piece body has two joints, the side seam of the wall and the joint between the wall and the bottom—both are interlocking folds. The side seam is formed by interlocking the fold on each edge and 'bumping' the interlock (Figure 3.9). The bottom joint is made by 'double seaming' (Figures 3.10, 3.11). The common food can of course is produced by securing a third component (an end) by double seaming after filling.

The rigidity of any built-up body is determined by the security of the joints and the stiffness provided by the side wall which is dependent on its shape. A fundamental difference in strength exists between round and all other shapes. The even stresses and the strength of cross-section of round bodies permit the closure of open-ended bodies (e.g. slip lid types) to be made with both the lid and the body in a state of even tension and compression. With square bodies, for example, the corners, which are relatively more rigid, transmit the compression forces to the flat sides, which (since they are restrained by the lid from moving outwards) bow inwards. The influence of material consumption on the

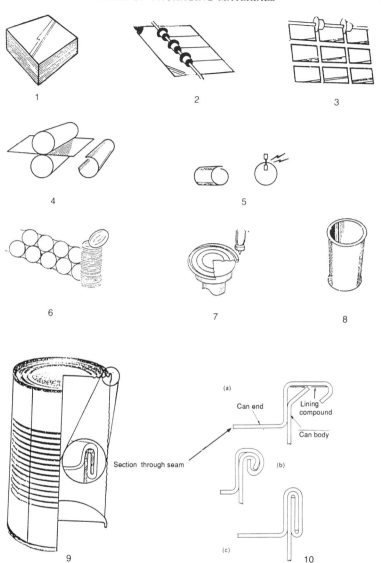

Figure 3.5 The three-piece electrically welded food can. (1) Starts from a stack of tinplate sheets which are coated with a lacquer, dried and stoved in ovens for 20 min. Different lacquers are used for different products, (2, 3) the lacquered sheets are first cut into strips and then the strips are cut to the correct blank size for the can bodies, (4) the body blanks are now rolled into cylinders and (5) the two edges are welded together electrically. The area adjacent to the join is again coated with lacquer and oven dried, (6) other sheets have meanwhile been cut into circular blanks for the ends of the can, (7) these can end blanks are curled at the rims and a sealing compound flowed into the curl, (8) a lip (or flange) is now formed at both ends of the welded body cylinder and (9) the end is now seamed to the body to give a can ready for filling and closing by the packer who seams another end on to the body after filling, (10) formation of a double seam, (a) end placed on body, (b) seam part-formed, (c) finished seam.

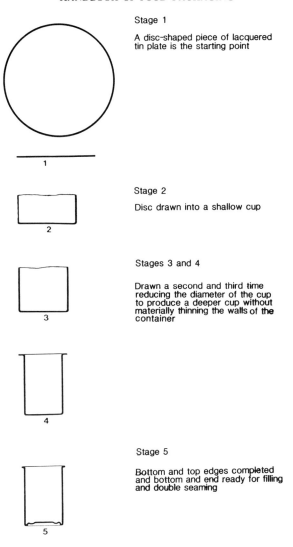

Figure 3.6   The drawn and redrawn (DRD) can.

choice of shape is significant. Built-up bodies are preferable from the point where a seamless or locked corner body is too expensive or does not provide the depth required. The effective size limits are dependent on the shape, the type and the function.

Built-up body cans may be divided into open top cans and general line cans. Both are used in the food industry.

The essential characteristic of the *hooked or seamed corner body* is the

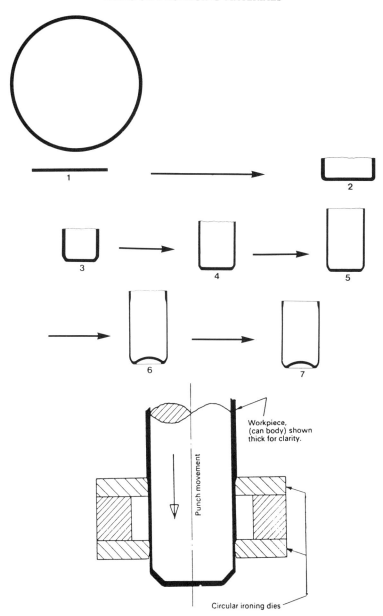

Figure 3.7   The two-piece drawn and wall-ironed (DWI) can. The cup-shaped component is first drawn from a round blank. The cup is forced through the ironing rings by the punch. Only two rings are shown, but three or more may be used in practice. Each ring reduces the wall thickness by between 20% and 35%. Stage 1, circular metal blank drawn into a shallow cup shape; stages 2, 3 and 4, the cup is wall ironed into the can shape; stages 5, 6 and 7, top trimmed and bottom finished, can cleaned and printed and walls finished; now ready for filling and closing.

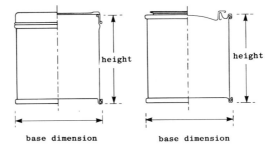

Figure 3.8 The built-up body can.

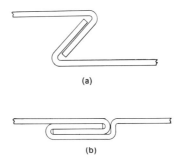

Figure 3.9 Interlocked side seam: (a) before hammering; (b) finished seam.

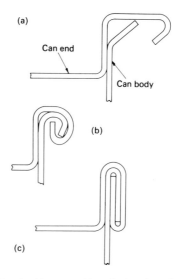

Figure 3.10 Formation of a double seam: (a) end placed on body; (b) seam part formed; (c) finished seam.

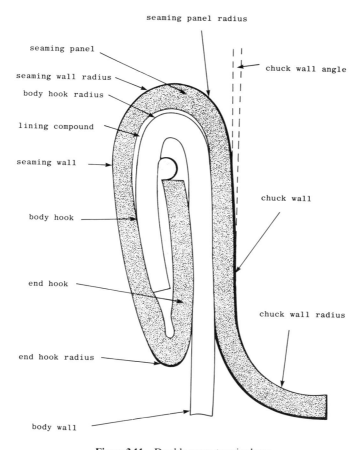

seaming panel radius

seaming panel

seaming wall radius

body hook radius

lining compound

seaming wall

body hook

end hook

end hook radius

body wall

chuck wall angle

chuck wall

chuck wall radius

Figure 3.11 Double seam terminology.

ability to form an almost square corner and for this reason it is popular for beef cubes, etc. Although the corner is sharp, it is not truly square.

Seamless (or solid drawn) body containers are pressed into shape and have no joints. The body is usually finished at the open mouth by trimming, beading or curling to receive the lid, which is frequently of the slip-on type.

The depth to which a seamless body can be pressed is related to the radius of curvature over which the pressing is performed, the thickness of the material and the ability of the surface finish to remain anchored to the base stock. Round, oblong and square shapes are all possible but above about 100–110 mm (4–4.5 in) in diameter the round seamless tin tends to be both uneconomic and functionally poor. All these containers are relatively flat with a large surface area relative to the depth.

Closures

There are five basic methods of closure:

(1) Frictional engagement
(2) Screw thread engagement
(3) Permanent mechanical interlocking of the lid and the body
(4) Plastics closures
(5) Atmospheric (differential) pressure on the lid (often called vacuum seals).

All methods find uses in the packaging of foods (Table 3.14) but the most widely used form of tin box closure in the food industry is the permanent mechanical interlocking of the lid and the body. The top of the can is joined to the body by a rolled permanent joint which is unique in two senses: (a) because of the close control possible, its performance can be accurately predicted; (b) it will withstand internal pressures, external pressure or balanced pressures. Such a joint, the 'double seam', is very reliable and this is what is required in food canning (Figure 3.11).

The *open top can*, defined as a one piece closure type tin filled through an unrestricted orifice, and particularly associated with thermally processed, canned foods, is used throughout the processed food industry. The older 3-piece can is still widely used, but the drawn and wall ironed can (DWI can) has made great advances for carbonated beverages. The processes of making both types were outlined in Figures 3.5 and 3.7 and the associated terminology is shown in Figure 3.12.

Table 3.15 gives the dimensions for open top cans, the metric diameters agreed by ISO and their relationship to the older nomenclature. Open top cans are lacquered internally if the nature of the contents requires it. There is no universal lacquer, each type of product has its own requirement, and the

Table 3.14 Typical can sizes used in food technology, and characteristic features

Type of can	Capacity	Typical uses	Features
Open top can	100 g to 4.5 litres	Fruits, vegetables, meat products, fish products	Pilferproof, easy to fill and handle automatically
Key-opening cans with reclosure	200–1000 g	Coffee, nuts	After key has been used to open the can, a push lid (frictional) is released for subsequent reclosure
Key-opening cans with no reclosure	Wide range of both shape and size	Sardines, large hams, poultry, processed meats	Contents can be removed with minimum of damage
Slip-lid cans	Wide range	Cocoa and other dry powders	Simple reclosure

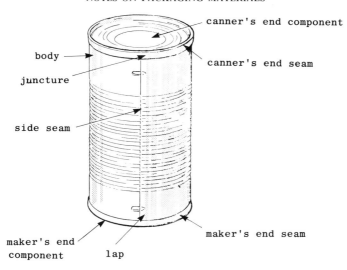

Figure 3.12 Can terminology. Open top processed food cans have three seams, drawn aluminium and drawn tinplate and have only one seam around the top end. Cans may be round, rectangular or irregular in shape.

Table 3.15 Diameters of open top cans: International Standards Organisation list for thermally processed food can diameters, together with British diameters

ISO diameters (mm)	British diameters	
	Old nomenclature	New nomenclature
52	202	52
60	—	—
63	—	—
66	211	66
73	300	73
77	—	—
84	307	84
99	401	99
105	404	105
127	502	127
154	603	154
189	—	—
230	—	—

lacquering is intended primarily to preserve the colour and appearance of the food, not to protect the tinplate from attack.

General line cans are cans used for foods and other products and are distinct from open top cans. The principal types used in the food industry are as follows.

- *The collar can*: a re-closable type of round tin incorporating a removable tear strip and internally fitting collar which provides a seating for the lid.
- *The paperboard composite*: a container with side walls based on board and having end components of metal, plastic, board or combinations of these.
- *The seamless tin*: the body is drawn in one piece.
- *The lever lid tin*: usually a round, built-up tin, the ring component being secured to the body and having an orifice into which an inverted hat-shaped lid is pressed.
- *The slip lid tin*: the body is usually seamless, and closed by a separate lid which fits over the mouth of the body.

Natural materials

When man first began to pack things, he used materials which were naturally to hand. Apart from broad leaves which could be used for wrapping purposes, and the skins of animals which could be made into suitable containers for liquids, such as water and wine, he also constructed baskets from reeds, withies and similar materials, while from mud and clay he made vessels such as pots to hold materials requiring longer term storage. Some knowledge of these materials is useful in understanding their use in the past in particular industries.

Natural materials were also exploited in the provision of protection. Here, straw, mosses, wood shavings and sawdust were used long before man-made cushioning materials were available.

Round-stick joinery

The pottery industry was very concerned with protection and achieved it, not only by use of straw to separate the pots, etc., but also to some extent by what was then called 'round-stick joinery'. This is an old system used by crate makers in North Staffordshire utilizing hazel wood for making 'round-stick' crates. Similar techniques are still employed in less developed countries.

Straw

Even when the round-stick method was abandoned casks, tea chests and crates made of wood or metal were still employed and straw was used up to a relatively short while ago for separating items and preventing impact. The use of straw as a general 'cushion' ceased when the governments of certain countries, notably Australia and New Zealand, announced that they would only accept such packing material if it was accompanied by a veterinary certificate as to its cleanliness, and later on, refused to accept it under any circumstances and insisted on its being burnt at the dockside.

Waxes and bitumen

Two other materials which have been used for packaging from ancient times are waxes and bitumen. The principal waxes employed today are paraffin wax and microcrystalline wax but in earlier times considerable quantities of materials such as beeswax and other vegetable and animal waxes were used for certain purposes, notably in closures for glass containers and for making paper and similar materials water-resistant. Bitumen has long been a favoured material, particularly where water and water vapour proofing of paper is concerned, because of its relatively low cost compared with the cost of paper pulp. The earliest use of bitumen was in a material first known as 'tar paper' and sometimes referred to as 'ocean paper' but more properly described as bitumen union kraft paper. This consists of two sheets of paper laminated together with a layer of bitumen applied molten, which, after it has set, forms a water-resistant barrier between the two sheets of paper. Water-resistant board has also been produced in this way. Most of the problems relating to the use of bitumen have been due to the cutting problems involved; cutting knives and similar equipment rapidly get dirty since bitumen, being black and sticky, tends to transfer from one place to another. Bitumen has also been used to wind the plies of fibre drums and similar canisters where water-resistance is required. Unfortunately, bitumen cannot be cross-linked, and hence always remains liable to soften, particularly if exposed to the sun's rays, and so tends to be unsatisfactory for use in outside conditions or in warm climates. It has been used, however, with extenders to make it less susceptible to softening, and several types of natural resin can also be admixed with it for this purpose.

Stoneware

There have, during the course of packaging history, been several periods when stoneware has been used extensively. At one time in the chemical industry many chemicals were packed in stoneware jars and closed with natural bark corks, particularly where the chemical was regarded as highly hygroscopic, potentially corrosive and liable to give off odours. Such uses have almost entirely been superseded today, although in certain parts of the world stoneware jars are still employed.

Textiles

Other materials which have been used are textiles, including cotton and linen, which are still used in certain parts of the food industry. Cotton scrims and muslins have been, and still are, extensively used to pack fresh meat and, to some extent, in the packaging of poultry. Many of these nowadays, however, are being replaced by synthetic scrims and the use of natural materials is dying out.

Wicker baskets

Wickerwork in one form or another has been with us for centuries and is still used for certain operations, particularly for packing glass, pottery and building materials, as well as collecting agricultural and horticultural produce. Baskets are generally employed as returnable containers.

References

1. F.A. Paine (ed.), *The Packaging User's Handbook*, Blackie, Glasgow (1990).
2. E.V. Southam, *Packaging Technology* **15**(104) (1969), 17.
3. J.M. Montresor, H.M. Mostyn and F.A. Paine, *Packaging Education*, Newnes-Butterworth (1974), p. 105 *et seq.*
4. D.J. Hine, Paper presented at 9th IAPRI Symposium, St. Gallen (1977); *Pira Report Pk 14* (R)/1977.
5. M.E.K. Styles, D.J. Hine and F.A. Paine, The machine/material interface in flexible package production.
6. Celerynova M. Obaly, 1st February (1986), 22–24.
7. E. Morgan, *Tin Plate and Modern Can Making Technology*, Pergamon Press, Oxford (1985).
8. M. Sodeik *et al.*, Fundamentals of modern can making and materials developed for 3-piece can manufacture, *Trans. ISIJ* **28**(8), (1988), 663–671.

4 Packaging machinery

A package is designed to protect and to sell the product it contains, and this generally requires a mechanical process on a packaging line selected to carry out efficiently those operations necessary to put the product into the package.

The majority of the operations on a packaging line are concerned with the package itself, such as making or forming sachets, erecting or closing cartons, feeding and seaming cans, and presenting bottles to filler heads and capping them. Secondary operations, such as coding, labelling, detecting metal, check weighing and collation for despatch, also involve the package in the main. However, it must be obvious that the ability of the packaging line to handle the product itself is of paramount importance. The line must put the product in the package economically, in the desired condition, at the required speed, and to the stated quantity.

A wide range of products has to be considered, from water-like liquids through various creams, carbonated beverages, powders, granule and piece products such as confections, vegetables (fresh and frozen), meat, household goods and toiletries and cosmetics. The nature of the product will have a more profound effect on the performance of the packaging line than any other factor.

Production and packaging line requirements

This section could well have a subheading of 'machineability and the material properties influencing it'. Until the decision as to the style of package to be used has been decided, work on the packaging line engineering cannot start. However, once this decision has been made, the choice lies between the various machine types available for producing the selected package structure.

It should be recognized that the machinery, the product and the package are part of an integrated system. If the machine is well made it is the most precise part of the system while both the product and the packaging are likely to be more variable. Hence the machinery must be selected to accommodate the variations (in dimensions and in critical properties) that will inevitably occur in both product and package.

The principal factors which affect efficiency and utilization of a packaging line may be considered under three headings:

(1) The suitability of the machine for the purpose

(2) The output speed required
(3) The likelihood and frequency of stoppages and the time taken to clear them

The manufacturers of the machines available are not usually in a position to know much about the variability of the particular product(s) to be packed and although they will be aware of the variability of general package types, they may not have a knowledge of the particular packaging specified. Information on both of these subjects must be supplied by the product maker and the packaging supplier (preferably in written specifications).

The machine functions can be broken down into a number of subsystems. Under these circumstances, although the subsystems overlap, it becomes easier to analyse the possibilities. The subsystems are:

(1) The product handling system—filling, weighing, loading, etc.
(2) The packaging handling system—unreeling, erecting, closing, etc.
(3) The basic machine framework
(4) The power transmission
(5) The control system
(6) The timing system
(7) The lubrication system.

Once identified, each subsystem can be analysed for both purpose and effectiveness. The timing system is one of the most important and in many machines there is a drive shaft which makes one revolution for every package produced. Relating machine actions to the angular position as recorded by a circular disc marked in a 360° sector can be of great assistance in determining correct machine setting. Getting the timing right is a first priority and seeing that it remains correct is vital.

Output speeds have been and are still growing in automatic packaging operations and in many instances line speeds are 10 or 20 times as fast as they were 25 years ago. The speed is very much related to the effects of stoppages.

A stoppage, which takes 1 min to clear, happening on average every 1000 packs means that at a machine speed of 10 packages/min the line is 99% efficient, i.e. 1 min of production (10 packages) is lost in every 100 min. The package delivery rate will be 594/h instead of 600 and this would not generally cause difficulty. However, at a machine speed of 200/min there would be a stoppage every 5 min (12 in each hour), 2400 packages plus the 12 causing the stoppages would be lost and the delivery rate will fall to 9588 in 1 h from the rated 12 000—an unacceptable level.

Stoppages during production are due mainly to three sources:

(1) Malfunction of parts of the machine due to wear, fatigue, bad adjustment or failure to clean heat sealers or adhesive applicators, etc.
(2) Variation in the product beyond the limits which the machine can

handle, e.g. size variations in blocks of product, inadequate weighing or volumetric control of quantity, etc.

(3) Packaging material problems such as the failure to replenish reels of material or reload magazines, and variable properties of the packaging materials outside the limits specified.

Good maintenance should virtually eliminate malfunction especially if preventive rather than breakdown maintenance is practised. Efficient quality control of the product should reduce the second factor to an acceptable minimum.

The minimum internal dimensions of a package must be specified so that the smallest size container is capable of accommodating the maximum permitted dimensions of the product. The average size of any rectangular block means that 50% of them will be bigger and 50% smaller than the mean. Specifying the average value of the internal package dimensions is incorrect.

Stoppages due to packaging materials out of specification are more difficult to control and good communication between packaging suppliers, machinery makers and the product manufacturer is essential. All products have inherent variations and both the packaging and the machine must be capable of handling these. Equally important is the identification of those properties of the packaging material which affect the running of the machine and their incorporation, with tolerance limits, in a workable specification for the packaging supplier.

The property profile

An important factor in trouble free running of machines is that the material has the right property 'profile'. This is the term used to define the limits of each critical property of the material within which the machine can operate successfully.

The various operations performed on a packaging material may involve different properties. For example, the unreeling of a plastic film is influenced by the static electricity generated causing blocking between the turns on the reel; it is not affected very much by the stiffness or the frictional properties of the material. Both of these properties are, however, involved in the bending and sliding movements as the material passes through a machine to be wrapped round a product or be formed into a pouch.

Studies to determine property profiles need a stable machine in a state of adjustment which can be defined. Initially, this means that the setting and critical dimensions of the machine production rate, temperature setting, etc., must be recorded and kept as constant as practicable. However, even though settings are maintained, some factors may drift with time. Secondly, a range of generally similar packaging materials, with differing property details, for example coated papers, may be required. Thirdly, an analysis of the machine motions to determine the properties of the packaging material likely to influence the machine's efficiency will be needed. Each operation must be studied in detail—feeding of material into the machine, conveying it through the machine, forming it into a pack and closing it. Lastly, a means of quantifying the efficiency of the packaging line under the headings already noted is required.

The material outlined above will assist in the design considerations of any package engineering line and may be summarized in the series of steps below once the nature of the packaging to be used has been established:

(1) Decide on the number of packages to be produced daily, both at start up and as anticipated when in full production.
(2) Outline a preliminary packaging line in terms of operations, e.g. package forming, filling, closing, collating, etc.
(3) Consider each operation separately in terms of performance.
(4) Discuss the requirements with own company engineers, etc., machinery makers and packaging suppliers.
(5) Prepare preliminary specifications.
(6) If possible, see similar machines in a working environment.
(7) Decide on which machinery to use, get quotes and write detailed specifications.
(8) Discuss with packaging suppliers, own engineers, etc. and machinery makers.
(9) Place orders.

Bottling

The word 'bottling', like its partner 'packaging', means little unless related to a product, for it is apparent that the problems related to bottling haircream, for example, are different from those associated with bottling milk or instant coffee. At one time, bottling lines handled only glass bottles, whereas today the container can be not only of glass, but one of a variety of plastics with shape and stability as varied as marketing may demand. Different-shaped containers need handling in different ways, and their contents are filled according to their characteristics.

A bottling line can be a collection of automatic machines connected by a conveyor belt (Figure 4.1) or a group of highly sophisticated, fully automatic units completely integrated by a synchronized drive arrangement. Semi-automatic lines depend to a large extent on the dexterity of individual machine operators for their maximum output, and operating speeds on these vary from 30 to 60 units per minute. Above this speed, a fully automatic bottling line is usually necessary, but before changing from semi- to fully automatic equipment there are several points to consider, most important of which is the container.

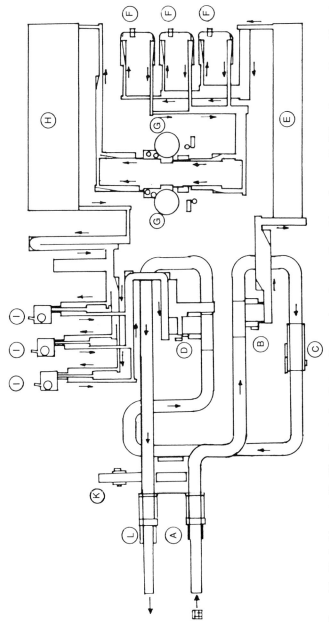

Figure 4.1 Typical layout of bottling line; A, depalletizer; B, emptying of crates; C, washing of crates; D, filling of crates; E, washing of bottles; F, bottle inspection; G, filling and closing of bottles; H, pasteurization; I, labelling of bottles; K, stock of pallets; L, palletizing.

Although it is possible on the most modern bottling lines, with synchronous drive and positive control, to handle containers which will barely stand up on their own, it is best to have a container design which will give sales appeal but at the same time incorporate stability in both the vertical and horizontal planes. A stable container has a relatively large base area, a low centre of gravity and parallel contact points on each side, so that there is no tendency for it to topple if a temporary hold-up brings it into contact with others. Similarly the cross-section of the bottle should be such as to avoid the possibility of jams and breakages which can occur when containers of oval or similar section twist and jam in conveyor guide rails (Figure 4.2). It is good sense when considering bottles of this shape to flatten the side parallel with the major axis and to give as large a contact area as possible between bottles by blunting the sharp radii at each end.

Having established that the proposed bottle shapes and sizes can be handled on an automatic line the individual bottling processes can be considered. These are:

(1) Bottle feeding
(2) Bottle cleaning
(3) Filling
(4) Closing
(5) Labelling
(6) Collating and packing for transport.

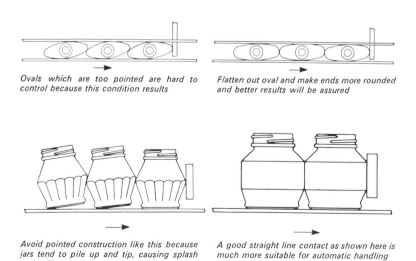

Ovals which are too pointed are hard to control because this condition results

Flatten out oval and make ends more rounded and better results will be assured

Avoid pointed construction like this because jars tend to pile up and tip, causing splash and spillage

A good straight line contact as shown here is much more suitable for automatic handling

Figure 4.2 Effect of bottle shape on filling lines.

Bottle feeding

Apart from specialized operators such as dairies, breweries and a decreasing number of carbonated soft drinks manufacturers who use de-crating machines for removing incoming bottles from metal, plastic or wooden crates, many bottlers arrange with their glass manufacturer and fibreboard case suppliers for their new glassware to be packed from the annealing conveyor directly into the cases in which the filled bottles will eventually be shipped. This saves space, labour and the necessity of tying up capital on returnables and means that relatively simple devices can be used to facilitate bottle feeding. A rotary unscrambling table is probably the simplest, but where shapes other than round are used a container feeder might be advisable. Once the filled cases have been inverted and the bottles pushed onto a moving belt, they are conveyed away, a line at a time, to the main slat chain conveyor.

Bottle cleaning

The traditional method of cleaning glassware is to wash it on one of a range of bottle washers available, and such washers are used extensively in dairies, breweries and allied industries using returnable bottles.

Washers are generally of two types, either 'hydro' or 'soaker-hydro', the former employing liquid jets and the latter complete immersion of the bottles, sometimes with the addition of rotary brushes. Operating speeds can be from 20 to 600 per minute and a variety of washers can be supplied for various conditions. However, where the one-trip bottle is used, a much simpler means of cleaning is often employed, utilizing compressed air.

Assuming that, as mentioned in the section under 'bottle feeding', the glassware has been packed ex-lehr into clean, new fibreboard cases and the flaps have been folded over for transit, the glassware which was commercially sterile when packed is likely to have been contaminated only by paper dust, straw or, perhaps, the occasional glass splinter which can be removed by the air-cleaning operation. It must be stressed here, however, that air cleaning is only a step towards obtaining clean glassware, and close cooperation with glass manufacturers must be maintained at all times to ensure that they are fully aware of the user's intentions and are therefore able to give the regular deliveries so necessary for the success of this system.

Particular attention must be paid to the conditions under which the bottles are stored, for humidity or temperature variations may create condensation, causing dust to adhere to the container wall and thus nullify the air cleaning process; 60 psi is the minimum air pressure for efficient cleaning, and instrument quality air, free from oil and water vapour, is essential.

Air cleaners of the fully automatic variety usually invert the containers prior to the insertion of air nozzles which release a blast of air, usually of two or three seconds' duration. The dust is thus blown out into a conveniently placed trap. Some air cleaners employ an exhaust or vacuum system as well as compressed air, but it should be remembered that this is not intended to improve the standard of cleaning, but merely to provide a method of conveying the foreign matter to a dust box or filter. Speeds of operation of automatic air cleaners range from approximately 60 to 600 per minute.

Filling liquids

Filling machines for putting liquids into bottles can be divided into four basic types: vacuum-filling, measured-dosing, gravity-filling, and pressure-filling (Figure 4.3a). Most liquid-filling machines are specifically designed to use only one of these four methods. But there are one or two multipurpose machines which can use a combination of gravity-, vacuum-, or pressure-filling methods. Filling by *vacuum* is the cleanest and most economical way to handle many products. In spite of the care which is taken in making bottles and cleaning them, there is always a percentage of defective bottles—ones with holes, chips and cracks. These are not easily detected in the pre-handling of bottles before

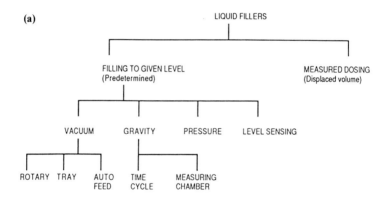

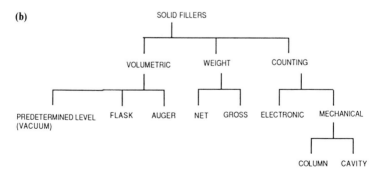

Figure 4.3 Types of fillers for liquids and solids.

filling, but vacuum-filling machines automatically avoid such bottles. Moreover, with vacuum filling there is no drip or other mess. There is little loss of product and it is unnecessary to wash or wipe the bottles before labelling.

Vacuum fillers are of three types: rotary, tray and automatic feed. On a rotary machine every bottle is handled individually. It is centred under a filling stem, raised, and then filled as it travels around the machine, independently of all the other bottles. On a tray-type machine, bottles are placed abreast in trays and rolled on conveyors under the filling head which may consist of one to eight feeding stems. The automatic feed type will operate by means of a lever that discharges the group of filled bottles and moves the empty bottles into position under these stems.

In general, the vacuum system requires a supply tank which is below the level of the bottles to be filled, and from each supply tank are supply pipes or lines to which is attached a filling nozzle. Also connected with the supply tank is an overflow receptacle. Once the machine is started, vacuum is fed into the overflow receptacle which is airtight. As soon as this happens a vacuum is created at the lower end of the air line or at the suction end of the nozzles. Around the suction nozzles is a gasket and the bottle is pushed into close contact with this gasket to make a positive airtight seal. This immediately creates a vacuum in the bottle unless the bottle is imperfect. If the bottle is perfect, the vacuum created draws the liquid from the supply tank, through the tube and into the bottle. When the liquid reaches the end of the overflow or suction airlines, it automatically breaks the vacuum causing a cessation of the flow of liquid into the bottle. The filling stem is then withdrawn and the bottle passes to the closure plant.

In *measured dosing* the height of fill is not constant. Here, the sequence of operations is as follows. Each individual filling unit consists of a calibrated cylinder and a piston. As the piston begins its downstroke a valve opens, allowing free passage of liquid into the cylinder. At the end of the stroke the cylinder is charged with a measured quantity, and when a container is correctly presented to the filling point, the delivery valve will open and the supply valve closes. Immediately this happens the piston discharges the liquid into the container on its return stroke and the whole sequence repeats.

There are two types of *gravity-filling* machines, one of which fills on a controlled time cycle, and the other using a measuring chamber. In the first, the correct presentation of the container to the filling head causes a valve to open which permits the liquid to flow for a predetermined time. The valve then closes and the container is taken away. The open time is determined by the viscosity of the product and the diameter of the filling orifice and control may be mechanical, by time clock or electronic. In the second type of gravity filler a supply valve opens to admit liquid to a calibrated chamber. When a container is correctly presented at the filling head, the supply valve closes and a delivery valve opens, thus charging the container. Finally, *pressure-filling* is basically similar to time cycle gravity-filling. An artificial head pressure is induced to the

Table 4.1 Comparison of liquid fillers in the food field

	Type of filler			
	Dosing	Vacuum	Gravity/low vacuum	Pressure level system
Suitable for	Thin and thick liquids of most types	Thin liquids and low viscosities	Thin liquids only	Thin and thick liquids
Food industry usage	Soups Sauces Nut and olive oils Flavourings	Sauces Flavourings	Essences Fruit juice	
Spirit and wine industry usage		All spirits	Wine Brandy All spirits	
Limitations imposed on container	None	(a) Stable between 100 and 650 mbar vacuum (b) Volume should be accurate	(a) Stable for vacuum of up to 50 mbar (b) Volume should be accurate (c) Neck opening not less than 7 mm	(a) Volume should be accurate
Limitations imposed by product	In practice none	(a) Not suitable for foaming products (b) Product will be aerated on suck-back	(a) Thin liquids only	(a) Strongly foaming product could give trouble
Accuracy of filling	± 0.1 to 0.5%	Dependent upon volume accuracy of bottle—usually ± 2%		

liquid by a pump or, sometimes, by air pressure within a closed tank. Gravity- and pressure-filling are really only suited to the moderately fast filling of low-viscosity liquids, such as fruit juices. They are capable of handling from one to eight bottles at a time at rates up to about 35/min. For really fast filling, vacuum filling is the best method. Methods of liquid filling are compared in Table 4.1.

Filling dry goods (powders and granular material)

Apart from counting (which may range from sorting equipment with electronic counters to devices consisting of a series of pockets, each holding a single item mounted around a rotating drum), there are two basic methods of filling powders and granules into containers. These are volumetric filling and filling after weighing. The nature of the product normally determines which method is used (Figure 4.3b). Developments are currently taking place in the field of counting devices using checking systems which use photocells, check weighers and sensing devices, all of which can stop the line if underfilled containers arrive.

Volumetric filling. The main methods are filling by auger, flask (or cup) fillers and vacuum filling. In *auger fillers*, an auger is fitted into a sleeve mounted below a hopper (Figure 4.4) containing the product, which should be granular and not too powdery. The diameter of the auger, D, and the pitch of its helix, P, are designed to suit the product being filled. Quantity delivered is controlled by the number of turns the auger makes in one cycle (Figure 4.5).

Cup or flask fillers will handle powders as well as granular materials. A flask of known and adjustable volume (the length is usually adjusted telescopically) accepts the product from a supply hopper, and when brimful the supply cuts off and the flask discharges its contents down a suitably designed chute (Figure 4.6).

The *vacuum-filling* method is essentially similar whether liquids, powders or granules are filled. The containers are raised to the filling head, making a vacuum-tight seal. The filling nozzle projects through the sealing ring, the depth of penetration governing the height of fill. When the seal is made, the container is evacuated and the product flows until the vacuum is shut off. With containers which could collapse under a vacuum, the filling head can be fitted with an airtight shroud and the vacuum drawn in both the container and the shroud, thus equalizing the pressure and preventing collapse.

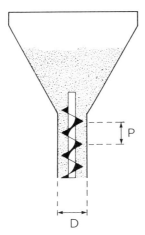

Figure 4.4 Volumetric auger filler (after D. G. Gray, Autopak Ltd.). D = diameter of auger, P = pitch of helix.

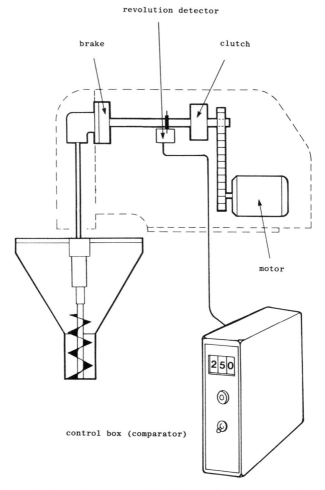

Figure 4.5   Controlling the auger filler (after D. G. Gray, Autopak Ltd.).

With all volumetric fillers, the actual weight filled is dependent on the bulk density of the product. Users have therefore been looking for mechanical assistance for manual check-weighing techniques. At a speed of 300 containers per minute, an operator engaged in manually check-weighing a predetermined number of containers from particular filling heads can only cope with one or two machine adjustments per minute. A better result can be obtained by the use of electronic weighing equipment which accepts the content of four containers manually tipped on to its scale, calculates the average weight and the variation and gives the operator an indication of the adjustment required.

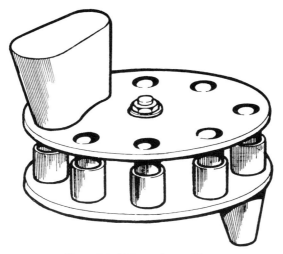

Figure 4.6 Volumetric cup filler.

Filling by weight. Weight filling is obviously the most satisfactory way of meeting the requirements of any weights and measures regulations. There are many different weighing techniques (Figure 4.7) which use single- or double-action scale beams, or heads which are operated by compressed air, and, most recently, electronic and microprocessor-controlled systems. It is also possible to fill products from a weight-balanced conveyor. In all, the basic principle is to divide the supply of product into a main feed and fine feed above the weighing head (Figure 4.8). At the beginning of the weighing cycle, both the bulk and the fine feed are operating to fill the weighing pan until about 80 or 90% of the required material has been added. When this is reached, the bulk feed stops and the dribble feed continues until the exact balance is reached. At that moment the fine feed ceases and the load is discharged, usually by tipping into the container.

It is important to remember that there is always the problem of reducing waste. All machines have an inherent variability, and package-filling machines are no exception. Because of this, if any machine is set to fill containers with a specified weight, some will be underfilled, and some will be overfilled. Overfilled packages waste material by unnecessarily supplying the customer with extra product. Insufficiently filled containers are undesirable and can lead to complaints if they are sold. To guard against too many below-weight packets which have to be removed, packaging operators have to set a target weight for the filling machine greater than the stated net weight. The problem always is to find the optimum target weight so that the number of overfilled packets can be minimized without producing too many containers with contents appreciably less than the correct weight.

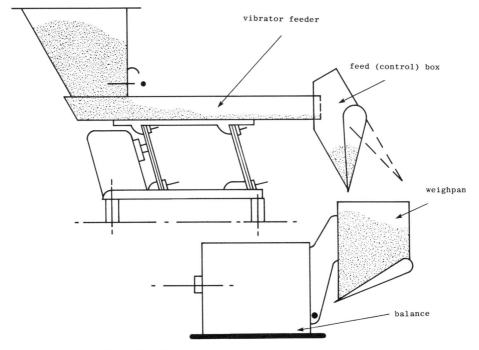

Figure 4.7 Filling by weight (after D. G. Gray, Autopak Ltd.).

Computers can now be programmed to find the inherent variability of a machine used in packaging, and to calculate the optimum target weight for any given allowed percentage of underweight packets. Such an analysis can be carried out on any type of machine. Hence a computer can predict the alterations in target weight which would be possible if new or better machinery were purchased, and hence determine whether the resultant savings due to reductions in overweight packages might justify its cost.

Statistical recording—the legal requirements

Packers in many countries are required to comply with regulations which include the requirement to maintain adequate statistical records of the packed product weight.

In the simplest form of automatic filling machine, a portion of product is measured into the container and the filled container is then passed on for subsequent operations (Figure 4.9). In this simple system, manual quality control must be applied. This usually amounts to taking selected samples during the production run, recording the net weight of product and then carrying out calculations to determine that the requirements are being met.

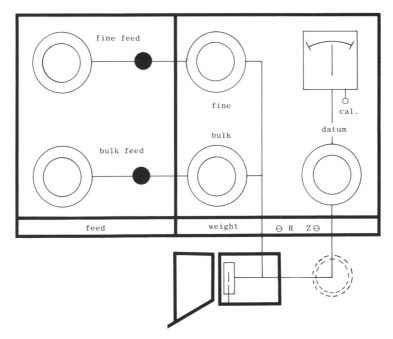

Figure 4.8 Use of bulk and fine feeds.

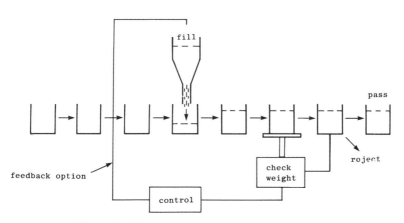

Figure 4.9 Simplest form of automatic filling operation.

The first stage in automating this quality control check is to provide a simple check-weighing device with reject facilities to the outfeed side of the automatic filling machine. Unless the container is consistent in its tare weight, allowance must be made for its weight variation. Therefore, the addition of an outfeed check-weigher can only check the gross weight of product plus

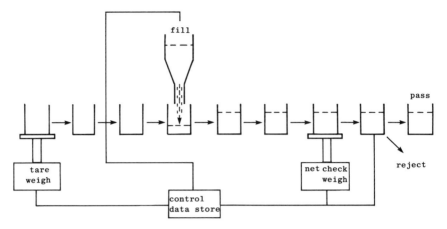

Figure 4.10   System modified to overcome the problem of variation in container weight.

container, with the attendant increased product 'giveway' to compensate for container weight variation.

To overcome the problem of container weight variation, it is possible to add a further check-weighing machine to the infeed side of the filling machine, and, by cross-linking the infeed and outfeed check-weighers, the net weight of product in each container can be determined (Figure 4.10). This eliminates the need for a product 'giveway' allowance against the container weight variation and offers a greater degree of control over the net weight of the contents. The system also has an option of a trend feedback system.

Both systems have a marginal disadvantage in that the control over the filler head is based upon historical information on the filler head performance and, therefore, the trend will only exert a limited influence on weight variations caused by changes in bulk density. However, both systems have the advantage of filling at relatively high speeds since they do not have a severe rate-of-weighing limitation.

To achieve more accurate control over the net weight of product within the container, it must first be weighed and its weight recorded, before being passed to a bulk-filling head where most of the product is filled into the container at high speed. The partly filled container is then transferred to a station comprising a smaller filler head mounted over another weigh cell. The remaining small amount of product is then filled into the container until the required net weight is reached. Any containers over or under weight are rejected (Figure 4.11). Machines of this type are in operation throughout the world. They have proved particularly useful in applications for high-cost products, such as instant coffee, milk powder and other food products.

Such systems have considerable value within the market, but, by weighing at the point of final filling, some constraint is placed on the output. To overcome

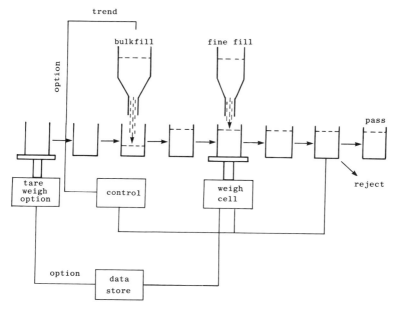

Figure 4.11 More sophisticated system using two filling heads.

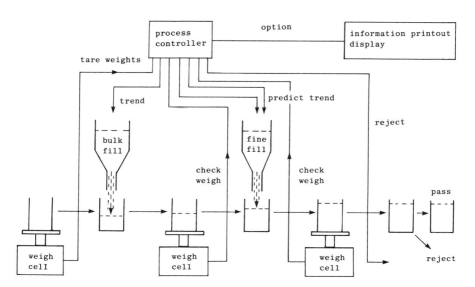

Figure 4.12 Fully automated control system.

this constraint we can remove the final weighing point from the final filling point. The container is first tare weighed and then transferred to the bulk-filling station where most of the product is filled at high speed. The partly filled container is then transferred to a second weighing station where, by use of a computer system, the amount of product required to attain the final net weight is calculated. This calculation is then transmitted to the fine-filling head. When the relevant container arrives at the fine-filling head, the required amount of product is dispensed. Because of changes in bulk density, the amount of product dispensed at the fine-filling station may vary marginally from that predicted and, therefore, a subsequent final check weigh station follows the final filling to ensure that the correct net weight has been achieved (Figure 4.12). Thus, by removing the weighing station from the fine-filling station, considerable increase in output can be achieved at a marginal expense of accuracy.

Semi-automatic weight controlled machines. Not every packer requires the outputs given by automatic machines, and in some applications the size of container is too great for automatic operation. Even when the output requirement is not high, however, there is still a need for accurate weight control. Therefore, similar weight control technology to that applied to the automatic machines has been adapted to semi-automatic machines.

(1) It is possible by placing a weighing system under the filling head to fill the product directly into the container, either rigid or flexible, and when the desired gross weight (product plus container) is reached, a signal is given for the filling to stop. A visual indication of the final weight is given to enable the operator to determine whether the desired weight has been achieved.

(2) Where a greater degree of filling accuracy is required, the container may be manually placed at a bulk-filling head where the bulk of the product is filled at a fast rate; the container is then manually transferred to the fine-filling station which comprises a filling head and weigh cell. Fine fill is initiated and the balance of the product is slowly filled until the desired gross weight is attained.

Capping of bottles and jars

For speeds of up to 60/min, cap tightening alone is usually found to be adequate. For this process the filled containers, with preformed screw caps loosely applied, are presented either manually or automatically to a rubber-lined spinning chuck which has a clutch set to operate when a set resistance is reached. For speeds of 60/min and over, a fully automatic capping machine is required, and this is normally supplied with a cap unscrambler for feeding caps to the machine in a steady stream and in a uniform manner. Transfer of the caps from cap chute to bottle is achieved by one of two methods:

(a) the bottles pick up their caps as they pass under the end of the chute, or
(b) transfer arms attached to a rotary turret strip the caps from the chute and feed them to the capping chucks individually.

Automatic capping machines. These fall into two main categories: in-line and rotary. *In-line cappers* are simpler, incorporating the direct cap pick-off principle, the bottles being held by travelling belts, while another set of belts, operating sometimes on the top of the cap and sometimes on the skirt, spin the cap home. *Rotary cappers* are more sophisticated, holding the containers individually while rotating chucks screw home the caps. The chucks on most rotary cappers operate with three or four self-centring jaws which are adequate for tinplate or other robust caps. However, where plastic caps or closures with a delicate finish are to be handled, a pneumatic chuck is used. This grips the caps by means of a flat neoprene 'doughnut' which is compressed by an air piston and thus grips the cap gently but firmly. Speeds of up to 350/min are usually possible on rotary cappers.

In-line cappers are available for speeds of up to 450/min on simple bottle shapes, but where precise cap torque control is necessary or where the container design is unusual, a rotary model is recommended.

Roll-on capping. The roll-on cap is in wide use for many bottling applications and, as the name suggests, the cap threads are formed by rotary heads whilst the cap is in position on the bottle. This type of cap can also incorporate a pilferproof device (or ROPP) so that the action of removing the rolled-on cap breaks a series of perforations and leaves a 'tell-tale' ring on the bottle neck. Roll-on cappers may be leased by the cap manufacturers to the users, and are of the rotary type with direct cap pick-off.

Corking and plugging. Corking and plugging machines are similar in construction to rotary cappers, the corks or plugs being fed from an unscrambling hopper to a chuck via a transfer arm or disc. The chucks do not rotate but push the closure home with a downward thrust. Some corking machines give one or more blows to the cork through the corking head to ensure that it remains firmly in position.

Crown corks. The crown cork, widely used for beer bottles on filling lines operating at up to 600/min, is applied by a rotary capper, usually an integral part of the filling machine. Pressure is exerted on the crown to compress the wad or liner, while clinching heads crimp the corrugated skirt into a groove on the bottle neck. The finished closure can withstand internal pressures of up to 100 psi.

Labelling bottles and jars

Up to speeds of 60/min, semi-automatic labelling machines are usually adequate. These require an operative to present a bottle to the machine which, in turn, glues and applies the label. Apart from the method of handling the container, there are three main methods of adhesive application used on all automatic labellers:

(1) Adhesive is applied directly to the labels while they are held on a vacuum pad or by grippers.
(2) The adhesive is applied to a plate or turret which, in turn, applies the adhesive to the label.
(3) The bottles receive a pattern of adhesive slightly smaller than the label itself and the label is then applied to the bottle.

The method adopted would probably be dictated by the type of application, i.e. the number and position of labels, whether or not they were to be washed off, etc.

High-speed labelling of round containers can usually be achieved effectively on one of a wide variety of rotary labelling machines, but where shapes other than round are to be handled, or where a round bottle is to be labelled in register with a stop or 'blip' on the bottle itself, an in-line labeller is required.

Case packing and sealing

After the bottles have been filled, capped and labelled, they are often packed into fibreboard outers for transit to the wholesale depot or retailer. (Quite frequently this process is preceded or replaced by shrink-wrapping). The case packing process is often performed manually, coupled with final inspection of the finished bottles; however, for high speeds of 120/min and over, automatic case packing is justified.

Where dividers are necessary, as with many glass bottles, a 'drop-packer' is required, which collates the bottles and then drops them, guided by spring or plastic guides, into the waiting cases with dividers already fitted. When this operation is completed the filled cases proceed to the case gluer which glues the flaps, ploughs them into position, and then places the cases in a compression section of suitable length to allow the adhesive to set before the case is released.

An alternative method of case packing is the 'wrap-around' method which, as its name implies, is a system of wrapping a fibreboard case blank around a collation of bottles or other containers, thus giving a tighter package and often board savings of up to 15%. Dividers can be inserted, if required, but these are not necessary on many containers due to the tightness of the wrap.

Shrink- or stretch-wrapping

Jars of jams, etc., were the first primary containers to be shrink-wrapped instead of being packed in fibreboard cases. The use of shrink- or stretch-wrappers has now extended to practically all forms of foodstuffs, whether in bottles, cartons or cans, as well as to many other types of consumer goods. The choice between case packing and shrink- or stretch-wrapping is not a subject to be dealt with here, but where shrink-wrapping is selected, the collation of containers is usually wrapped in a loosely fitting tube of shrink-film, normally polyethylene, which has been biaxially oriented, and the pack is subsequently passed through a heated tunnel which softens the film, causing it to shrink and tightly envelop the contents. A cooling section normally follows the heated tunnel so that the packs may be handled immediately. With stretch-wrapping the film is tensioned around the collation and no heat is needed.

Palletizing

Pallets are universally recognized as a convenient aid to stacking and storing regular-shaped packages. They are flat platforms, frequently made of timber, with a space between the upper and lower faces for the insertion of forks from fork-lift trucks. Pallet loads are roughly cubic in shape, being formed by layers of cases or goods which have been accurately placed to give a good bond or interlock between layers. While manual loading of pallets is common, high outputs require automatic palletizing equipment which receives filled cases from one or more lines, collates them a layer at a time, and loads them on to pallets ready for collection by fork-lift trucks.

Shrink- or stretch-wrapping of pallet loads is now widely used. This process obviates the necessity of strapping, gives a firmer bond to the load, is an aid against pilferage and a means of protection from the elements.

Canning operations

The basic format of the metal food can has not altered in over 50 years, but since 1970 a number of technical improvements have been introduced. Almost all three-piece cans now have welded side seams rather than soldered and some are now of a two-piece construction, a single piece body and makers end, either drawn and wall ironed (DWI) or draw redraw (DRD) and the closing end will frequently have a ring-pull easy-open feature.

The canning of processed foods may be divided into eight unit or basic operations:

(1) Handling and storage of empty cans
(2) Cleaning empty cans
(3) Product preparation

(4) Filling
(5) Closing
(6) Processing
(7) Cooling
(8) Handling and storage of filled cans.

Handling and storage of empty cans

Tinplate containers do not have unlimited resistance to physical abuse, nor can they be expected to withstand indefinitely conditions which could promote corrosion. Cans should therefore be transported in a manner that will avoid damage to the rims, excessive denting, or rupture of soldered laps. When transport conditions are unavoidably severe, empty cans can be packaged in paper wrappers or fibreboard cases. They must, of course, be kept clean and hygienic during transport. Empty cans should be stored in the dry and protected from changes in temperature that will result in moisture condensation and subsequent rusting. In coastal areas, even small amounts of sea salt promote corrosion. Storage facilities should be arranged so that the first cans received are those that are first used (see Table 4.2).

Cleaning empty cans

Although cans are delivered to the user in a clean condition, it is nevertheless often necessary to wash them before they are filled. To be effective, the washing

Table 4.2 Handling and storage of cans (adapted from information supplied by The Metal Box Co. Ltd.)

Can packaging	Cans are best delivered on pallets with an overcover of board, the layers of cans being separated by layer pads; the outer covering should not be removed until the cans are required
Can storage	Ensure warehouse building is weatherproof and maintained at a constant temperature (16°C), the humidity of the warehouse atmosphere being kept as low as possible
Can handling	Cans should always be handled with care, avoiding denting, rim damage or scratching
	Conveying is best done by plastic-covered wire rope, magnetic conveyors or slat conveyors
	Plastic-covered wire rope reduces 'cable burn'; 'buildback' must still be avoided
	Use an accumulator for feeding machines; magnetic conveyors used for elevating, lowering and dividing cans
	Alpine conveyors allow 'buildback' as well as height change, accumulator function.
	Slat conveyors, used for entry and exit to machines, are tolerant to can slip
	Dividers are used to produce or combine two or more can streams; selection must be made with care and related to can speed
	Can collators or unscramblers group cans into units for packing into retort baskets or cartons

should be done by sprays of hot water with the cans in an inverted position. A jet of steam is insufficient to ensure proper cleaning.

Product preparation

A first and important step in the sequence of canning operations, is that of cleaning and preparing the food before filling it into cans. This may be carried out in a variety of ways, depending upon the product. From an aesthetic or utilitarian viewpoint the purpose of such preparation steps as trimming and slicing is clearly evident. It is also obvious that adequate washing will diminish substantially the extent to which the food is contaminated with spoilage bacteria which could reduce the effect of processing that follows filling and closing.

Filling

Cans should be filled uniformly and with the proper amount of contents. Proper filling serves to expel undesirable gases, especially oxygen, and at the same time aids in creating an internal vacuum after processing and cooling. The formation of a vacuum in the container is accomplished by filling with a hot product or by heating the contents after filling but before closing. Underfilling the can to provide a head space, usually in the range of 6–9 mm, varies somewhat with the nature of the product and the size of the can. There

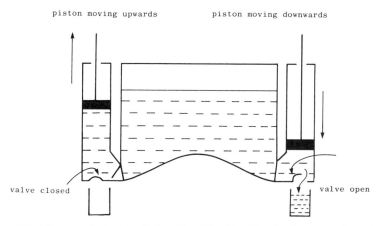

Figure 4.13   Diagram of external cylinder piston filler. The filling head consists of a tank with external measuring cylinders into which the product is drawn as the tank rotates. While the product is being admitted to each cylinder, the filling value is closed. It is then opened by the action of an externally operated cam and the product is delivered through the open valve into the can by a piston. The valve is closed by the valve operating cam. Products: jams, fruit juices, soups, baby foods, cream, milk products, sauces, purées, edible-oils, etc.

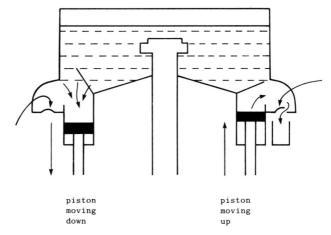

piston   piston
moving   moving
down   up

Figure 4.14 Diagram of internal piston filler. The filling head consists of a tank with internal measuring cylinders into which the product is drawn by a stationary flight as the head rotates. While the product is being admitted to each cylinder, the filling valve is closed. The valve is then opened by the action of an externally operated cam, and at the same time the product is forced by a piston through the open valve into the can. Products: viscous products such as some types of meat products and pet foods, as well as the medium to heavy range of products not handled by the external piston filler.

are several reasons for creating a vacuum in the container:

(a) To preserve flavour characteristics and nutritive components that are vulnerable to oxidation
(b) To leave room for gases that may be liberated during heat processing that might otherwise cause the can to distort in a manner similar to a can containing a product spoiled by microbiological action
(c) To avoid strains due to internal pressures
(d) To minimize or eliminate corrosion caused by oxygen.

To avoid a vacuum so great that panelling or collapsing of the can occurs after the processed cans are cooled, the magnitude of the vacuum produced is controlled by choosing suitable filling temperatures and head space dimensions.

Two types of piston filler are generally available, using either an external or an internal cylinder (Figures 4.13 and 4.14).

Closing (seaming)

The canning process is based on heating the sealed containers until the contents are commercially sterile. Clearly, then, it is essential that the sealing operation, also called closing or double-seaming, be such that recontamination of the food by microorganisms is precluded during the subsequent

Table 4.3 Closing (seaming) machinery (derived from information supplied by The Metal Box Co. Ltd.)

Definition	Closing or seaming machines are used for the insertion and double-seaming of the end component (processor's end 'lid')
Principles	There are two operations; (a) first roll operation and (b) second roll operation; (a) causes end component curl to turn inwards slightly to clinch on to the body component, (b) causes flattening of the clinched seam to the body component
	Each of the two rollers has a different profile to carry out its operation
	Seaming rolls may be in duplicate for each operation depending on manufacturers
Types	Semi-automatic or automatic for round or irregularly shaped cans; atmospheric, vacuum or steam-flow closing
	Infeed of cans may be straight or circular; can may remain static with rolls moving around seam or rolls may be fixed with can rotating; can size can be varied by diameter or height or both; the easiest change is by height
	Machines will include end component coding, usually up to 12 digits
	Machines operating over 160 cans/min are best linked mechanically to the filler which gives a smoother can feed transfer and operation
	'No-can-no-lid' feature is standard on all machines
Can seam criteria	Standards vary from country to country; generally it is accepted that an adequate seam tightness consistent with a satisfactory overlap coupled to minimum wrinkle formation relative to can diameter is satisfactory
	'Butting' is replacing overlap and additionally can seam dimensions are being related to can diameter and plate thickness

cooling, handling and storage of the cans. In modern canning, closing may be carried out at high speeds, 1000 or more cans per minute, by highly efficient closing machines. Alternatively, closing may be done on relatively simple, but powered, equipment or even with the aid of manually operated devices. Regardless of whether sophisticated or simple closing equipment is involved, it is essential that operators fully understand the principles of seam formation. Leaking through the seams of cans can occur after processing; therefore seam examination is very important (Table 4.3).

Processing

The thermal processing of canned foods, commonly termed cooking, retorting or processing, is the application of heat at a specified temperature for a specified time. This operation has two fundamental purposes. The first is to produce a commercially sterile product. This means that the product has been subjected to heat treatment at a temperature and for a time sufficient to destroy not only organisms which might adversely affect the consumer's health, but also those organisms which can produce spoilage under prevailing storage conditions. The second purpose is to cook the material to a point where minimum further preparation is necessary by the consumer. It must be

Table 4.4   Processing (derived from information supplied by The Metal Box Co. Ltd.)

Definition	'Process' in retort operations means the application of heat to foods in containers for a period of time at a temperature scientifically determined to be adequate to achieve commercial sterility
Process evaluation	The determination of a process is dependent upon obtaining accurate, reliable heat penetration data, that is obtaining information about how fast or how slowly the product will heat to a desired temperature; in addition, the thermal resistance (the amount of heat required to kill) of microorganisms in each specific product must be known
Equipment	Can be divided into three categories: (1) Batch retorts (2) Continuous rotary retorts (3) Hydrostatic retorts
Batch retorts	Vertical or horizontal; horizontal retorts may have a rotating basket Heated by steam or hot water or steam/hot water; controls may be manual, semi-automatic or fully automatic *Critical points* (1) Steam alone is good medium of heat transfer (2) Important to exclude all air, i.e. proper venting and bleeding (3) Steam inlet in retort must be adequate (4) Steam bleeders must be open during whole process (5) Retort *must* have mercury-in-glass thermometer which must be checked regularly and used for controlling the process (6) Accurate recording thermometer must be installed (7) Divider plates must have perforations 25 mm (1 in) diameter on 50 mm (2 in) centres (8) Steam spreaders are essential on horizontal retorts (9) Vents must not be connected to a closed drain system (10) Retort must be fitted with pressure gauge and safety valve
Continuous rotary retorts	Horizontal with two or three shells: (preheating, holding, cooling) Cans rotate by means of a reel guiding them through a scroll on the internal periphery of the shell Cans continuously rotate and agitate contents *Critical points* (1) All air must be excluded (2) Retort must have mercury-in-glass thermometer for process control (3) Accurate recording thermometers must be fitted (4) All other items are automatically fitted by supplier, e.g. safety valve, bleeds (5) Critical factors in the can such as minimum headspace, initial temperature, cooker speed, maximum consistency, vacuum and can specification must be carefully controlled (6) Can sizes must be restricted to that specified
Hydrostatic retorts	Basically a static retort operating at a constant temperature, through which the containers are conveyed in the required process time by a conveying system; the term 'hydrostatic' refers to the fact that the steam pressure is balanced by the weight of water in the feed and discharge legs, i.e. hydrostatic pressure. The range of can sizes is dictated by the can carriers but these allow a considerable degree of flexibility *Critical points* (1) All the features listed under continuous rotary retorts (2) Conveyor speed must be carefully checked

remembered that in accomplishing these two purposes, care must be exercised that processing does not destroy quality factors such as flavour, texture, colour and nutritive value. All reliable processing specifications are based on time–temperature conditions adequate to destroy *Clostridium botulinum*, a micro-organism that produces extremely potent toxins. Without diligent attention to processing conditions, it would not have been possible for the canning industry to have obtained such a remarkable performance record for the safety of canned foods, especially in tropical climates.

The destruction of microorganisms by heat, and their ability to multiply themselves, is highly dependent upon the acidity of the product. In general, products having a pH below 4.5 can be processed without recourse to temperatures higher than that of boiling water. Products that have a pH greater than 4.5 require a higher temperature in order that the processing time is not excessively long. Of course, the physical texture of the product and the size of the container also determine the processing time. Heat penetration through viscous or semi-solid products such as meat is slow. Accordingly, for solid products and larger cans a longer time is required before the temperature at the centre of the can rises to the necessary level. Regardless of the product, processing specifications are always based on the temperature at the coldest

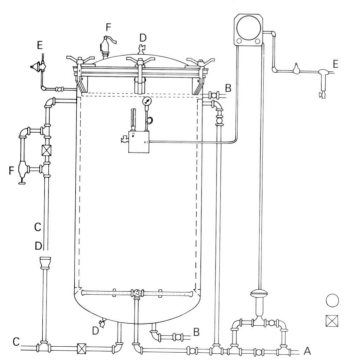

Figure 4.15   Vertical retort: A, steam; B, water; C, drain, overflow; D, vents, bleeders; E, air; F, safety valves, pressure relief valves.

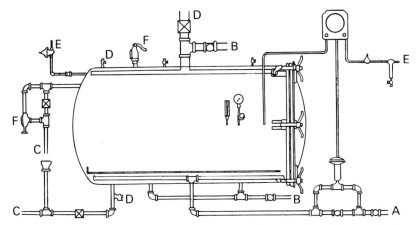

Figure 4.16 Horizontal retort: A, steam; B, water; C, drain, overflow; D, vents, bleeders; E, F, safety valves, pressure relief valves.

point within the can. All installations for processing canned foods under steam pressure must comply with local regulations for piping and necessary safety features.

The various methods of retorting, and the critical factors for each method, are summarized in Table 4.4. Two types of batch retort (vertical and horizontal) are shown in Figures 4.15 and 4.16.

Cooling

The last important operation before cans are ready for labelling, storage and marketing is to cool the sealed cans after processing. Clearly, the objective of cooling is to halt the deleterious effects that over-cooking might have on the

Table 4.5 Cooling methods and machinery (derived from information supplied by The Metal Box Co. Ltd.)

General principles	Cooling must always involve chlorinated water with an adequate level of free chlorine
	A 'wet' can is a dangerous can and must not be handled manually
Batch retorts	Cooling water may be used once only or recycled through a cooling tower with adequate control of chlorine level; additional cooling is sometimes used externally, i.e. open tank with baffles; can temperature should reach around 38–40°C and will then retain sufficient heat to dry off
Continuous retorts	One shell is used for continuous cooling; water flow must be carefully controlled to avoid overcooling; this would lead to drying problems and the danger of corrosion
Hydrostatic retorts	Final leg contains cooling water usually in countercurrent flow and spray system; flow rate is governed by water temperature
Can dryers	Used after cooling; may be simple air-jetting system directed at seams and/or passage through hot air drying tunnel

product, as for example excessive softening of the food or objectionable changes in flavour or colour. Small cans for certain products may be air-cooled, but cooling is mostly done in water in several ways. The hot cans may be cooled by admitting water into the retort in which they were processed, or the cans may be removed from the retort and conveyed through a tank or shower of cold water. Large cans and certain irregularly shaped cans must be cooled under pressure in order to avoid excessive strains on the container. This may be accomplished in the retort, using either air or steam to counteract the pressure developed within the can during the processing operations. Water for cooling must be free from microbiological contamination, or else it should be chlorinated. Critical points in the cooling process are summarized in Table 4.5.

Handling and storage of filled cans

Strictly speaking, the handling and storage of canned foods are not a part of the canning operation *per se*. Nevertheless, they are important to a successful operation. Rough handling and contaminated runways can lead to spoilage

Table 4.6   Post process can handling (derived from information supplied by The Metal Box Co. Ltd.)

General	Food preservation depends on: (1) destruction of bacteria by heat, and (2) prevention of recontamination of the product by means of sealed container
Process factors	Trend towards double-reduced tin plate and aluminium cans plus higher speeds calls for great care; in cooling, tins go from pressure to vacuum; small deformations in high-speed automatic equipment may be more significant than low-speed low-impact conditions
Spoilage factors	(1) Quality of can double seams (2) Presence of bacterial contamination in cooling waters or on runways (3) Excessive abuse due to poor operation or adjustment of filled can handling equipment.
Operating precautions	Inspect can seams regularly Ensure seamer adjustments have not exceeded tolerances Inspect can handling system from seamer to case packer Do not allow cans to drop freely into crates from seaming machine discharge tables Do not overfill retort crates Prevent sharp impacts between filled crates Chlorinate all cooling waters to give free chlorine level Thoroughly clean and sanitize all belts and runaways in contact with can seams at frequent intervals Dry cans in can dryer; remove all surplus water as soon as possible Keep handling to a minimum Avoid rolling can conditions where can-to-can contact may occur When cans are rolled, slopes, can spacing and can speeds must be carefully engineered to avoid can bumping Can seams must never contact the runway surface At palletizers and other take-off and transfer points, make provision for continuous and gentle deceleration of the cans Plant must be designed for ease of cleaning

due to the introduction of microorganisms; storage of cans at excessively high temperatures or under conditions favourable to corrosion may ruin an otherwise satisfactory pack. The storage of canned foods in temperate climates presents no serious problem so long as recommended practices are followed and providing cans are protected from moisture. Clearly the storage of filled cans in a damp warehouse or the filling of wet cans into cartons should be avoided.

In tropical climates, and especially at locations near salt water, the protection provided by conventional tin coatings may be inadequate to prevent corrosion on the outside of cans placed in storage. When cans are stored at temperatures as high as 50°C (120°F), internal corrosion will be very substantially accelerated, and the growth of certain spoilage organisms (thermophiles) that normally remain dormant is possible. This condition is encountered even in temperate climates when cans are stacked in large blocks before they are adequately cooled. A summary of the important points is given in Table 4.6.

Wrapping operations

For wrapping mass-produced articles in a constant flow, automatic wrapping machines replace the manual operator. The speed of packaging is greatly increased, and in the case of small objects such as toffees which are convenient

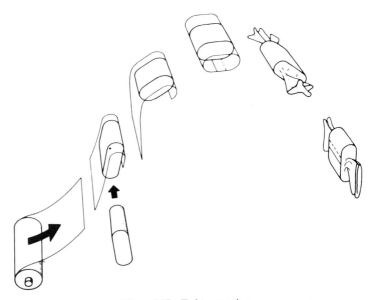

Figure 4.17 Twist wrapping.

to feed and wrap, speeds of up to 600 pieces per minute may be achieved by cutting a piece of film, forming it into a tube around the object and twisting the ends of the tube (Figure 4.17); this is known as *twist-wrapping.*

Rectangular objects lend themselves to a mechanized version of the *parcel wrap* (Figures 4.18 and 4.19). There are several versions which can be selected

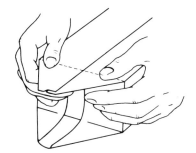

Figure 4.18   Parcel wrapping.

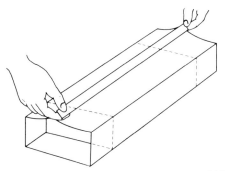

Figure 4.19   Parcel wrapping using a 'grocer's fold'.

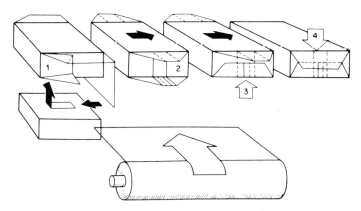

Figure 4.20   Mechanized 'parcel' wrapping.

according to the size and shape of the object to be wrapped. In each the principle is broadly similar: a length of wrapping material is drawn or fed from a reel and cut off, the object to be wrapped is pushed into it and the ends folded around the object, forming a 'tube', with an overlap of 5–25 mm. The open ends of the tube are tucked in appropriately, and the overlap and tucks are sealed in place by heat seals or adhesive. Neatly rectangular packets of cereals or cigarettes can have one end tucked in by grippers after which the package is then pushed through pairs of ploughs which fold the other flaps in order (Figure 4.20). When the objects to be wrapped are not constant in size, or are soft and compressible, the ends are better folded in by grippers as in Figure 4.21. In order to secure each fold as it is made, the ends are folded down in a style more suitable for large loaves of bread.

This principle of the bread-wrapping machine produces a direct wrap using material drawn from the reel. The product is fed by a flighted conveyor through a curtain of heat-sealable material on to an elevator. The wrapper is gripped between the product and a keep-plate on the elevator, and the first end-fold is made. As the elevator moves upwards, the wrapper is pulled from the reel, formed around the loaf and the second end-fold is made. By using the product to pull-feed the wrapper, the length of film used is determined by the product girth. This is particularly useful for bread which varies somewhat in size from loaf to loaf. The wrapper is separated by a serrated knife as the loaf is moved forward by a reciprocating pusher. The third and final end-folds are

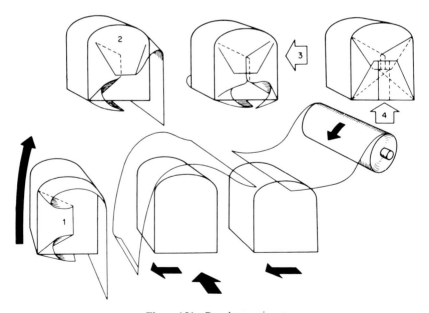

Figure 4.21 Bread wrapping.

formed and the base longitudinal seam is made. This and the end-folds are then heat-sealed and the wrapped loaf is discharged by belt conveyors.

Waxed paper and heat-sealable cellulose films can be used with the standard machine and a version is available to use polyethylene and cast polypropylene. Sealing of the plastic films is achieved using heated discharge belts.

A machine (Figure 4.22) for wrapping crusty bread (French sticks, bloomers) uses the horizontal form-fill-seal principle. This produces a pillow-pack with crimped seals at each end if required. However, a twist-closure can be formed on one end to produce an easily opened bag-like pack. This machine uses heat-sealable film and product feed is by hand. Bread may also be sold in preformed bags and both automatic and semi-automatic equipment is available.

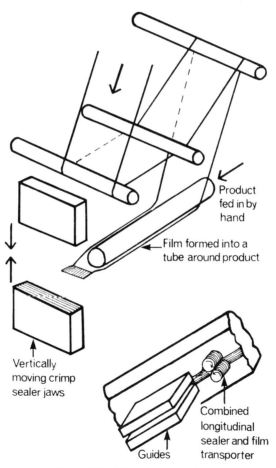

Figure 4.22 Wrapping crusty bread.

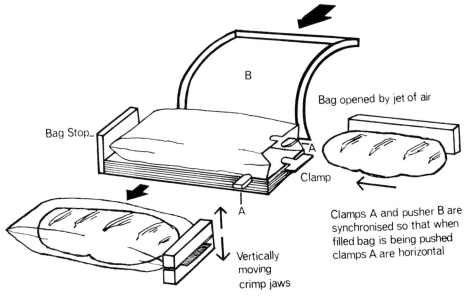

Figure 4.23 Bread bagging.

Bread can be bagged automatically by using 'wickets' of pre-made polyethylene bags. A bag is blown open by air and a scoop enters the open mouth, expands, grips the bag, and pulls it from the wicket over a suitably-positioned loaf. The open end is then closed by a twist-tying machine. This is achieved by gathering the open mouth of the bag together to form a neck and twisting a plastic-covered wire round it.

For small bakers, suitable semi-automatic bagging machines are available (Figure 4.23). Wickets of polyethylene bags or cellulose film bags with a thumb-cut can be used on this equipment. Again, air is used to blow the bag open, and the product is inserted by hand, this action separating the bag from the stack or wicket of bags. Closure can be effected by semi-automatic tying, by adhesive-tape applicators, by stapling or by a straight heat-seal where applicable.

Bags: manufacture, filling and closing

Any form of bag consists simply of a prefabricated wrapper which only needs the product and the closure to complete the pack. They have always been useful where hand-filling operations are involved, and recent developments in automatic bagging machines have raised the potential yet again. Bags have been used for many years for such foods as flour, sugar, and rice, and the

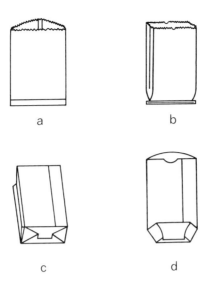

Figure 4.24   Main types of paper bag: (a) flat; (b) gussetted; (c) self-opening satchel (SOS); (d) rose bottom.

improvement in bagging equipment coupled with the wider choice of materials, notably polyethylene, has increased the potential.

Paper bags

There are four standard styles of paper bag: flat, gusseted, self-opening satchel and rose bottom (Figure 4.24). They are made in various types and weights of paper according to the size, weight and nature of the contents, and the purpose for which they are intended. Sulphite papers, particularly the machine-glazed (MG) variety, are used for general purposes, and kraft (made from sulphate pulp) paper bags are used where strength is essential. If resistance to grease or fatty foods is necessary, bags are made in greaseproof, vegetable parchment, or glassine, all of which materials can provide some resistance to the loss of flavours. Paper bags made in wet strength or waxed paper are also available, but it must be remembered that such materials will only give resistance to the passage of water vapour if special attention is given to the closure and to the seams.

Film bags

These are used extensively for packaging. In general they all provide more protection than a paper bag while some films give visibility to the product. Flat, satchel and gusseted styles are available, as with paper bags. In some instances, film-fronted or -windowed bags using both films and paper may be desirable. Films such as LD or HD polyethylene, which may be extruded in tubular form, can be produced both with or without gussets and contain no seam in the length direction of the bag. The availability of this type of bag and the great improvement in its resistance to moisture, water vapour transmission, and grease, and an easy, efficient closure has led to an upsurge in the use of such bags for foods. HD polyethylene bags, for example, have virtually eliminated waxed paper, greaseproof and vegetable parchment from the field of packaging fatty and wet foods such as bacon, meats, fish, etc.

The principal film materials used for food bags are low and high density polyethylene and cellulose film, in the plain, moisture-proof and heat-sealing moisture-proof grades, although polypropylene, polyester, vinyl and saran-coated type films are also used. Cotton bags also have limited uses for packaging larger quantities of foods. They are obtainable in bleached and unbleached qualities and may be printed if required.

Open mesh bags

Products such as fresh vegetables that require complete ventilation in transport and storage are frequently packed in open mesh bags. In smaller quantities, up to about 3 kg weight, polyethylene film bags with perforations may be used for the same produce, but above this weight film bags are not so good. These open mesh bags were first made in hessian with a very open weave, and were then produced in yarn twisted from special kraft paper, but now most are made in a mesh of tough resilient plastics. Maximum ventilation with good product retention and handling characteristics, plus a certain attractiveness, is obtained for products such as carrots, sprouts, etc., in weights up to 15 or 20 kg.

Bag filling and closing equipment

There are four operations involved in bagging:

(1) Feeding the flat bag to the loading point
(2) Opening the bag and keeping it open
(3) Loading the product
(4) Making the closure.

The initial opening is usually achieved by an air-blowing device, often assisted by the incorporation of lips on the bag. Mechanical fingers are also used either alone or to assist the air-opening device.

Originally the bag was then kept open and the product guided into it by means of a scoop or other means down which the product slid. Now the alternative principle of inserting scoops into the bag and (holding it under slight tension) pulling the bag over the product, is employed particularly for the bagging of sliced bread. The principle of moving the bag rather than the product enables bagging to be performed on products with little resistance to distortion or crushing

A wide choice of closures is available. Ties, clips, staples, tapes and heat seals are all employed.

Bag-in-box packages

Given a wrapping material inherently water vapour resistant and heat-sealable (whether the seal is a weld right through or only a surface seal), it is always possible to make a bag more efficient than an overwrap for a box of a given size. Moreover, many film materials are also odour-resistant and therefore can protect food products better when in immediate contact, and thus avoid locking up potential sources of odour within an overwrapped and usually printed carton. Cheaper grades of boxboard may be used for the carton if an inner bag is employed.

For such reasons there has always been interest in bag-in-box packages. There are three basic types. In the first, the bag is loose within the box. This type is usually produced during the package-forming and filling operation. The second type produces a bag and then spot-glues it into the box or carton during the making operation and delivers the folded flat bag-in-box to the user. In the third style, which is also delivered as a flat unit, the bag liner is secured firmly to the walls of the carton of which it forms an integral part, being produced from sheet or tube during the finishing operations of carton manufacture. These styles usually extend the bag beyond the top and bottom flaps of the carton and the bags are frequently heat-sealed and secured in position by the closure of these flaps.

Such bag-in-box packages are not only used for the packaging and protection of dried and powdered foods, but also for liquid packaging such as fruit juices and even wines in quantities of up to a gallon or more. Obviously careful evaluation of both the product requirements and the handling involved is essential.

Cartoning

A cartoning system combines a special carton with the machinery to erect it from a flat condition, fill it with a product, and close it. The machinery varies from simple hand-fed machines to automatic stations coupled with means for packing the cartons directly into cases for dispatch. Whatever the system employed, three main operations are performed.

(1) *Forming or erecting the container*: material may be fed to the carton erection point as a continuous web, as a flat carton blank, or as a folded carton flat with a manufacturer's joint secured.

(2) *Loading or filling the container*: in the continuous web-fed and top-loading systems, the container has only one face open at the moment of filling; with an end-loading system it is possible with certain products to load the carton and close both ends afterwards.

(3) *Closing or sealing.*

In addition to these main operations, cartoning systems may be required to carry out secondary operations such as handling paper liners, embossing codes and inserting leaflets. All these can be performed along with the three main operations on manual, semi-automatic or fully automatic lines.

When forming, filling and closing must be carried out on one machine in a single operation, fully automatic machinery is usually used. The division between semi-automatic or fully automatic machines relates principally to how the filling or loading of the carton is carried out. If loading is done directly into the cartons, even though an operator inserts it into the infeed conveyor by hand, the system is classed as automatic. If the rest of the operation is automatic but the load is inserted directly into the carton by hand, the system is described as semi-automatic. Most systems requiring high speeds use continuous-motion machines; lower-speed systems usually use intermittent-motion machines. The latter can also be of advantage where the nature of the product demands a stationary carton at the moment of filling.

Cartons for liquid products

Requirements with respect to impermeability and hygiene are particularly stringent when packing liquid food products. It is therefore often necessary to use other methods than those already indicated in order to get satisfactory results. Sealing must naturally be very tight, and there are problems to be solved in connection with filling and how to seal or avoid open-cut edges. The most important liquids at present are milk and milk products and fruit juices. The machines described in the following section may normally be used for all these products. There are two main groups of machines, those which work from a reel, form the package and fill it in a continuous operation, and those which start from a pre-manufactured blank.

Figure 4.25 shows the 'TetraPak' principle of forming from a reel, the different stages being combined in one machine. This machine permits carton-filling under fully aseptic conditions. Developments in this system have led to the 'TetraBrik' package which is oblong (Figure 4.26) and not a tetrahedron in shape. Figure 4.27 shows the 'PurePak', which starts from carton blanks, heat-sealed along one side; output is about 60/min.

Machines for the packaging of liquids are frequently not purchased outright but rented, because of the special working conditions and the importance of perfect servicing. In such instances a down payment plus a certain sum per filled carton, and/or an rental charge is made. Important features of such

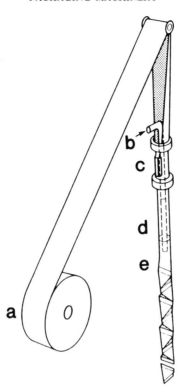

Figure 4.25 The principles of making tetrahedral packages from a reel. (a) Packaging material on a reel is fed to the machine and formed into a tube by a longitudinal seam (b). The liquid enters the tube from (b); filling is below surface level (d). Finally, the package is transversely sealed (e) and given its final shape.

packages for liquids—apart from economy in manufacture and handling and absolute leakproofness even under difficult transport conditions—are the necessity for aseptic filling, simple transport containers to permit easy stacking and easy opening and closing of the units.

Cartons for solid products

As already mentioned, cartoning systems can load the product through the end of the carton or fill the product vertically, and both of these may be performed as continuous processes (Figures 4.28 and 4.29). Cartons may also be supplied for systems which will erect, fill and close either a prelined carton (Figure 4.30) or one which makes its own liner on the machine (Figure 4.31). Both these types of lined carton can be used for vac/gas and vacuum packaging at speeds of up to 60 or 70 cartons a minute.

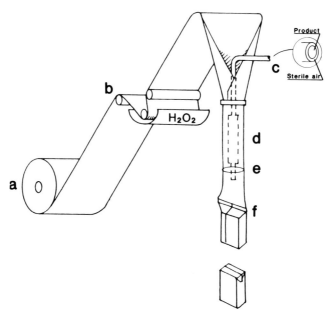

Figure 4.26   The TetraBrik aseptic concept is shown. (a) Packaging material on a reel is fed to the machine passing through a hydrogen peroxide bath (b), and formed into a tube by a longitudal seam. The tube heater zone is indicated by (d). The liquid is admitted through (c), and filling and sealing is performed below surface level (e). Finally, the package is given its final shape (f).

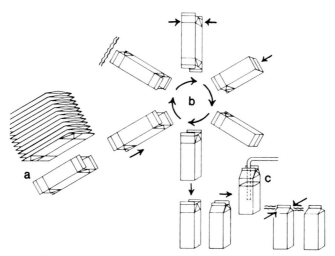

Figure 4.27   The filling and sealing procedure of a gable-top carton from a prefabricated blank is shown. (a) The blanks are fed to the machine from a magazine, unfolded and introduced into a mandrel (b) where the bottom is heated, folded according to scores, and heat-sealed. Now an open box, it is removed from the mandrel, filled (c), and top-sealed by means of heat and pressure.

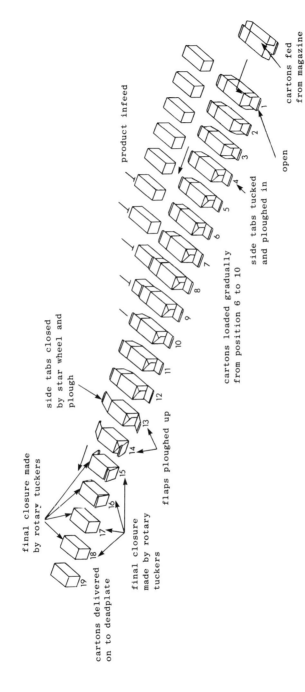

Figure 4.28 Constant-motion cartoning machines—horizontal (courtesy of Baker Perkins Ltd and Baker Perkins (Export) Ltd).

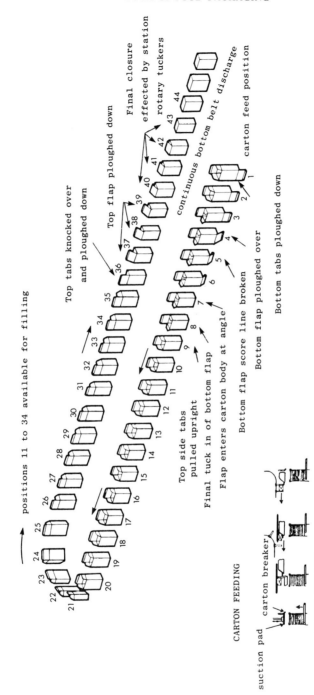

Figure 4.29 Constant-motion cartoning machines—vertical (courtesy of Baker Perkins Ltd. and Baker Perkins (Export) Ltd.).

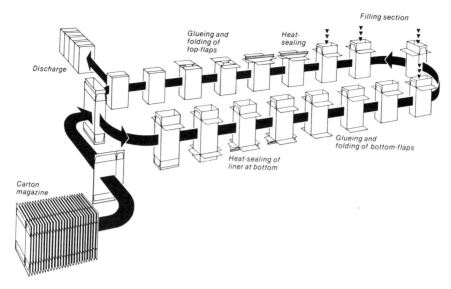

Figure 4.30   Carton system for prelined cartons (courtesy of Pembroke Cartons Ltd.).

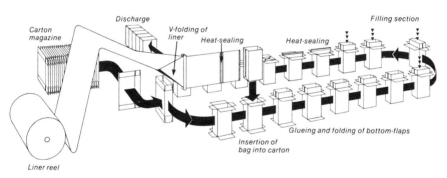

Figure 4.31   Carton system for lining cartons with bags made in line (courtesy of Pembroke Cartons Ltd.).

Cartoning systems

From the point of view of the package user, there are several advantages in utilizing a cartoning system. Firstly, the administration operation involved is simpler. One company is responsible for the supply and servicing of both machinery and cartons. Since the particular supplier of cartons will normally be producing them for the type of machine used in the system on a relatively large scale, he should be familiar with the detailed requirements of the carton blanks or flats and the limitations of the carton handling machinery concerned. The package user need not, however, be completely tied to only one carton supplier as most of the systems available are licensed for more than one production unit. In some instances, there may be two or three sources of supply within one group of companies, and with other systems competitive companies who are all licensees for a particular system can provide alternative sources of supply. The tied cartoning system is principally applicable where versatility and flexibility rather than full automation is needed, and this is frequently the requirement with many food lines.

There are several factors to be considered when selecting a cartoning system.

Machinery considerations.   These include:

(1) The required production rate of the machine both immediately after installation and at the expected peak of production.
(2) The number of package sizes which may be involved.
(3) The frequency of size changes.
(4) Whether the system will have to cater for alterations of the product type or number within the size variation anticipated.
(5) The availability of labour for the machine lines. For example, if in the initial stages a team of five or six operators is needed for one line and the expected production peak would mean a total production of four times the volume catered for on one line, will the system require four times as many operators or can it operate on only seven or eight? If four times the number of operators are involved, are they obtainable in the particular area concerned?
(6) The space available for putting in the machinery, bearing in mind a possible increase in production later on.

Product factors.   These include:

(1) The variations likely to be expected in the product itself, how easy it will be to control the size of the product and within what limits.
(2) The method available for handling the cartons after they have been filled. For example, are they required to go into a deep freezer, or are they to be over-wrapped with a barrier material to provide protection from moisture in an export market? Such factors will determine the nature of any protective barriers that may be required on the board itself.
(3) Coupled with the requirements of the last question, the protection the product itself requires from moisture, from oxygen, from outside odours, etc. Is it greasy, or wet, or otherwise able to affect the board from which the carton is made?
(4) If similar products have been packed in the past, the experience with the method used. For example, has it been found that top loading is better than side loading in terms of product handleability? If there have been preferences in the past, are there new reasons which would nullify such results now?

Board and carton requirements. These include:

(1) The board requirements in terms of the decoration and selling effects required
(2) The protection the carton must provide to its contents, and whether these will have repercussions on the ease of functioning of the carton system from the feeding, erecting and closing standpoints.

General considerations. These include:

(1) Whether the proposed new equipment must link up with any other equipment already in existence, such as overwrapping machines, case packers or filling heads
(2) Whether advice can be obtained from the carton supplier on such other ancillary equipment with which he may have had experience on his cartons on other lines.

Form, fill and seal machines

These machines use a reel of flexible material (paper, film or laminates of paper/film/foil) and either form it into a tube and then seal and fill it at regular intervals, or fold it lengthwise and seal it at right angles to the fold to form a series of pockets (sachets) which are filled and closed.

Machines of the first type form pouches with a seal at each end and down the centre of one face, while the second type of operation produces sachets with three or all four edges sealed. Notice that the difference between the first type and the TetraPak operation (Figure 4.25) is that in the latter, alternate seals are made at right angles to one another, while in the pouch machine they are all in the same plane.

Form-fill and seal machinery may be conveniently divided into three types, which we shall examine in more detail:

(1) Vertical machines, in which the material is formed into a circular section tube over a forming collar (Figure 4.32).
(2) Horizontal machines, in which the material is formed into a rectangular section tube through a forming box (Figure 4.33).
(3) Sachet-forming machines which of themselves are of two types. The first uses a single web folded in half and then cross-sealed (Figure 4.34) and the second utilizes two webs which are brought together and initially sealed on three sides. In each instance, after filling, the remaining side is sealed to form the complete pack.

The seals in sachets are always made between different areas of the same face of the web, but with pouches either face-to-face or overlap sealing is possible on the long face seal produced in forming the tube.

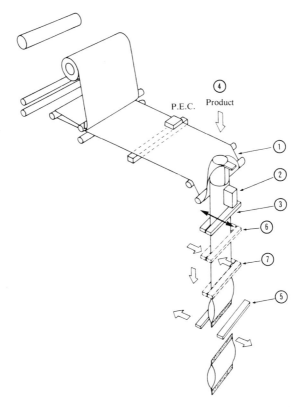

Figure 4.32 Vertical form-fill-seal machine; 1, film from reel made into a tube over forming shoulder; 2, longitudinal seal made; 3, bottom of tube closed by heated crimped jaws which move downwards drawing film from reel; 4, predetermined quantity of product falls through collar into pouch; 5, jaws open and return on top of stroke; 6, jaws partially close and 'scrape' product into pouch out of seal area; 7, jaws close, crimp heat seal top of previous pouch and bottom of new one. Crimp sealed container cut off with knife.

Vertical form-fill-seal (f.f.s.) machines

Vertical f.f.s. machines can produce three styles of pack.

(1) *Pillow packs* (Figure 4.35). These characteristically have fin or overlap seals on the base of the pack and transverse seals at either end. The most common applications are 'solid' preformed items or multipacks, e.g. candy bars or chocolate enrobed biscuits, and collections of smaller solid preformed items, e.g. a given weight of pre-wrapped sweets or a given volume of non-wrapped sweets.

(2) *Sachet packs* (Figure 4.36). These have a four-sided fin seal (occasionally only three) around the edge of the pack. The most common applications

(a)

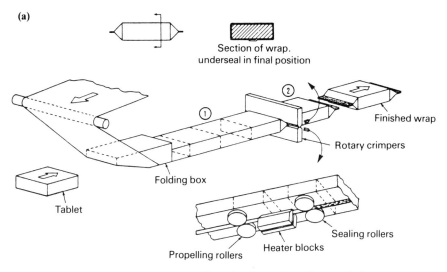

Section of wrap.
underseal in final position

Finished wrap

Rotary crimpers

Folding box

Tablet

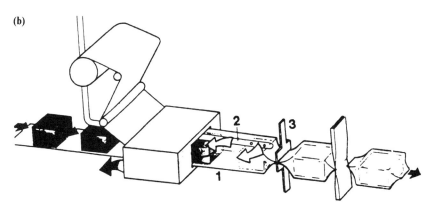

Sealing rollers

Heater blocks

Propelling rollers

View showing sealing of longitudinal seam

(b)

Figure 4.33 (a) Horizontal form-fill-seal machine: 1, film drawn from reel, formed into horizontal tube around product with continuous longitudinal seal underneath formed by heater blocks and crimping rollers; 2, rotary heaters make the crimped end seals and a cut-off produces individual packs. (b) Gas flushing. Purging the air from the package is accompanied by continuous gas flushing. The packaging machine provides a film tube (1). Gas is injected into this tube through an injection pipe (2), which extends to a point just before the sealing jaw (3).

are powdered and granular or similar products (e.g. instant soups, instant potato, instant desserts).

(3) *Strip packs* (Figure 4.37). These consist of two layers of material sealed together to contain the product between them in individual pockets. The most common applications are in pharmaceuticals: pills, capsules, suppositories.

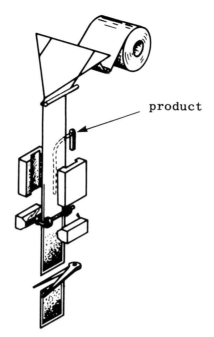

Figure 4.34 Three-sided sachet making.

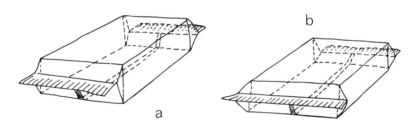

Figure 4.35 Pillow packs: (a) with gusset; (b) without gusset.

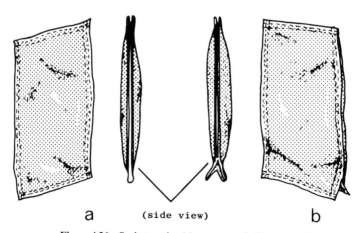

a   (side view)   b

Figure 4.36 Sachet packs: (a) ungussetted; (b) gussetted.

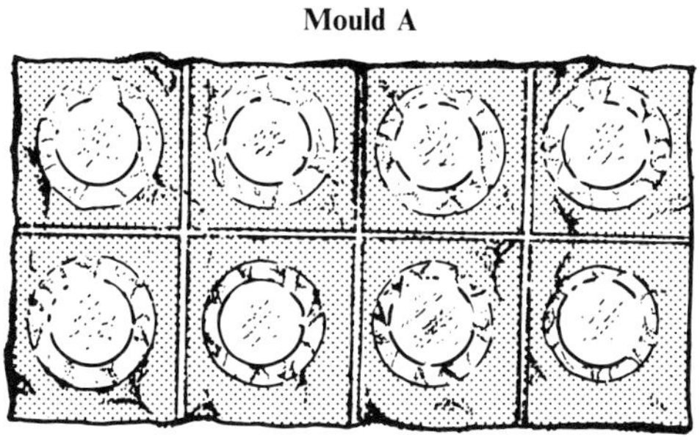

Mould A

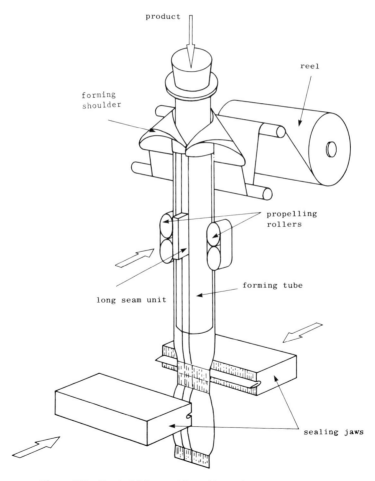

Figure 4.37   Typical strip pack.

product

reel

forming
shoulder

propelling
rollers

forming tube

long seam unit

sealing jaws

Figure 4.38   Vertical f.f.s. machine with special long seam sealer.

Packaging machinery for vertical f.f.s. pillow packs

Sequence of operations of a typical basic machine (Figure 4.38). Most vertical f.f.s. machines make use of a forming shoulder to convert the flat web drawn from the reel into a tube. In most machines the horizontal sealing system contains a driven sealing carriage. The sealing jaws come together to make the cross seal, and then move downwards pulling the required length of packaging material from the reel. At the bottom of this stroke, the sealing jaws open and return to begin the next pull off/sealing cycle. The vertical seal may be either an overlap seal or a fin seal depending on the pack presentation required and on the packaging material used. Figure 4.38 shows a conventional long seal system, where a heated element reciprocates in the horizontal plane to make the vertical seal.

There is one other important alternative system used on certain vertical pillow-pouch machines. This is the principle of the fixed horizontal seal carriage, operating in conjunction with a friction driven film feed. Here the sealing jaw carriage does not reciprocate vertically; only the sealing jaws open and close with a timed sequence. The flexible caterpillar-type rollers drive the film by friction against the forming tube (Figure 4.38).

Overlap seal or fin seal? The overlap seal is more economical in wrapping material usage than the fin seal. It is suitable for most monofilms and for laminates which have a sealing medium on both sides. One-sided films and laminates require a fin seal system.

Sealing techniques. The packaging material used will be determined by a number of factors, including the nature of the product, its marketing

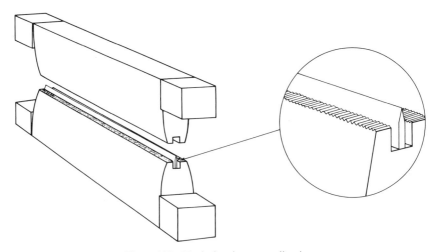

Figure 4.39 Typical resistance sealing jaws.

parameters and the distribution system used for that product. The material selected will have a vital influence as to the sealing system employed on the packaging machine. This may be one of two types.

Resistance sealing (Figure 4.39) is used where the film consists of a carrier or body material with a heat seal coating. The carrier material is not normally affected by the heat and on its own would not adhere with heat and pressure (e.g. aluminium foil, paper or cellulose film). Such materials must be coated with, or laminated to, a heat sealable layer. Continuously heated sealing jaws are mostly corrugated or grooved, and the corrugations on the jaws should be in mesh to obtain a good seal. Such sealing systems incorporate a knife to separate the seal in two (one-bag top seal, one-bag base seal).

Impulse sealing is used for unsupported materials, where the material can be sealed to itself on either surface. The components of an impulse sealing unit are shown in Figure 4.40. The nichrome resistance strip is heated (often to red heat) by an accurately timed, low voltage, electrical current impulse. The radiant heat wave melts the polymer films clamped in the sealing jaws. The short duration of the impulse is followed by a cooling time (to give good seal strength).

Three-sided and four-sided sachet packs. Typical applications are for liquids, instant soups, instant desserts, instant potato and tea bags. The sequence of

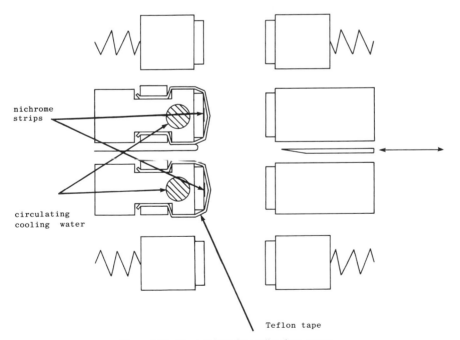

Figure 4.40 Typical impulse sealing jaw system.

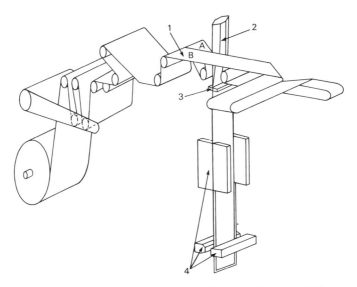

Figure 4.41   Four-sided sachet making: 1, feeding of the wrapping material from two webs, A and B; 2, filling tube; 3, shoulder plate; 4, longitudinal sealers; 5, top and bottom sealers and cut-off.

operations for four-sided sachets is shown in Figure 4.41 (for three-sided sachets, see Figure 4.34). Advantages and disadvantages of the vertical f.f.s. system are summarized in Table 4.7.

Horizontal f.f.s. machines

These have much in common with their vertical counterparts. Like the vertical machine, the horizontal machine combines the three separate operations of pack forming or shaping, product introduction into the pack and pack closure. The process also uses a reel or reels of wrapping material; the principal difference is that material is pulled from the reel into a *horizontal* plane where the subsequent operations take place.

Pack styles can be conveniently grouped into two main types. *Pillow packs* are virtually similar to the vertical packs. The most common *horizontal* applications are for 'solid', preformed single items or multipacks (e.g. candy bars, biscuits). *Sachet packs* typically have three- or (more usually) four-sided fin seals around the edges of the pack. The most common applications are virtually the same as for the vertical sachet machine—powdered, granular or similar products and liquids, e.g. instant soup, instant potato, instant desserts, etc.

Sequence of operations of a typical basic machine (Figure 4.42).   The products are loaded into a continuously moving infeed conveyor. The type of conveyor

Table 4.7 Advantages and disadvantages of vertical f. f. s. pillow and sachet packs (adapted from material presented by A.P. Benson, Institute of Packaging Education Course, February 1980)

Advantages	Disadvantages
Pillow pack	
Very wide range of materials, from inexpensive coated film and papers to complex laminates (resistance seal); wide range of plastics-based materials (impulse seal)	Pack filling and sealing take place in the same location, so fine powder products can interfere with the seals
Relatively simple machine construction and comparative low cost of packaging machinery	Multicomponent filling not usually possible
Easy adjustability for a wide range of sizes	Most machines are intermittent-motion, and to gain high outputs, a more complex multiple tube arrangement must be used
Can handle a wide range of powder, granular and small sized, regular-shaped products using a variety of product feed systems: auger fillers, volumetric feed cups, weighers (also piston pumps from liquids)	
Various machine executions available, from low output, inexpensive single-tube machine (60 pm) to high output multiple-tube machine (240 pm)	
Numerous options available for special applications (gas flushing, vacuumizing, simulated block bottom bag, etc.)	
Comparatively good pack size/product volume ratio	
Sachet pack	
Generally more popular for marketing of smaller quantities of powdered products	Comparatively poor pack size/product volume ratio
Very good containment of fine powder products (perhaps less potential leak areas than with pouch packs)	Only suitable for 'flat' packs—more bulky products excluded
Can use relatively inexpensive wrapping material (e.g. heat seal cellulose film, coated polypropylenes and polyethylene) and certain laminates	Pack filling and sealing take place in the same location, so fine powder products can interfere with the seals

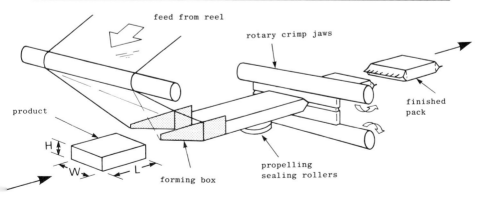

Figure 4.42 Horizontal f.f.s. machine: sequence of operations.

is dependent on the product, but a 'chain-lug' infeed is the most commonly used. The wrapping material is formed into a tube inside a folding box. The edges of the tube are fin-sealed together as they pass between a pair (or several pairs) of horizontally mounted propelling and sealing rollers. The products are pushed into the formed tube by the advancing infeed. A pair of rotating sealing jaws seals the tube transversely between the products, and separates the packs with an integrally mounted knife.

Automatic product feeding. For pouch pack machines these may range from relatively simple units for the more common applications, to custom-designed systems to suit a particular individual requirement.

Alternative sealing systems. The type of folding box and the layout of the long seam propelling/sealing unit depend upon the packaging application and the materials used. Additionally, different types of transverse sealing units may be required.

High integrity pack sealing For pack seals of high integrity on heavier gauges of material, it may be desirable to have a reciprocating sealing head instead of the conventional rotary crimp jaw assembly (Figure 4.43). Such a head is used for pharmaceutical packs or for long shelf life packs, as required by the bakery, biscuit and other food industries. The unit increases both the sealing pressure and the time during which the sealing jaws are in contact with the wrapping material. It comprises a reciprocating carriage with combined cross sealing jaws and cam-operated knife. At the start of the cycle, the high-pressure jaws come together to make the transverse pack seals. The meshed jaws then move forward at the same linear speed as the wrapping material. At the end of the horizontal stroke, the cut-off knife operates and the jaws separate vertically. The return stroke then commences, ready to begin making the seals on the next pack.

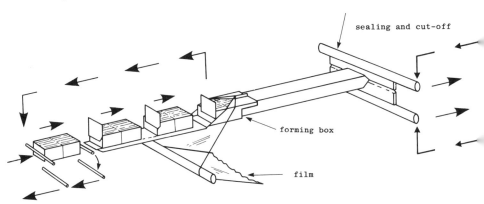

Figure 4.43 Pillow pack making with a reciprocating sealing head.

The sealing of unsupported polyethylene or similar plastic-based materials.
Unsupported plastic-based materials are in nearly all instances unsuitable
for handling on conventional horizontal pillow pouch machines. On these
machines, heat is conducted through the wrapping material to the inner
coated surface which 'melts' to form the seal under pressure. With an unsupported plastic material, the outer surface of the material would also melt,
and stick to the heated rollers and crimper jaws, before the inner surfaces were
sealed together.

The most common method of overcoming this utilizes a combination of hot
air blasts and heated-wire units to form a bead-seal (Figure 4.44). The long
seam is made by a hot air blast which welds the *overlapped* edges of the
wrapping material. The pack ends are sealed and cut off by a cam action which
controls the passage through the material of a heated wire, forming a bead-
style end seal. For wrapping *biscuits on edge*, in single column, machines are
equipped with overhead transport-fingers on the infeed and with special side
belts to hold the biscuit column in position after passage through the folding
box (Figure 4.45).

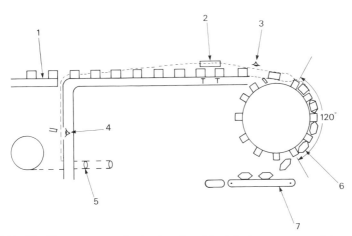

Figure 4.44   The Hayssen system for bead-sealing. The infeed conveyor (1), if fitted, automatically positions the product on the moving film web. If no infeed conveyor, the product
is placed directly on the web of film. The film is gradually folded into a tube around the product
and both move forward to the fin or lap sealing system (2). The girth of the package is adjusted
to the product girth as film is pulled around the product and the fin or lap seal is heat sealed.
As the film and product advance towards the constantly rotating die wheel, an electric eye (3)
registering off the product releases the end sealing dies. With printed film an electric eye (4)
registers off an eye mark. If a cone timer (5) is fitted, this unit automatically makes an ink mark
on the film to assist in product placement and also releases the end seal dies. The product is
carried around the die wheel through a 120° sealing arc (this gives a long dwell time) when the
die cut-off wire (6) is operated and the package is released. The discharge conveyor (7) carries
packages away and up to normal working height.

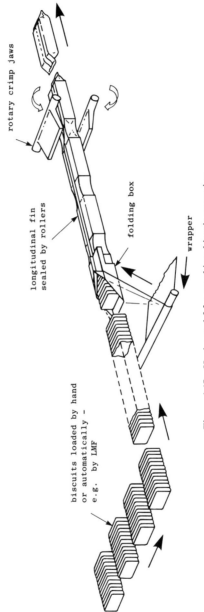

Figure 4.45 Horizontal f.f.s. machine: biscuit wrapping.

Packaging machinery for horizontal f.f.s. sachet packs

Sequence of operations of a basic twin web machine (Figure 4.46). Wrapping material is drawn, in a horizontal plane, simultaneously from two reels. Material from one reel overlays that from the second reel. The product is introduced between the two layers. The edges of the two layers are then sealed together, cross seals are then made between each product, and the individual packs so produced are cut off.

Common applications are for greetings cards, rubber gloves, bandages, and similar products (generally flat objects). Special applications are in tea bag making (Figure 4.47).

Sequence of operations of a basic single web machine (Figure 4.48). Material is pulled off from the reel in a horizontal plane, over a forming shoulder ('plough' or 'nose'). The shoulder guides the material in such a way as to fold it in half, with the 'open' portion towards the top. The folded material is drawn through a series of sealing stations, forming a three-side seal. Individual packs are then cut and the material advanced for product filling and closure.

Advantages and disadvantages of horizontal f.f.s. machines are summarized in Table 4.8.

Table 4.8   Advantages and disadvantages of horizontal f.f.s. pillow and sachet packs (adapted from material presented by A.P. Benson, Institute of Packaging Education Course, February 1980)

Advantages	Disadvantages
Pillow pack	
Good product volume/pack size ratio	Unsuitable for powders or granular products
Very wide range of materials from inexpensive coated films and papers to complex laminates	Limited size range versus conventional overwrapping machines
Relatively simple machine construction and comparative low cost of packaging machinery	
Easy adjustability for a wide range of sizes	
Smooth continuous motion action, giving options ranging from high output (600 ppm lines) to low output (40 ppm lines) high versatile units	
Sachet pack	
Short product drop (compared to vertical sachet machines) means reduced filling time and high line speeds (up to 400 ppm)	Product volume/pack size ratio is not as good as on pillow pouch machines
Pack forming and sealing is performed away from the filling station(s) so that sealing efficiency is not impaired by product trapped between the seals	Inexpensive, 'unsupported' wrapping materials cannot generally be used, since the operating principle demands the use of more rigid, laminated materials
Pack rigidity superior to pouch-style pack	Packaging machinery tends to be significantly more expensive than pillow pouch packaging machinery
High degree of seal efficiency from four-sided fin-seal system	
Automatic filling of multiple product loads is possible	

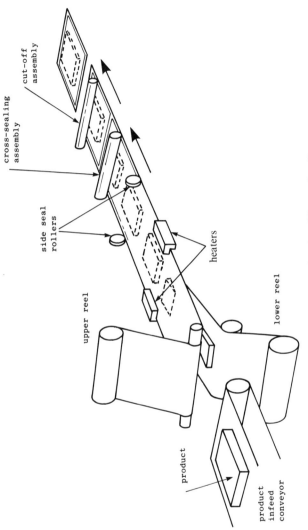

Figure 4.46 Twin reel sachet forming, filling and sealing.

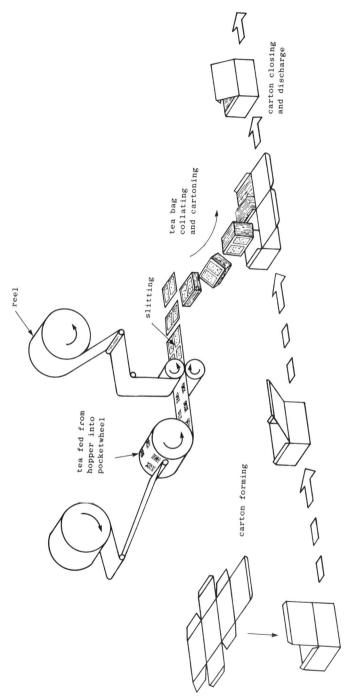

reel

tea fed from
hopper into
pocketwheel

slitting

tea bag
collating
and cartoning

carton forming

carton closing
and discharge

Figure 4.47 Making and cartoning tea bags.

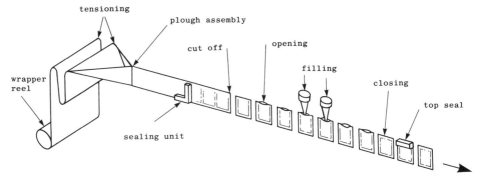

Figure 4.48   Single web sachet packaging.

Thermoformed f.f.s. packs

Containers may also be produced from reels of material by thermoforming a series of trays in a web, filling them with the product and then feeding another web over the open tray to just cover the flanges of the trays to permit a heat seal closure (Figure 4.49). The web of filled and closed packages is then punched out to form individual packages, or may be slit at intervals to give a number of units joined together. Such machines operate at speeds of from 6 to 20 cycles per minute, and the number of packages produced will depend on the number formed per cycle which in turn is generally dependent on the area they occupy.

Figure 4.50 illustrates the steps in producing the thermoformed (deep-drawn) package. Thermoforming the film is especially important. Figure 4.51 contains a summary of the possible forming methods, of which the most commonly used are negative vacuum forming, with or without plug assist and negative compressed air forming, with or without plug assist. In these methods, the heated film is formed in a mould (negative forming). In vacuum forming, the force is provided by the difference in pressure between the evacuated mould and the atmosphere of 1 bar (1 kg/cm^2). In compressed air forming, forming pressures of 6–8 bars are common.

Vacuum forming results in irregular wall thicknesses in a deep-drawn cavity (Figure 4.52). The most uniform wall thickness distribution is produced by means of compressed air forming with plug assist. Compressed air forming without plug assist also results in more uniform wall thicknesses than with vacuum forming. However, the machinery required for this method is more complicated (and more expensive) than comparable machines for vacuum forming.

The requirements for processing laminated films (PVC-PVDC, PVC-PFC, PVC-PE-PVDC) and for producing more complicated shapes, are basically

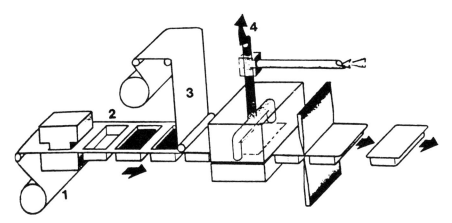

Figure 4.49 Thermoforming packaging machine fed from two film coils. One inner thermo-formable film (1) is formed into a tray (2). The food product is placed in this tray covered by an upper film (3). A vacuum is created in the tray (4), and broken by the gas mixture just before the upper film is sealed.

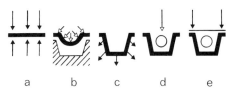

a   b   c   d   e

Figure 4.50 The steps in making thermoforms: (a) heating; (b) forming; (c) cooling; (d) filling; (e) sealing.

different. As a result of the required high forming force in the partially limited forming temperature range, only the compressed air forming method, either with or without plug assist, can be employed. Mechanical pre-stretching is always required for complicated shapes. Compressed air forming alone does not provide uniform distribution of the material in the deep-drawn cavity. There are six main materials used for thermoforming.

(1) *Rigid PVC film* (polyvinyl chloride). This is produced primarily by calendering. It has good thermoforming properties, and is generally processed clear, coloured and printed. It can be sealed by HF, ultrasonic, radiation, heat impulse and heat contact methods.

(2) *Polystyrene film*. This is produced almost exclusively by extrusion. It has good thermoforming properties, similar to rigid PVC film. Clear, stretched polystyrene is brittle, but non-transparent copolymers have good impact strength. Polystyrene cannot be sealed by means of HF, but is well suited for heat contact sealing.

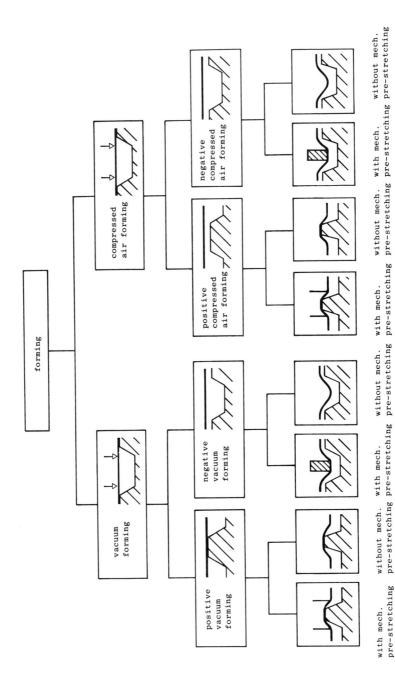

Figure 4.51   The possible ways of thermoforming.

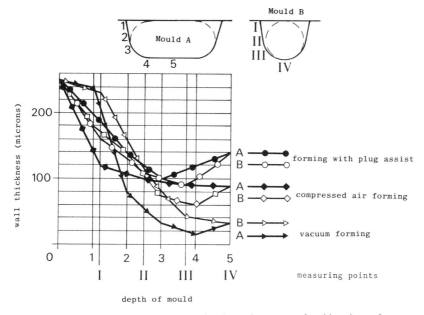

Figure 4.52 Variation in wall thickness for the various ways of making thermoforms.

(3) *Polypropylene film.* This is produced by extrusion and generally employed in a laminate with other films, e.g. polyamide.
(4) *PVC-PVDC* (polyvinyl chloride-polyvinylidene chloride). The polyvinylidene chloride is applied as a dispersion or emulsion and dried.
(5) *PVC-PVF* (polyvinyl chloride-polyvinyl fluoride). The two films are laminated together. The laminate is currently employed for products which are extremely hygroscopic to be sold in areas with tropical climates.
(6) *PVC-PE-PVDC* (polyvinyl chloride-polyethylene-polyvinylidene chloride). This triple laminate has been on the market for several years.

Labelling

Labelling machines for packaging combine three components: the container to be labelled, the adhesive, and the label. They may be divided into two main categories based on the general nature of the adhesive system: machines using 'wet' adhesive (provided separately from the labels), and those using pre-adhesed labels.

In selecting a machine we must consider which system is best suited to the particular application—wet adhesive, heat activated or pressure-sensitive adhesive. Factors to be considered are:

(1) Economic:
   (a) basic label cost
   (b) adhesive cost
   (c) setting up, cleaning down and routine stoppages
   (d) inventory cost
(2) Marketing:
   (a) label material and print requirements
   (b) container material
   (c) container shape
   (d) container reuse
   (e) security considerations
(3) Handling:
   (a) storage conditions
   (b) transit hazards
   (c) retail conditions
   (d) in-use conditions

The details of the label specification and the adhesive type will depend to a great extent on the machine selected. Wet-gum machines generally operate with cut labels, thermosensitive labels are supplied cut or reel-fed, while pressure-sensitive labels are invariably reel-fed. Cut labels can usually be replenished while the machine is operating, while reel-fed stock requires a machine stop to recharge. As reel diameter is limited, the frequency of stoppages will depend on label length and thicknesses of labels and backing material. Thermosensitive labels require no backing material.

Four principal operations are performed by a labelling machine:

(1) Feed the label from the magazine or roll.
(2) Pick up the label generally by suction cups, compressed air or by secondary adhesive.
(3) Apply adhesive either with full coverage, or in vertical or horizontal stripes (generally from rollers), on to either label or container.
(4) Press the label to the container by pressure pads, compressed air, belt or brushes, during which process the container will be moved into position, held firmly while the label is applied, and then removed. This may be achieved by a rotary movement, the containers being held in a rotating turret: or by a 'straight-line' movement, by conveyor star wheel or screw mechanism. Straight line operation is generally more versatile, particularly where containers of unusual shapes are concerned. Most bottle-labelling machines hold the bottles vertically, but some machines, particularly for cans, hold the containers horizontally.

Table 4.9 Check list of details for consideration when selecting and ordering labelling machinery

Containers to be labelled
Shape
Sizes
Dimensions
Materials

Labels: user's requirements
Number and type
Dimensions
Materials including treatments, e.g. whether embossed

Adhesive and application system
Cold glueing   Full coverage
Hot glueing   Stripe application
Heat-activation Machine maker's
Self-adhesion   recommendation on
               adhesive supply

Payment and delivery
Cost
Method of payment
Weight of machine
Point of delivery (machine site? delivery bay?)

Installation
Services needed:
  electricity (voltage, frequency, power, phase)
  compressed air (pressure and flow rate)
Weight and dimensions of machine
Height of IN and OUT feeds for containers

Responsibilities: supplier/user for:
Preparation of floor
Provision of electricity, air, etc.
Move machine to site
Install machine
Conveyors to and from machine
Provision of safety devices (switches, guards, etc.)

Labelling machine capacity
Maximum/minimum label widths
Maximum/minimum label heights from base of bottle
Maximum/minimum width and depth of rectangular bottle to be labelled
Maximum/minimum diameter of cylindrical bottle

Operation
Roll or magazine (hopper) feed
Normal, maximum and minimum speeds required
Left-to-right or right-to-left operation
Labour skills and training arrangement for:
  operation
  maintenance
  change-over
Change-over times
Change-over parts:
  provision
  identification
No bottle/no label, or no label/no bottle devices
Reload capacity during operation

Servicing
Labour skill and training required
Recommended spares holding by user
Terms of machine manufacturers' repair service

Records
Layout plan
Electrical wiring diagram
Operational manual
Maintenance instructions
Parts list

Attachments needed
Coding
Counting
Wrong label detector
Over-printing
Special fittings for odd-shape bottles
Logical sequence control to keep output speed at optimum, influenced by number of bottles on line before and after labelling machine

Security
Label machinery manufacturer to maintain security on any information concerning the product to be bottled

Acceptance
Bottle/label combinations to be used on acceptance trials
Length of trial run for each bottle label combination

Purchasing, installation and operation of labelling machinery

The first step to successful labelling is to ensure that the machine supplier is aware of the exact requirement, and to this end he should be supplied with samples of the containers and labels to be used, throughput required and other details. An acceptance trial should be conducted, and a careful record made of the details of the label, adhesive and containers that were successfully used. The machine must be operated and maintained strictly in accordance with its operating manual, which should be readily available. Control should be kept of the labels, adhesives and containers, to ensure they are as close as possible to those successfully used in the acceptance trial. Finally, when the same machine is to be used for a new design of labels and/or container, sufficient time must be allowed for trials to find a new, probably different, adhesive. A full check list is given in Table 4.9.

Container handling

This can be by hand if labels are applied by hand or by semi-automatic machines. In fully automatic machines where containers are continuously supplied, it is necessary to provide some form of separation. The two most common methods are star separating wheels or infeed worms. Passage through such machines may be intermittent or continuous, straight line or rotary, or cans may merely roll through. Discharge may be to a collecting table (stationary or rotary), direct to end packaging (shrink-wrap, etc.), or to trays.

Standard case sealing, wrap-around case sealing and tray erection

Standard case sealing

The process of erecting, filling and sealing cases is as follows. The case blanks are manually or automatically fed to an erector which glues the base of the case or simply folds in the flaps of the base without glueing. The formed case then passes to a filling machine which puts the product into it. The filled case enters the case sealer, which either seals both the top and the bottom flaps in one pass or the top flaps only where the case erector had glued the base previously.

Adhesive application systems

There are five types of adhesive application systems used for case sealing.

(1) *The open pot roller system*. This is the traditional method. Water-based adhesives are applied from open overhead and bottom applicator pots. Adhesive patterns can be varied to obtain strips, spots, stripes or overall

coverage of the flaps by varying the design of the rollers. Such a system can handle the widest range of viscosity and adhesive type.

(2) *The closed extrusion system.* This system applies lines of aqueous adhesive from nozzles in the applicator head. Either gravity flow or light air pressure feeds the adhesive from a closed central reservoir to the applicator nozzles. The equipment is designed so that adhesive flows out of the nozzles only when the case flaps are in contact. As the case passes under the applicator heads, a controlled pattern of adhesive is applied to the flaps. The adhesive flow is then automatically shut off until the next case arrives. A skipping mechanism may be used to provide an intermittent pattern on each flap. By applying a closely controlled adhesive pattern from a closed extrusion system, we can reduce compression time, adhesive consumption and clean-up costs.

(3) *The closed spray system.* Adhesive is fed under pressure from a closed central reservoir to a spray head applicator. The spray nozzles apply a fine mist of tiny adhesive droplets in a pattern which can be varied from a broad band to a thin stripe depending upon the distance between the spray head and the case and various controls available on the spray head. Spraying requires much higher line pressures than extrusion to give effective atomization. Adhesive consumption is further reduced by using a closed spray system. Both extrusion and spraying require a lower viscosity adhesive than the open pot roller.

(4) *Ball pen application.* The ball pen applicator is normally used for top sealing only. The adhesive is fed either by gravity or a low-pressure closed system to the applicator head. The adhesive is applied in continuous or intermittent lines. The ball raises a valve system when pressed which allows adhesive to flow past the ball. A low viscosity water-based adhesive is used to allow sufficient adhesive to flow past the ball valve.

(5) *Hot melt jetting systems.* A hot melt jetting system consists of a heated reservoir fitted with a mechanical or air-operated pump. The molten adhesive is pumped through a filter and heated hose to the applicator gun. The gun is fitted with a small orifice nozzle and a solenoid or air-operated needle valve which controls the flow of adhesive. Several hoses and guns can be fed from the same pump and tank. This type of equipment is used for hot melt case sealing primarily because the gun can be mounted in any position for applying the hot melt.

Wrap-around case sealing

In recent years has been a significant increase in the use of wrap-around case sealing. There are several reasons for this, namely the significant board saving (up to 30%); one machine does the work of three; the tighter pack can eliminate

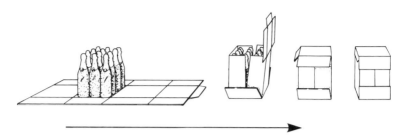

Figure 4.53 Operations in wrap-around case forming.

the need for divisions for bottles; and warehouse space saving results from the smaller case size (up to 20%).

The conventional regular slotted case has a manufacturer's joint (also known as the glue lap) which is normally made with a resin emulsion adhesive. In wrap-around case sealing, this joint is bonded with a hot melt adhesive during the case sealing operation. Figure 4.53 shows the wrap-around process, which takes a case blank, places the collated product on to the case blank, wraps the case tightly around the product and seals the end flap and side seams with hot melt using jet applicators.

The hot melt adhesive is applied to the ends of the case and under the side seam from three guns jetting horizontally. Multiple orifice guns may be used depending upon the size of the case. Horizontal jetting requires an adhesive of similar viscosity to that used in jetting upwards, to prevent the adhesive running down the case before sealing and compression. Wrap-around case sealers are available for speeds in the range 10–60 cases per min.

Tray erection

The main materials used for trays are corrugated and solid fibreboard. Blanks are individually fed from a stack, adhesive is applied as dots or stripes to the corners, the tray is erected and the corner joints compressed to bond them (Figure 4.54).

Adhesive systems for tray erection.   There are two primary adhesive systems: hot melt jetting and spray application of resin emulsion. A secondary ball pen application system is sometimes used. The adhesive requirements and end use performance of *hot melt jetting* are similar to those for case sealing. The hot melt is applied by means of a jetting unit. *Resin emulsion* normally has a longish setting time, typically around 20 s on these types of board. When such adhesives are spray-applied, less adhesive and therefore less water in the adhesive is used, and the setting speed will be faster, say 8 s.

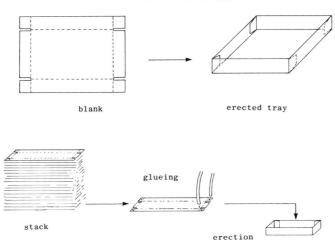

blank   erected tray

glueing

stack   erection

Figure 4.54 Making trays.

In tray erecting, higher pressures can be applied to the bond since it can be compressed from both sides at once. Using pressures of up to 200 psi, the adhesive will set in 1 or 2 s. Thus, production speeds of up to 30 trays per minute can be obtained on machines with short compression times using resin emulsion adhesives.

Some machines are fitted with *pen applicators*. These typically use fast-setting emulsions or latex adhesives. Because of the lower setting speed of latex and emulsion adhesives, lower production speeds result, typically 5–20 trays per minute.

Organization of packing lines

We have now arrived at the conclusion that any packaging line starts a long way before the end of the production line for the product and only ends when the product and the package reach the consumer. All the operations between these two very important points have so important a bearing on the packaging that they should be regarded as part of the packaging process.

Moreover, it is not important whether the operations we are concerned with are carried out by hand or by fully automatic machinery: the same basic principles apply. Additionally, the type of package is not particularly impor-tant, as once again essentially the same operations are set in train. The containers to be used are either formed or erected or presented to a filling point, the product is filled either by hand, by semi-automatic operation or fully automatically, some form of closing is carried out, the packages may then require to be collated and put into larger outer cases, and these in turn are closed and sealed by some means or other. The principles of layout, dictated by good materials handling, have to be followed so as to achieve the maximum efficiency in terms of labour and machinery cost.

Further reading

F.A. Paine, *Package Design and Performance*, Pira International (1990).
Encyclopedia of Packaging, Wiley, New York.

5 Packaging for physical distribution

Introduction

The principal shipping containers are made of wood, fibreboard and metal. Glass is still sometimes used for carboys for corrosive liquids such as acids, etchants and other chemicals, but the quantities are small and the carboy must be well protected in another pack. Wood is normally used when the package is large or the product is of high density. Thus timber cases and crates are used extensively for weights above 100 kg (220 lb); below this weight, fibreboard (both solid and corrugated) is the favoured material. Timber is also used for wine and beer casks, but there is a trend toward its replacement by metal (stainless steel or aluminium), either alone or with inner liners of plastic.

Plastic is also used for shipping containers and these are often returnable because the cost of the basic material is high. Plastic crates, long established in the dairy industry, now find more exacting use for bottled beers, minerals, and soft drinks. As beer crates are usually stacked higher and for longer periods than those for milk, polypropylene is preferred for beer, while high-density polyethylene has proved adequate for milk crates. Polyethylene has also replaced glass carboys for many liquids eliminating the need for extra protection. Polyethylene drums and jerricans are widely used for bulk foodstuffs. Expanded polystyrene is employed as a shipping container for tomatoes and grapes, and cured and fresh fish. Its insulation properties are useful for keeping the product cold with a minimum of solid coolant.

Solid and corrugated fibreboard cases are the most widely used, convenient and economical shipping containers, particularly for foods and other household products. The principal uses of fibreboard in food distribution are as regular slotted and other cases and trays, the latter often in combination with shrink-wraps. An increasing percentage travel as palletized or unit loads, handled mechanically rather than manually, and this affects the performance requirements in distribution and storage. Automated packaging lines running at speed also impose stringent requirements on the fibreboard packaging if it is to operate satisfactorily on the line.

Solid fibreboard is made of paperboard (often waste pulp board) lined on one or both sides with kraft (strong paper, usually brown, made from sulphate pulp) or similar material of between 0.1 and 0.3 mm thick. The total thickness of the combined board ranges from 1 to 3 mm. It is often coated with plastic film to give water resistance. Corrugated fibreboard includes both double-faced board and double-wall board. Double-faced board consists of two flat sheets separated by a fluted sheet made from straw or a special hardwood pulp; different grades and types of board can be produced by varying the materials, the thickness, and the weight of the liners and the fluting medium. Double-wall board is made from two fluted sheets separated by a flat sheet and faced on both sides with a further flat sheet. Flute configurations vary in number of corrugations to the metre and hence in thickness of the combined board.

The conventional case is the one-piece or regular slotted container, although open-tray and wrap-around styles are used extensively. The normal range of weight that corrugated and solid fibreboard cases can carry lies between five and 50 kg (110 lb) without any special fittings being used, while reinforced containers are capable of carrying loads of powdered or granular material of up to 500 kg.

In the 1960s, important innovations in the movement of goods appeared, including palletization, modular packaging, and freight containers. In some instances the pallet has now superseded the wooden case or crate, and in others the shipping container has been eliminated since the goods are loaded directly into a freight container and secured. This last is particularly applicable to heavy machinery. The concept of unit loads is now fully accepted worldwide.

Functions of a shipping container [1, 14]

The essential purpose of a shipping container is to contain and protect its contents during transport from manufacturer to consumer. Where an industrial package is involved, the consumer could be another manufacturer, and not the ultimate user of the finished goods. The function of the consumer container or retail unit is to provide for a convenient quantity or number of articles in one unit, which will be purchased by the ultimate user in a retail shop or store. It is obvious, however, that to some extent the outer or shipping container will be complementary to the inner or retail units.

So far, we have referred to what is inside the container as its 'contents', and it may have appeared that the whole of this *contained material* was the article or product to be transported. In reality, the contents of a shipping container may include cushioning material, blocking and bracing, water vapour barriers, or other ancillary materials which may have a prime function in the protection of the article or product to be packaged. The shipping container should be defined as the unit which holds the product and any other protective materials.

A *shipping container* has several functions to perform. The first and prime one, of course, is to *contain* the article and other protective material in such a way as to prevent spillage during the journey between manufacturer and ultimate consumer. It may also be required to keep out dirt, dust, moisture, heat, insects, rodents, or other foreign bodies. Additionally, it may need to be designed to prevent pilferage. A third important property of the shipping container is that it shall be *compatible* with the materials contained within it. For example, where foodstuffs are concerned, the container must be of such a nature that it does not contaminate the product, either by introducing toxic material into it or, alternatively, by providing a source of odour which may flavour the food.

A shipping container must possess sufficient *strength* to withstand stacking hazards during transport and storage. It is possible, however, to use certain containers which are not strong enough on their own to withstand the stacking loads to which they will be subjected, by providing internal fitments which are capable of increasing the stacking resistance. In many instances the unit packages inside the shipping container will perform this function, and a relatively weak outer container can be considerably reinforced by the presence of folding cartons, as well as by cans and jars. The shipping container may also have to provide some degree of *ventilation* between its interior and the outside conditions. This is particularly true in the case of perishable foodstuffs and livestock. It must also provide *information* on its outer surface which will enable the material within, the manufacturer of the product, and the destination of the package, to be identified. This information must be in a form that will not be obliterated either by the mechanical hazards of transport or by climatic hazards during transport and in store. Finally, a shipping container should be so designed that *access* to its contents can be achieved relatively easily without recourse to a book of instructions.

For the remainder of this chapter, we shall concentrate on distribution packages made of solid and corrugated fibreboard, as these are the types most widely used in the transport of foods.

Primary, secondary and tertiary levels of packaging [2]

The requirements for packaging must be integrated. The retail unit, the shipping container and the unit load should be complementary to one another and the packaging specification must cover all the levels of packaging that the distribution pattern requires. As a minimum, there must be the primary package that holds the basic product—a bag, can, carton, bottle, tube, or other form of container. For small packages distributed in large quantities, a shipping container is then specified to hold a group of primary packages. The corrugated box that holds one or two dozen primary packages is the most common example of a secondary package or shipping container. Large appliances, such as refrigerators and washing machines, are shipped individually, often in double-wall corrugated boxes, fastened to a wood platform or skid. In such instances the primary and secondary packages are therefore one and the same.

There is also a third level of packaging, which comes into use when a number of primary or secondary packages are assembled on a pallet or slip sheet, for handling as a unit load by materials-handling equipment in ware-

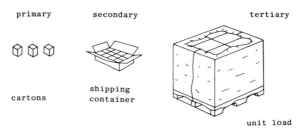

Figure 5.1 Examples of three levels of packaging (after E.A. Leonard).

housing, and in loading into or out of trucks and railcars. The specification for a unit load, or tertiary package, may include a requirement for strapping the secondary packages in place, or for sheathing the unit load with plastic film for protection against the weather, as two examples of several options. Examples of the three levels of packaging are shown in Figure 5.1 [2].

Fibreboard case performance

Probably the two most important tests which a fibreboard case has to meet are those related to dropping and to stacking performance [3,14]. Good stacking performance, of course, may be required both in warehouse conditions and in transport where vibration from the vehicle is likely to affect the compression strength.

Drop performance.   Let us look briefly at the drop hazard, and in some greater detail at the stacking performance hazard. Figure 5.2 and Table 5.1 show the relationship between the number of drops to spillage of a case containing cans,

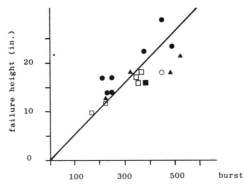

Figure 5.2 Drop versus burst. ○, solid kraft; ●, B-flute kraft; ■, solid test; □, B-flute test; ▲, A-flute test.

Table 5.1 Drop versus burst data for a carton $18\frac{5}{8} \times 12\frac{5}{16} \times 7\frac{1}{4}$ in, 40 lb in weight, containing six no. 10 cans (after Kellicut and Landt [4])

Liner material	Burst (psi)	Flute	Average drop height (5 boxes) at failure (to nearest inch)
Test	167	B	9
	351	B	15
	221	B	11
	366	B	17
	346	B	16
	481	A	17
	224	A	12
	319	A	17
	522	A	20
	377	Solid	15
Kraft	206	B	16
	242	B	13
	234	B	13
	244	B	16
	376	B	21
	439	B	27
	489	B	22
	446	Solid	17

and the bursting strength of the material from which the case was made [4]. The correlation looks extremely good, but the variation about any one point, each of which is the average of dropping five cases, is considerable, so much so that if every result was plotted the figure would not look like a straight line but a scatter diagram. Hence, although there is a general relationship between the bursting strength test and the drop hazard, it is not a very good one and we certainly should not regard bursting strength as a relevant property in this sense.

Stacking performance. Since their introduction, corrugated cases have been used for an ever-increasing number of applications in food packaging. The contents they carry, the severity of the distribution system, and the height to which they are stacked, have all increased over the years, and the continued success of the corrugated fibreboard industry illustrates the way in which demands for improved cases have been met. Several factors focus attention on their stacking performance (among other aspects), and these include:

(1) The use of higher stacks to offset warehouse costs
(2) The use of lighter boards to reduce cost
(3) The incidence of industrial accidents
(4) The conditions of acceptance imposed by carriers
(5) The implementation of statutory requirements

An example of a contrary trend is the greater use of pallet racking which can reduce the effective stack height.

The relation between safe stack loadings and the box compression failing load.
It is first necessary to appreciate the difference between a long-term stack loading, where packs are subjected to a constant sustained weight, and a laboratory compression test, carried out using machine-driven platens till failure occurs in a matter of minutes.

In the laboratory the test takes place in a very short time under an increasing load at a constant rate of deformation. Furthermore the platens are rigid and arranged so that the force is applied evenly around the perimeter of the case, and generally the point of failure is sharply defined.

Stacks may exist for an appreciable period of time (days, weeks or months) under a constant load with uncontrolled rate of deformation. Corrugated fibreboard is a visco-elastic material and therefore exhibits 'creep' under such conditions; that is, there is a continuing deflection of the material when subjected to a constant force. This characteristic has long been recognized and Figure 5.3 shows the approximate relationship [5] between the applied load (expressed as a percentage of the compression failing load) and the time required for the pack to fail. It can be seen that a pack subjected to a load of 55% of the compression failing load (CFL) will take 100 days to collapse. Increasing the load to 63% of the CFL will cause failure in 10 days, while a load of 75% of the CFL is likely to cause collapse in a single day.

It is also necessary to correct for differences in moisture content of the case between test conditions and field conditions. Figure 5.4 shows how moisture content affects the failing load. Other factors, such as stacking pattern, vibration under load, and pallet bearing areas, must also be taken into account. It is normal therefore to apply a suitable correction to compression failing load data to allow for these factors. This correction factor (safety factor) has a value generally lying between 1.5 and 5.5 depending upon circumstances. It is advisable to assume the worst stack loading condition in individual circumstances and make the best estimate available of the various factors which could reduce the stacking strength compared with the compression failing load.

The quality of the conversion of the corrugated board into a case can have an important effect upon the strength of individual cases, as also can subsequent damage to the cases during transit to the user, storage in his warehouse and handling during packing operations. When determining the CFL of a given type of case, therefore, it is advisable to test at least 6 (and preferably 10) cases withdrawn at random from stock so that some indication of the possible variation of the CFL is obtained. It is possible to calculate the probability of failure of the weaker cases by a statistical calculation, but it is probably sufficient to use a safety factor of 1.3 to allow for deviation of compression strengths from the mean value.

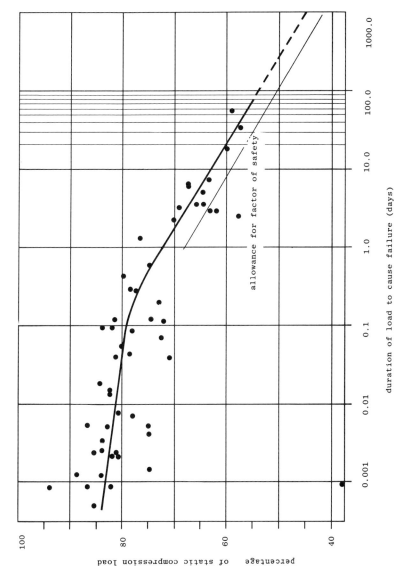

Figure 5.3 Duration of load to cause failure in stacked fibreboard cases (redrawn from Kellicut and Landt [5]).

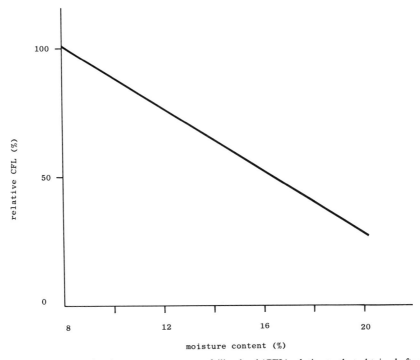

Figure 5.4   Effect of moisture content on case failing load (CFL) relative to that obtained after preconditioning at 23°C and 50% r. h.

Estimation of safe stacking loads.   In a simple bulk stack, it is generally assumed that the load is evenly distributed. Under these conditions, the maximum load will be experienced by the cases in the bottom layer of the stack, and the load on any individual case in this layer will be equivalent to the gross weight of a single case multiplied by the number of layers supported by the bottom layer. Thus if a stack is comprised of *n* layers of cases, each having an overall weight of *W* kg, then the load on a case in the bottom layer will be *W*(*n* − 1) kg. This figure can be compared directly with the CFL of the cases and the appropriate safety factor applied directly.

It is the magnitude of this safety factor and how it is determined that provides most of the uncertainty. Simple modifications such as Martin's formula suggest that

$$\text{Required } P = \frac{W(N-1)}{F(1 + T - (S/B))}$$

where *W* = weight of one case; *N* = number of layers in stack; *F* is a fatigue factor; *T* is a layer factor; *S* is a deviation factor; *B* is the number of cases in one layer.

Table 5.2 Factors affecting stacking performance

Contents factors	(1) Number, arrangement, weight and dimensions
	(2) Nature, strength and deformability
	(3) Interaction with case
Distribution factors	(4) Stacking method, pattern, type of pallet, etc.
	(5) Stacking height and loading
	(6) Stack duration
	(7) Climatic conditions
	(8) Handling damage before stacking
	(9) Case attitude in stack
Case factors	(10) Size, shape and style
	(11) Materials used, construction and tolerances
	(12) Fittings?
	(13) Printing effects, cut-outs etc.
	(14) Closure method
Other factors	(15) Relevant statutory regulations; company specifications, carriers's conditions of acceptance, etc.

The values for the various factors are then inserted as appropriate. The importance of stacking resistance and its relation to the cost of the case warrant more detailed consideration, however.

Board properties affecting stacking performance

Table 5.2 lists the factors affecting stacking performance and these are divided into three main areas: those related to the contents, those related to the distribution system, and those related to the case. A great deal of research [5–10, 12] has attempted to relate the compression failing load of a case to the relevant factors—the properties of the material, the temperature, and so on. Two board properties are found to be significant, edge crush and bending stiffness.

Edge crush is measured by the edge crush test (ECT). In this, a narrow strip of board (with, for corrugated board, the flutes parallel to the short side), is crushed between parallel platens. The force to initiate crushing is measured; the dimensions of the test piece ensure that the board is crushed without bending. The result is expressed as force per unit length of the test piece, which in the FEFCO standard is 25×100 mm.

Stiffness is measured by the force needed to deform the board into a curve. For fibreboard, two tests are possible: 4-point bending in which the test piece is supported near the ends on two supports, and then deformed by applying force at two inner points, or 3-point bending, where the force is applied at one central point. Four-point bending has the advantage of giving pure bending between the two centre supports, but needs more complex measuring equipment since the relative positions of the outer and inner supports do not give the deflection of the central section. Three-point bending is simpler, but should not be used to compare boards of differing thickness.

The relative importance of bending stiffness and edge crush on compression vary with the ratio of the panel dimensions and the board thickness. In the case of corrugated fibreboard, a number of major investigations have shown that the compression failing load, P, is related to case perimeter, Z, bending stiffness, S, and edge crush, E, by formulae of the type

$$P = KZ^{1/2} \cdot E^{3/4} \cdot S^{1/4}$$

where K is a constant and S is the geometric mean of the bending stiffness in the two principal directions. This formula [8] applies over most of the sizes and aspect ratios of corrugated board cases. As the stiffness of the board relates to the panel dimensions, the compression failing load becomes directly proportional to the perimeter and the edge crush value.

Using regression analysis Kainulainen and Toroi [13] derived the following empirical formulae for single and double wall board cases respectively:
For single wall:

$$P = 17.7 \cdot (L + W)^{0.31} \cdot E^{1.06} \cdot d^{0.85}$$

For double wall:

$$P = 17.7 \cdot (L + W)^{0.38} \cdot E^{0.85} \cdot d^{0.65}$$

where E is measured in kN/m; L and W are the length and width of the case in metres and d is the total caliper of the board.

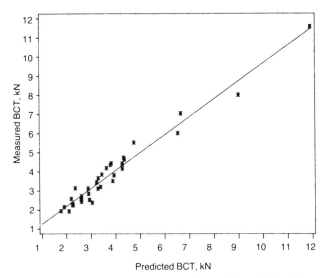

Figure 5.5 Measured versus predicted BCT. Correlation coefficient 0.981. Estimated ECT and bending stiffness values have been used.

Using these formulae the measured and predicted results for 34 different boards are compared in Figure 5.5.

Stacking performance has also been related to the various use situations, but all are lacking in some respect. Table 5.3 gives a list of the significant variables [11] in stacking, but does not necessarily decide how they are related. Commonly, all the factors are multiplied together.

Let us examine these factors [13] in greater detail. The duration of the stack has been covered by the pioneering work of Kellicut and Landt [5] (Figure 5.3). The effect of moisture content is far more serious. Figure 5.4 shows the effect of moisture content on case-failing load and it can be seen that we get a 50% loss of compression strength if the moisture content rises from 8 to 16%, a level not impossible to reach, particularly if one is packing fresh fruit or vegetables.

Filled handling can obviously be a most important factor also, and Table 5.4 shows the reduction of case-failing load as the filled handling goes up. Note that we can lose as much as 43% by relatively low impacts. It is fairly obvious that case attitude in transport—whether it is standing on the base properly, or whether it is closed properly—will have a major influence, but in general the

Table 5.3 Significant variables in stacking performance

(1) Duration of stack	T
(2) Moisture content of board (including rate of change)	M
(3) Filled handling damage	D
(4) Case attitude and closure method	A
(5) Overhang and type of pallet	O
(6) Stacking pattern	S
(7) Interaction with contents	I
Required CFL = P is maximum stack load × $T \times M \times D \times A \times O \times S \times I$	

Table 5.4 Reduction of CFL with filled handling

Case style 0201–300/B/300
    (437 × 311 × 242 mm)
Contents—Plastic bag filled with powder to gross weight of 15 kg
    Headspace greater than maximum deflection during compression test

Handling	CFL		
	kN	lb	% reduction
(a) Nil	3100	700	—
(b) 15 in drop on LBE	2800	630	10
(c) 4 × 15 in drops onto each base edge in turn	2800	630	10
(d) 4 impacts from 6 ft on inclined plane onto each VE	2800	630	10
(e) (d) plus (c)	1800	400	43

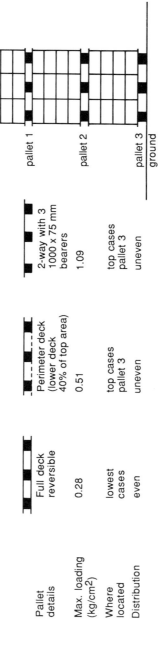

Pallet details	Full deck reversible	Perimeter deck (lower deck 40% of top area)	2-way with 3 1000 x 75 mm bearers
Max. loading (kg/cm²)	0.28	0.51	1.09
Where located	lowest cases	top cases pallet 3	top cases pallet 3
Distribution	even	uneven	uneven

Figure 5.6 Variation in stack loading with type of pallet. Case details: 500 × 400 × 300 mm, g wt 50 kg. Pallet details: 6 cases per layer, 4 layers per pallet 1000 × 1200 mm, pallets weighing 30 kg. Stack details: 3 pallets high.

amount is difficult to estimate, although instances where 15–20% compression load is lost can be detected.

When cases are palletized into unit loads and these are then stacked, a different set of conditions will exist, and the assumption used for the simple bulk stack calculations may not be applicable. It is therefore first necessary to determine the points of maximum loading in the palletized stack; unless a full-deck reversible pallet is used, this will not be on the bottom layer of cases in the lowest pallet load, but upon the top layer of that load, because of the geometry of the bottom deck of the second pallet (see Figure 5.6). It is no longer possible to use the figures for the average loading on a complete case as the force is no longer evenly distributed. It is therefore necessary to consider both stacking loads and compression failing loads in terms of force per unit perimeter length of the case.

Consider the following example. Cases weighing 5.5 kg each when filled, of a size $500 \times 330 \times 250$ mm, are stacked on a standard 1200×1000 mm pallet in five layers of seven cases each in the pattern shown in Figure 5.7. The pallets weigh 15.5 kg each and the pallet loads are stacked three high. The lower decks of the pallets are full perimeter-boarded, using 100 mm wide boards, and this lower deck pattern is shown superimposed in Figure 5.7 by dotted lines. The base of the pallet can be found, by calculation, to be bearing upon

$$2 \times (1160 + 990) + 16 \times 100 = 5900 \text{ perimeter mm of case}$$

compared with the total of the 11 620 perimeter mm of case available in the seven cases of the layer.

The total force distributed over the bottom layer of the lowest pallet is provided by two complete pallet loads plus four out of the five layers of the

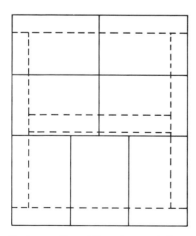

Figure 5.7 Diagram of pallet layout. Dotted lines show main bearers of pallet; full lines show case edges (7 cases per layer).

bottom pallet, that is

$$70 \text{ cases @ } 5.5 \text{ kg each} + 2 \text{ pallets at } 15.5 \text{ kg} = 416 \text{ kg}$$
$$\text{plus 4 layers of 7 cases @ } 5.5 \text{ kg each} = 154 \text{ kg}$$
$$\text{Total load} = 570 \text{ kg}$$

This load is spread over 11 620 perimeter mm, giving a force of 570/11 620 or 4.9 kg/per 100 mm of perimeter or 81.5 kg per case. The force upon the top cases of the lowest pallet is 416 kg distributed over 5900 perimeter mm, giving 7.1 kg per 100 mm perimeter.

Therefore the most severe loading is upon the cases in the top layer of the lowest pallet load and corresponds to an effective load of 117 kg per case. It is this latter value of stacking load which must be used in further calculations. (It is assumed that the case contents do not help to support the load.)

Calculation of compression strength required.   It is known that the cases remain in the stack for 100 days and Figure 5.3 shows that the stack load must therefore not exceed 55% of the CFL. Therefore required CFL = $\frac{117}{55} \times 100 =$ 213 kg under the atmospheric conditions during storage.

Assume that the relative humidity in the warehouse can be as high as 75% (i.e. a moisture content of the case of 12%). Figure 5.4 shows that the CFL of cases at this moisture content will be 75% of that at standard test conditions. Hence the figure at standard conditions will be 100/75 × 213 kg or 284 kg. The cases are 'brick-stacked' on the pallet as distinct from vertically stacked (all vertical case edges coincident) and a further factor must be allowed for this

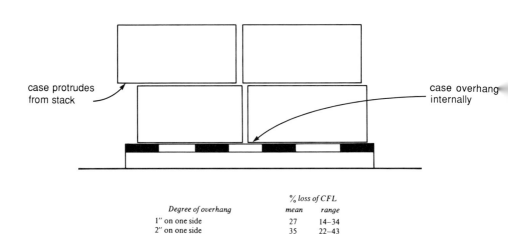

case protrudes from stack

case overhang internally

Degree of overhang	% loss of CFL	
	mean	*range*
1″ on one side	27	14–34
2″ on one side	35	22–43
1″ on one end	11	4–28
2″ on one end	26	9–46
1″ on one side and one end	34	27–43
2″ on one side and one end	12	34–46

Figure 5.8   Effects of overhang in pallet loads.

configuration. Allowing a safety factor of 25% this brings the required CFL at 50% r.h. up to 355 kg. The effect of overhang on pallets is illustrated in Figure 5.8. Note here, again, that 30 or 40% of compression strength can be lost quite easily.

It is fairly obvious that the several percentage reductions associated with the various factors concerned cannot be added up, and if we multiply them, probably unjustifiably, we see that there can be a considerable loss of performance. Also, the interaction of the case with the contents has been ignored: some classes of contents help stacking and some do not. It is, however, drops and compression forces which are the main hazards to fibreboard packing cases.

Unitizing methods

The virtues of unitizing—the art of making big ones out of little ones—have long been known [12]. Unitized loads yield savings in labour costs, simplify materials handling and often increase shipping and warehousing efficiencies. Pallet size loads shipped and stored as integral units make logistical and economic sense.

A list of the predominant unitizing methods (see Table 5.5) includes strapping, film wraps and pallet-stabilizing (non-skid) adhesives. Added to this are tapes, cords, netting, wires and a number of less important methods. Within each category, too, choice of materials must be made. If strapping is selected, a decision must be made as to whether steel or non-metallic straps will be used, if the choice is non-metallic, which will best serve (rayon or polyester cord, nylon, polypropylene, or polyester straps) must be decided. If film wrap is the chosen method, whether to use shrink or stretch, must be determined. Superimposed on all this is the application method. Every one of the unitizing methods can employ fully automatic, semi-automatic, or manual techniques.

The choice is simplified a little by concentrating on the end result. What will be the effect on the product's usability? The idea is to get the product into the hands of the ultimate consumer at the lowest overall cost. Examination of the nature of the load, and the total distribution and storage environment, will make the choice easier.

What exactly is a unit load? 'Unitizing' and 'bundling' tend to be used interchangeably, and this is correct in one sense, because whenever several packages are combined the result is a unit load. To be more precise, however, unitizing refers to a pallet-sized load bound together by some means. When referring to palletless loads, some unitizing equipment suppliers differentiate by saying that if the load can be man-handled, it is a bundle. As soon as mechanical materials-handling equipment becomes necessary, it is a unit load (really just another way of saying 'pallet-size load').

Table 5.5 Comparison of unit load methods

Method	Load stability	Benefits	Limitations
Strapping			
Automatic	Excellent	Handles cases in	Cannot be used if layers have gaps
Semi-automatic		columns	Not restackable
Manual		Reduces case cost	
Shrink-wrap			
Automatic	Excellent	Reduces case cost	Cannot be used if layers have gaps
Semi-automatic	Excellent	Cleanliness	Takes up much floor space
Hand gun	Good	Handles uneven loads	Not restackable
		Handles cases in	High cost per pallet
		columns	High labour cost
Stretch-wrap	Excellent	Reduces case cost	Cannot be used if layers have gaps
		Cleanliness	Takes up much space
		Handles uneven loads	Not restackable
		Handles cases in columns	High cost per pallet
Wet adhesive			
Automatic	Good	Low cost per pallet	Will not handle cases in columns
Manual	Good	Unitized load	Possible fibre tear when
		Restackable	dismantled
Hot melt	Good	Lowest cost per pallet	Will not handle cases in columns
		Unitized load	Requires conveyor system
		Restackable	
		Gap filling	
Tape	Fair	Lower equipment cost	Not restackable
		Portability	Marginal stability
Cord	Fair	Lower equipment cost	High labour cost
		Portability	Not restackable
			Marginal stability

Film wraps

Film wraps (especially in a fully automatic system) hold the load together, protect it from pilferage, and depending on type, seal the load against weather hazards. Since both shrink-wrapping and stretch-wrapping with film operate in much the same areas, it will be useful to look at those in which shrink-wrapping has made its biggest impact and then try to see if and where stretch-wrapping has particular advantages.

Shrink-wrapping. Awkwardly shaped articles were the first to be shrink-wrapped. The earliest film used was polyvinylidene chloride, and the close wrapping of food items such as hams and poultry was carried out by evacuating the air from a bag and shrinking by immersion in hot water. The use of tougher, cheaper films with higher softening points, such as low-density polyethylene, had to wait for the development of hot air shrink tunnels with reasonable control of temperature. The development of shrinkable low density polyethylene film led directly to applications such as the collation of cans, jars, bottles or cartons, and put shrink-wrapping into direct competition with the fibreboard case. For cans and bottles, it is usually necessary to use collating trays made from fibreboard or thermoformed plastics, such as PVC or expanded polystyrene. Shrink-wrapping was usually cheaper than a fibreboard case, but it was the additional advantages of shrink-wrapping which usually led to its adoption.

One advantage at retail store level is the fact that the total amount of packaging to be disposed of is smaller, and this is an important positive point from the environmental protection viewpoint. A further advantage is the fact that shrink-wrapped goods take up progressively less storage space as the cans and bottles are removed, whereas a fibreboard case takes up the full amount of space until the last of the contents have gone.

A later development in the use of shrink-wrapping was the overwrapping of complete pallet loads, displacing the use of paper or board coverings held in place by metal or plastics strapping. Shrinking of larger areas of film can be carried out in heated cabinets or tunnels, or by the application of hot air from hot air 'guns', similar to large hair dryers. Shrink-wrapped pallet loads are more stable than strapped loads (particularly when the objects being palletized are irregularly shaped) and the film wrap confers extra protection against ambient conditions.

Stretch-wrapping. The general principle is that the film is stretched around the object or objects to be wrapped and then heat sealed. Residual tension in the film provides the tight contour wrap. The simplest analogy is that of an elastic band. Like shrink-wrapping, one of the earliest applications of stretch-wrapping was for the packaging of poultry. Film bags were stretched over a former, the poultry inserted and the bag then allowed to retract to its original dimensions. Again, like shrink-wrapping, it is now competing in the fields of collating, pallet overwrapping and retail wrapping.

The main films used in stretch-wrapping are low density polyethylene, ethylene/vinyl acetate copolymer (EVA) and PVC, the choice depending on factors such as appearance (sparkle, clarity, etc.), protection required (gas barrier, water vapour barrier, etc.) and the susceptibility to damage by compression of the articles to be wrapped.

One way of stretch-wrapping pallet loads is by spiral winding so that a standard width of film can be used, irrespective of the pallet load dimensions. Basically, pallet stretch-wrap equipment consists of a turntable on which is placed the pallet load to be wrapped. To one side of this there is a vertical pillar, up and down which moves a carriage holding a reel of film. After placing the pallet load on the turntable, the film is secured to the bottom of the pallet, the turntable is set in motion and a suitable number of turns of film given to provide adequate anchorage. The film carriage then moves up and down again until the required number of layers has been applied.

Pallet stabilizing adhesives

These are rather peculiar in that they retain some degree of tack even after setting, and have good shear strength, but weak cleavage strength. This allows high adhesive qualities but at the same time results in low cohesive qualities. At first, all such adhesives were cold glues, applied by spray methods or by extrusion. The amount of glue applied is critical: too little will not improve the coefficient of friction, too much may cause fibre tear, and in some cases loss of package integrity.

Hot melt pallet stabilizing adhesives possess advantages in that they have:

(a) Ability to break and remake the bond: hot melts stay soft and tacky after they cool, permitting packages to be restacked
(b) Gap filling properties: because of bead height, hot melt helps bridge any gaps due to irregularities in the package
(c) Applicability under adverse conditions: hot melts are unaffected by moisture, and by airborne dust and dirt
(d) Cost: far less hot melt is used than with a comparable cold glue system, usually yielding cost savings, even though the cost per pound of hot melts is higher.

There are numerous hot melt formulations, designed both for general purpose and special purpose situations. The choice is dependent on:

(1) Weight of packages
(2) Nature of contents
(3) Whether cases will be filled or only partially loaded
(4) Whether fibre tear is allowed and if so, to what degree
(5) Pallet loading pattern, that is, interlocking or column stacks.

Strapping materials and methods

At one time strapping was relatively simple, and consisted of a number of steel bands, wrapped around a pallet load (usually vertically, or both vertically and horizontally), and secured with a metal seal. Nowadays, strapping choices consist of steel straps, rayon or polyester cord, and polypropylene, nylon or polyester straps. And, although some strapping still is secured via metal seals, closing methods now extend to buckles, punch and die seals, and friction welding as well. The tension/elongation characteristics of typical materials are indicated in Figure 5.9.

Which of these alternatives to choose depends as much on the nature of the

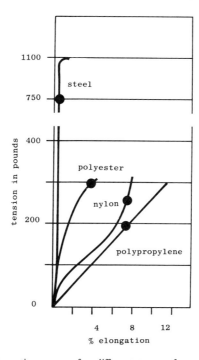

Figure 5.9   Tension/elongation curves for different types of strapping (courtesy Signode).
● = maximum normal working range.

load as it does on the materials themselves, but general properties of strapping can be summarized as follows.

Steel.   A high tensile strength, general purpose strapping material, especially good for very heavy loads, or loads which have sharp or abrasive edges. Steel has a unique elongation of about 0.1% which does not vary until just before the breaking point. It can be applied manually, semi- or fully automatically.

Cord.   First of the non-metallic strapping materials, it is noted for simplicity; it can be hand-tied, buckled, or sealed. A general purpose material for purely manual applications, with up to 25% elongation.

Polyropylene.   A general purpose product, recommended for light to medium loads but susceptible to tension decay. May be used in either manual or automatic systems, sealed by buckle, metal seal, heat, or friction weld. It has upto 25% elongation.

Polyester.   The non-metallic strap, with characteristics more akin to steel. It has low elongation (only 2–3%) and so can be tightened strongly around a load. It costs about 50% less than nylon and 30% less than steel.

References

1. F.A. Paine, *Fundamentals of Packaging*, revised edition, Institute of Packaging (1981), p. 20.
2. E.A. Leonard, *Managing the Packaging Side of the Business*, AMA Management Briefing, AMACOM, New York (1977).
3. G.A. Gordon, *How hazardous is your journey?*, paper presented to *Packaging Industries Workshop*, Wembley, Pira Report Pk 25 (R) (1979).
4. K.Q. Kellicut and E.E. Landt, *Fibre Containers*, **42, 44** (1955), 3334–3940.
5. K.Q. Kellicut and E.E. Landt, Basic design data for use of fibreboard in shipping containers, Report no. D1911, Forest Products Laboratory, US Department of Agriculture.
6. G.G. Maltenfort, Compressive strength of fibreboard containers, *Fibreboard Containers*, July–November (1956).
7. A.E. Ranger, *Paper Technol.* **1**(5) July (1966).
8. R.C. McKee, J.W. Gander and J.R. Wachuta, *Compression strength formulae for corrugated boxes*, report from Institute of Paper Chemistry Sept. (1963).
9. J.S. Buchanan, J. Draper and G.W. Teague, Combined board characteristics that determine box performance, *Paperboard Packaging*, September (1964).
10. L. Nordman and M. Toroi, paper presented at *FEFCO Conference* (1968).
11. H.P. Mostyn, Corrugated cases—development in the assessment of stack performance, paper presented at *9th IAPRI Symposium*, St. Gallen, Pira Report Pk 16(R) (1977).
12. *Paper och Trä*, **68** (1986) No. 9, 666–668.
13. F.A. Paine, *The Packaging User's Handbook*, Blackie (1990) Chap. 22.
14. F.A. Paine, *Package Design and Performance*, Pira International (1990).

6 Spoilage and deterioration indices

In the past when food was scarce man survived by developing methods of preserving foods, such as fermentation to prevent spoilage during storage. Nowadays society is even more dependent upon stored foods and it is estimated that in the UK alone the cost of food losses due to spoilage is more than £100 million per year [1, 2].

In considering the packaging of food, the way in which the food deteriorates must be determined and the influence of transport, storage and sales conditions on the rate at which this deterioration takes place must be estimated [3]. The role of water in foods is of major importance and the principal factors causing deterioration in any food [4, 5] can be divided between biodeterioration and abiotic spoilage.

Biodeterioration

The life forms that most commonly cause deterioration of food products or damage packaging materials are bacteria, moulds, insects, birds and rodents. Microbial attack requires the presence of moisture and warmth. Insects, such as moths and beetles cause damage to stored grains and other foods. Birds, rats and mice are well known as pests where food is processed, packaged and stored in warehouses or ships' holds but the control of these is more a matter of good housekeeping than of packaging.

Biodeterioration is also caused by the normal processes of ageing (senescence) which occurs in living organisms such as fruits and vegetables. Senescence may often be delayed by the combined effects of suitable processing and packaging, and by appropriate temperature control in store.

Effects of temperature on senescence

Even after harvesting, fruit and vegetables are still living products and metabolic processes continue. Senescence cannot be slowed down to any noticeable extent by packaging alone, although it protects the produce from soiling during handling and transport. The package must also permit the produce to 'breathe', to give off carbon dioxide and moisture, and to take in oxygen (see chapter 8). Cooling, however, can slow down these processes. For example, if the life of a particular fruit is 4 days at 20°C and it is marketed and consumed within 2 days at this temperature, only half of its possible storage life is used. Therefore before marketing the fruit can be held for 2 days at 20°C. However, if the storage temperature is lowered to 10°C the storage time of 2 days can be increased to 5, and at 2°C, it might even become 12, substantially increasing the time available to effect marketing, sale and consumption.

Microbial growth

The species of microorganism causing food spoilage depends on the surrounding conditions. Some microorganisms need an organic nitrogen source such as amino acids, whilst others grow only if sufficient glucose is present. All microorganisms need water. The water activity (a_w) of a food indicates the amount of water available. Thus, high water activity will mean more bacterial growth [6]. If water is less available there will be less bacterial growth. Acidity or pH is also important; most organisms reproduce less quickly in acid surroundings [1].

The temperature range within which microoogranisms can flourish is from $-10°C$ to 80°C and microorganisms are classified as thermophilic (40–80°C), mesophilic (10–40°C) or psychrophilic ($-10°C$ to 10°C). Mesophilic bacteria are mostly responsible for the spoilage of foods.

Aerobic bacteria grow in the presence of oxygen whilst anaerobic bacteria cannot. Those that can grow in both are called facultative anaerobes. Microorganisms are also affected by the presence of other microorganisms and antimicrobial agents such as preservatives [1].

The principal factors affecting the growth of microorganisms responsible for food spoilage are summarized in Table 6.1.

Bacterial growth in food passes through a succession of phases. If the numbers of bacteria are recorded periodically and the results plotted on a graph as in Figure 6.1 [8], a growth curve is obtained. This curve may be divided into phases as shown:

(1) an initial lag phase (A to B) during which there is little or no bacterial growth;

Table 6.1 Factors influencing microbial growth and spoilage (adapted from Gould [7])

Physical factors	Chemical factors	Microbial factors
Temperature	Available substrates	Substrates utilized
Water activity	pH value	End-products formed
Oxidation–reduction potential	Concentration of major solutes present	Numbers and types present
	Presence or absence of oxygen	Maximum growth rate
	Preservatives	

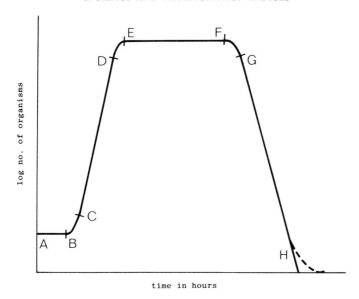

Figure 6.1 Growth curve of microorganisms [8]. A–B, lag phase; B–C, phase of positive acceleration; C–D, logarithmic or exponential phase; D–E, phase of negative acceleration; E–F, maximal stationary phase; F–G, accelerated death phase; G–H, death phase.

(2) a phase of positive acceleration (B to C);
(3) the logarithmic or exponential phase of growth (C to D) when the rate is at its most rapid;
(4) a negative acceleration phase when the rate is decreasing (D to E);
(5) the maximum stationary phase where the number of organisms remains constant (E to F);
(6) the accelerated death phase;
(7) the death phase or the phase of decline (G to H) during which the number of organisms decreases.

With many microorganisms, the number of organisms does not decrease at a fixed rate to zero as shown by the continuous line in the figure but tapers off gradually as indicated by the broken line, a few viable cells remaining. At this stage the food may well have spoiled but may still be a good source of nutrients for different bacteria which thrive in different conditions. Lengthening of the lag phase and the phase of positive acceleration is very important in food preservation and can be achieved by:

(1) The introduction of as few spoilage organisms as possible, i.e. by avoiding contamination and preventing the addition of actively growing organisms; such as those growing on unclean containers, equipment or utensils

(2) Making the environmental conditions (food supply, moisture, temperature, relative humidity, pH, etc.) unfavourable in at least one respect.

Figures 6.2 and 6.3 show the influence of temperature on the multiplication of putrefying bacteria on fresh meat [8]. Figure 6.2 shows that lowering the storage temperature not only prolongs the lag phase of the bacterial growth curve, but also reduces the slope of the second (logarithmic) growth phase. Similar results are also achieved by storage in carbon dioxide, but side-effects such as off-colouring may occur; the usefulness of carbon dioxide increases at lower temperatures [9].

As microbial growth depends upon the water content of the food or water activity (a_w), it is also affected by the relative humidity (r.h.) of the atmosphere. This is defined as the ratio of the actual amount of water vapour present to that which would be present if the air was saturated at the same temperature and expressed as a percentage. For every food there is a relative humidity with which it is in equilibrium and this is called the equilibrium relative humidity (e.r.h). Reducing the a_w of a food or its e.r.h. increases the lag period and decreases the growth rate during the logarithmic phase of growth.

Figure 6.4 shows that various types of microorganisms grow only when

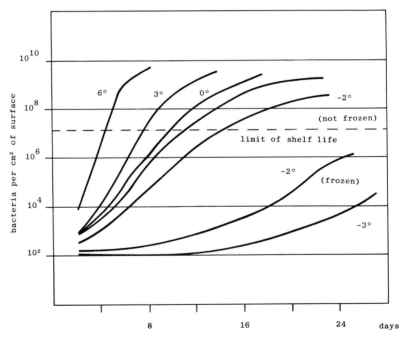

Figure 6.2 Lowering the storage temperature of fresh meat prolongs the lag phase of bacterial growth and reduces the slope of the logarithmic phase; r.h. 90%. After Heiss [5].

Food poisoning microorganisms: influence of temperature on the rate of growth in susceptible foods and the consequent risk.

Food spoilage microorganisms: influence of temperature on the rate of growth.

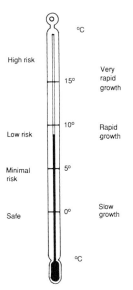

Figure 6.3   Effect of temperature on spoilage and poisoning.

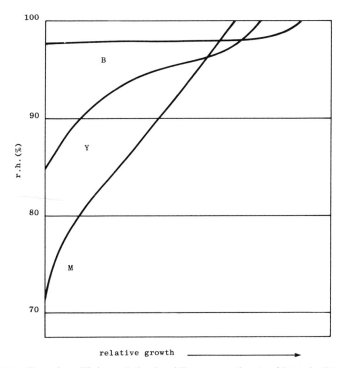

relative growth

Figure 6.4   Effect of equilibrium relative humidity on growth rate of bacteria (B), yeasts (Y) and moulds (M). After Heiss [5].

their relative humidity is above a minimum value dependent on the particular organism. Therefore, whenever biodeterioration plays a decisive role, shelf life can be increased by lowering the storage temperature.

Food spoilage and food poisoning

Some microorganisms form spores when the environment becomes less favourable and in this way they can maintain themselves but they do not multiply. Spores are not easily destroyed by many preservation methods.

Microorganisms may be divided into those that cause food poisoning (pathogens) and those that only cause spoilage. The main food spoilage microorganisms [10] are:

- *Lactic acid bacteria*: facultative anaerobes which develop in foods containing sugars which are then converted to lactic acid. Although the smell and taste of the food changes, this is not generally harmful to humans and some are used in food manufacture, e.g. in making yoghurt and cheese. They do not form spores.
- *Acetic acid bacteria*: aerobic and non-spore forming microorganisms which develop in alcohol-containing foods. The alcohol is totally or partly converted into acetic acid which affects the smell and gives a sour or acid taste to the food.
- *Proteolytic bacteria*: develop in and on non-acid protein foods. Toxins are produced which can be harmful to humans and unpleasant gases can be formed. The food becomes sticky and slimy.

The main bacteria which cause food infections or food poisoning either because of their number or because of the toxins they produce are:

- *Clostridium botulinum*: a spore forming anaerobe that produces dangerous toxins which can be lethal. Spores find their way into food through contamination from the soil. Growth of the microorganism itself takes place in low acid foods (those with a pH above 4.5) especially in meat and fish. However, the spores are destroyed by heating at 121°C for 3 min (the 'botulinum cook').
- *Clostridium perfringens*: occurs on meat and poultry contaminated by the contents of the intestine of the slaughtered animal. It is anaerobic and forms spores which can survive high temperatures.
- *Bacillus cereus*: a facultative anaerobe which also forms spores which are heat resistant. Contamination is from the soil and it occurs in cereal foods such as rice and bread.
- *Listeria monocytogenes*: develops in soft cheeses and patés. It is a dangerous bacterium which will multiply slowly even at low temperatures (1–8°C) but is destroyed by pasteurization.
- *Salmonella* bacteria: a facultative anaerobe which grows quickly at

lukewarm temperatures. Food can be contaminated through unhygienic preparation and through contamination at slaughter houses. *Salmonella* become a hazard to health when they have multiplied considerably. They will survive refrigeration and freezing but heating above 60°C destroys them.

- *Campylobacter jejuni*:   a non-spore forming facultative anaerobe which grows best between 42° and 45°C but not below 28°C. Pasteurization destroys *Campylobacter*. The main source of contamination is poultry.
- *Staphylococcus aureus*:   occurs in the mucous membranes of the nose and throat and is a non-spore forming facultative anaerobe. It develops in protein-rich environments and produces intestinal toxins. Heat kills the bacteria but does not destroy the toxins. Cool storage temperatures (5°C) are required to prevent growth and toxin production.

Moulds and yeasts

Moulds are aerobic and spore forming. They differ from bacteria in that when their spores grow, instead of one individual dividing into two, they send out thin root-like filaments (hyphae) which branch and grow to form an easily visible cotton-like mass (mycelium) which produces many more spores on the surface. The species most commonly found are *Penicillium* (the common green moulds) and *Aspergillus* especially the black variety (*Aspergillus niger*). The conditions necessary for moulds to germinate and grow are very similar to those needed by bacteria, i.e. a food supply, such as starch and protein, damp conditions (80–90% r.h.) and a suitable temperature (25 ± 5°C) are sufficient for rapid growth. Again low temperature has a retarding effect; growth at 4°C is very slow. Moulds do not need liquid water like bacteria and although some will grow at humidities as low as 70%, most prefer a moist environment. Mouldy foodstuffs smell and taste musty. Some moulds form mycotoxins which cause food poisoning or other illnesses.

Yeasts prefer slightly acid sugar-containing foods and convert sugars into alcohol and carbon dioxide. Fermentation continues until the alcohol content reaches about 15%. Yeasts are facultative anaerobes. Fermented products usually have a characteristic smell and gas bubbles may be visible.

Viruses can cause illness and foods can often play a part in their transmission. They are usually killed by heat but low temperatures can preserve them. Ice-cream can be a very suitable medium for viruses.

Preventing bacterial and mould growth

Since both bacteria and moulds only grow in the presence of water and/or high humidities, it follows that no damage will be done if the conditions are too dry for them to germinate and grow. The logical course is therefore to exclude moisture. This can be done by packaging in a sealed water vapour resistant barrier, provided the contents themselves have a low e.r.h. (below say 45–50%). Where high moisture contents mean that the e.r.h. of the product is liable to support moulds, a film must be used that permits moisture to escape sufficiently speedily to avoid condensation but not so fast as to promote too rapid drying out.

Another method was introduced about 1974 from Japan involving the use of oxygen 'scavengers' or 'deoxidizers', substances which absorb oxygen (see also chapter 11). Under cool conditions the lack of oxygen inhibits the growth of aerobic organisms. The most successful scavengers are based on iron (ferrous) powder, but hydrosulphite and organic reducers are also used. Since iron will set off metal detectors used during the packaging process the small container of iron powder is only introduced after the filled (but not sealed) package of food has been through the metal detection area.

Insect infestation

Many stored products (cereals, dried fruits, nuts, spices, cocoa and coffee beans) have a sufficiently low moisture content to practically preclude the possibility of mould growth. It is this type of product which is often stored for long periods that is most susceptible to attack by moths and beetles. The extent of such damage has been estimated to 20–30% of particular crops in some seasons.

Both moths and beetles lay eggs which hatch into a larval stage. The larvae feed on the food product before pupating into adult insects which then lay eggs and restart the cycle. Obviously, it is essential that the food itself contains no eggs before it is filled into the package or these will hatch out later and bore their way out.

Against such attack, scrupulous cleanliness in stores is essential and good housekeeping is paramount. Premises that are cool, dry and well ventilated, containing tidy stacks with adequate gangways to permit regular inspection and cleaning are essential as is strict rotational turnover of stock. When considering packaging to keep out insects it is obvious that, since the larvae are quite tiny, cracks and crevices which can develop into small holes are to be avoided. There is no material that is insect-proof in every circumstance but generally the thicker and stronger the packaging material the more resistant it will be. Smooth surfaces are better than rough ones and folds and creases in the package must be minimal.

Rodents and birds

Rodent damage is also best prevented by good housekeeping. Clean conditions and prompt destruction whenever traces are found are the best safeguards. It is difficult to keep food factories free from birds but again good housekeeping and attention to roof access will assist.

Abiotic spoilage

This is caused by chemical and physical changes in the product such as reactions between proteins and sugars (browning reactions), enzymic reactions, hydrolytic reactions, the oxidation of fats (producing rancidity), and the physical changes of swelling, drying out, caking, melting and so on. Some of this spoilage can be prevented by packaging alone, provided the food preservation process has been adequately carried out, but to give the required shelf life, temperature also must often be controlled. Some reactions are also affected by light.

The role of water in foods

Water [11] is the most abundant individual constituent by weight in the majority of foods, and is an important constituent even in those foods in which the proportion of water has been deliberately reduced during manufacture or processing either to change the properties or as an aid to preservation (see Table 6.2). It is hardly surprising therefore that water exerts an important influence on many aspects of food quality, and foods are often divided into three main categories, according to the proportion of water that they contain: dried foods, foods of intermediate moisture content and wet foods.

Table 6.2 Typical water contents of selected foods (adapted from DeMan [12])

Product	Water (%)
Tomato	95
Lettuce	95
Cabbage	92
Beer	90
Orange	87
Apple juice	87
Milk	87
Potato	78
Banana	75
Chicken	70
Salmon (canned)	67
Meat	65
Cheese	37
Bread (white)	35
Jam	28
Honey	20
Butter and margarine	16
Wheat flour	12
Rice	12
Coffee beans (roasted)	5
Milk powder	4

The amount of water present in many foods can vary over a limited range without causing much apparent alteration in the product itself. For example, some biscuits can absorb 2% more moisture than that present when they are freshly baked, and the consumer would not be able to detect a difference. However, a distinct lowering of quality would be noticed above this level. In many instances, therefore, a *critical moisture content* is defined for a product, with upper and lower limits within which the product is satisfactory.

The main changes which are brought about by moisture gain or loss can be summarized as follows:

(a) *Physical changes.* These include hardening or caking, caused by wetting and subsequent drying out of crystalline products, or the loss of crispness that can take place because of moisture gain.

(b) *Microbiological changes.* Essentially, these are due to the growth of moulds or bacteria which can occur if the moisture content of the product rises above a critical level, or if water droplets form within a package at any point, even if the moisture content of the bulk of the product is not high enough to promote growth.

(c) *Chemical changes.* These include Maillard reactions, solute effects and enzymic reactions which only occur in the presence of moisture and are very slow in its absence. Temperature is an important influence and generally the higher the temperature the more rapidly changes occur.

At low moisture contents autooxidation of lipids occurs but rapidly decreases as the a_w increases. Changes in nutritional quality can subsequently occur as a result of these reactions [13, 14].

Sorption isotherms

When the moisture content of a food is plotted against relative humidity or water activity at constant temperature, the curve obtained is called the *sorption isotherm*. Isotherms for different foods vary both in the shape of the curve and in the water present at each relative humidity [11] (see Figure 6.5). The water content of a food at a particular humidity is dependent on its water-soluble constituents and the presence of colloids. Any water-soluble material in a food reduces the water vapour pressure by lowering the amount of free water present. The degree to which this water activity is lowered depends to a large extent on the total concentration of all dissolved molecules and ions. Foods high in protein, starch or any other high molecular weight materials, have relatively high equilibrium moisture content at low humidities and this does not increase greatly at higher humidities. On the other hand, foods high in sugars have low moisture content at low humidities, but as the humidity increases above a certain value the water content rises sharply and continues to increase.

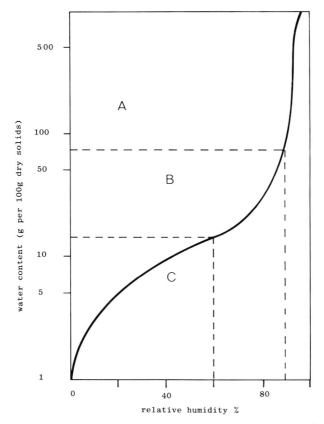

Figure 6.5 Moisture categories of foods. A, high-moisture foods (most or much of water usually relatively free of restrictive interaction with the solids). B, medium-moisture foods (most if not all of the water physico-chemically restricted by interaction with the solids). C, low-moisture foods (most if not all of the water severely restricted by interaction with the solids). After Duckworth [11].

Most sorption isotherms for foods show hysteresis, i.e. two curves are produced (Figure 6.6), one slightly above the other, the upper curve being obtained as the food is dried down from an initially high moisture (the desorption curve), while the lower curve is obtained as the moisture content increases from the dry side (the absorption curve) [12]. Frequently, changes brought about by moisture take place only when the moisture content reaches a critical level. As already stated the amount of water present in many foods can vary over a limited range without any apparent change. The water in a given amount of air will also vary, but the capacity of the air for water vapour is strictly limited. The amount it can hold varies with the temperature and the air becomes 'saturated' with water vapour when there is sufficient present to provide a vapour pressure equal to the vapour pressure of water

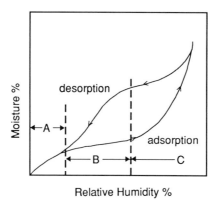

Figure 6.6 Adsorption and desorption isotherms [12].

at that temperature. The ratio of the actual amount of water vapour to that which would be present if the air was saturated, expressed as a percentage, is called the *relative humidity*. Water activity (a_w) is directly proportional to the vapour pressure and represents the ratio of the water vapour pressure of the food to the water vapour pressure of pure water under the same conditions, expressed as a ratio. Thus a water activity of 0.7 corresponds to a r.h. of 70%.

The water present in hygroscopic foods (those which absorb water) also exerts a vapour pressure, but this pressure is not in simple proportion to the amount of water present, even if the temperature is constant. If hygroscopic food is placed in a closed atmosphere, therefore, the water will transfer from the food to the atmosphere, or vice versa, until equilibrium is reached. If the atmosphere is not a closed one but has an unlimited capacity, then the final water content of the food will be dictated solely by the relative humidity of the atmosphere. Each water content will correspond to a given relative humidity and, under these conditions, it is known as the *equilibrium relative humidity* (e.r.h.).

Some materials, usually crystalline ones, undergo a change of form when they gain or lose more than a small quantity of water, and these changes take place only at specific relative humidities. Figure 6.7 shows typical curves for sugar, table salt and a non-food chemical as specific examples.

The most common deterioration here is the development of caking or lumpiness, produced by wetting of the product followed by subsequent drying out. For example, pure sugar does not absorb water from the atmosphere at all unless the relative humidity is above 85%. Above this, it absorbs moisture continuously and would finally dissolve in it if left long enough. The critical relative humidity for table salt is 75%. Consequently, in temperate climates both these materials can be packaged without moisture barriers. In

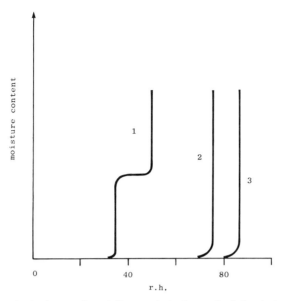

Figure 6.7 Stepwise isotherms of crystalline products: 1, non-food chemical; 2, salt; 3, sugar (sucrose).

tropical countries, however, they must be protected by good moisture barriers well closed to keep out moisture.

Hygroscopic materials like flour and flour confectionery, cereals, custard powders and other starch products are found to gain or lose moisture smoothly and continuously when exposed to any atmosphere with which they are not in equilibrium. They all have moisture isotherms of the S-shape, or sigmoid type shown in Figure 6.8. The moisture isotherm is the most useful representation of moisture relations and generally gives all the information necessary for determining the degree of protection required. The relative humidity in stores and shops in a temperate climate lies between 40 and 70%, and superimposing this information on the isotherm shows that, providing the critical moisture content of the products being packed lies between those limits, little or no moisture protection will be required.

If, however, the upper critical moisture content of a product is in equilibrium well below 40% r.h. or the lower critical moisture content is in equilibrium above 70%, then protection is needed. The farther from the expected storage atmosphere, the more effective the barrier required, Naturally for other markets conditions other than 40–70% r.h. may apply.

If the amount of moisture permeating a package under the expected conditions is known, as well as the filling weight, the initial and the permissible final moisture content and the average humidity difference between the exterior and the interior of the package, the permissible storage life can be

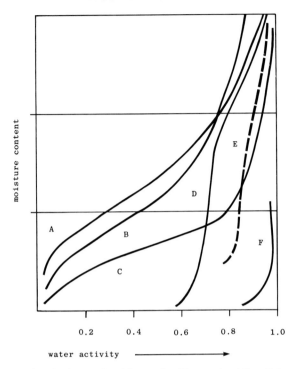

Figure 6.8   Absorption isotherms for (A) starch, (B) protein, (C) cellulose, (D) glucose, (E) sucrose, (F) lipid.

estimated. If the calculated shelf life is too short, a packaging material with higher water vapour resistance must be chosen. However, we can determine the permissible shelf life more accurately by storage tests with selected packages under expected storage conditions, but even here it is advantageous to know rough values in advance in order to make a suitable choice of package.

Whenever mixtures of different components are to be packaged, the equilibrium moisture of the less sensitive components must be altered in accordance with the permissible equilibrium humidity of the most sensitive component. Care must be taken that all components are fresh and have not suffered from unsuitable or over-long previous storage.

These measures do not ensure complete safety for a product packaged in a water vapour barrier if the ambient atmosphere undergoes fluctuating diurnal temperatures and humidities. The atmosphere of a non-air-conditioned store changes continually at rates that depend on prevailing winds and temperature. Sugar and salt can become as hard as bricks if they have not been thoroughly dried before packaging or if they are not enclosed in a highly moisture-resistant package.

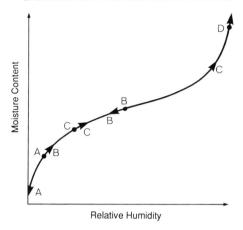

Figure 6.9 Different safety limits for one food [5]. A, for auto-oxidation; B, for Maillard reactions; C, for enzymatic changes; D, for microbiological damage. Direction of arrows indicates increasing danger for A–D.

The sensitivity of foods to water vapour at different water activities can be estimated from the shape of the isotherm as well as the critical points above which distinct deteriorative changes occur [14].

Figure 6.9 shows some safety limits that have been observed in practice [5]. These safety limits are to some extent dependent on storage time and storage temperature.

In summary, there are no foods in which water has no part in influencing quality. Its effect can vary from a minor effect to the predominant, but in most instances it has a major determining role. While temperature and moisture changes are the two most vital influences leading to the deterioration of foods, they are not the only ones: oxygen, physical damage, taint or flavour loss and the effects of light can also cause difficulty.

The more important deterioration indices for several typical groups of foodstuffs are listed in Table 6.3 [15]. It will be noted that different foods are affected by different factors, and hence have different packaging needs. The first step in all food packaging is to determine the relevant deterioration mechanism, so that optimum protection by packaging for the distribution system in use can be applied. In addition, techniques of food preservation are designed to prevent spoilage or adverse changes and to impart a shelf-life. Table 6.4 lists the major food preservation technologies available [7].

Packaging, then, can influence the biochemical and microbiological deterioration of foods, flavour, colour and texture, moisture and oxygen transfer, and the effects of temperature changes and light on foods. It is thus an integral part of the processing and preservation of foods. Indeed, without it, current methods of production, marketing and distribution would be impossible.

Table 6.3 Approximate order of importance of specific deterioration indices for classes of food. 1 = most important, 7 = least important.

Foods / Index of deterioration	Microbial changes	Inherent changes	Moisture changes	Oxidation changes	Taint, etc.	Light	Physical damage
Baked goods							
Bread, biscuits, cakes, etc.	4	4	1	2	4	4	2
Meat and meat products							
Raw, cooked and cured meats	1	2	4	2	6	4	6
Sausages	2	2	4	1	5	5	7
Poultry	1	2	5	2	4	6	6
Animal fats	—	—	—	1	2	—	—
Fish and fish products							
Wet, smoked and cured	1	3	4	2	—	—	—
Fish cakes	1	3	1	3	—	—	5
Shellfish (cooked)	1	4	1	1	—	—	—
Vegetables and fruits							
Brassicas	—	—	1	2	—	—	3
Legumes	—	—	1	2	—	—	3
Potatoes	2	—	1	—	—	3	4
Root crops	1	—	—	—	—	—	2
Top fruits	1	—	2	—	—	3	—
Soft fruits	1	—	—	—	2	—	3
Salads	2	—	1	—	—	—	—
Cereals and cereal products							
Grains (e.g. rice)	—	—	1	—	—	—	—
Flour	—	—	1	2	—	—	—
Breakfast cereals	—	—	1	4	3	—	2
Confectionery							
Hard and soft boiled sweets	—	—	1	—	—	—	—
Chocolates	5	—	1	2	3	—	4
Jams	1	—	1	—	—	—	—
Jellies	1	—	1	—	—	—	—

Table 6.4 Major food preservation technologies (adapted from Gould [7])

Objective	Factor	Preservation method
Reduction or cessation of microbial growth	Reduced temperature	Chill storage Freezing and frozen storage
	Reduced a_w/raised osmolality	Drying and freeze-drying Curing and salting Conserving with added sugars
	Nutrient restriction	Compartmentalization of aqueous phases in water-in-oil emulsions
	Decreased oxygen	Vacuum and nitrogen packaging
	Increased carbon dioxide	Carbon dioxide enriched 'controlled atmosphere storage' or 'modified atmosphere storage'
	Acidification	Addition of acids Lactic or acetic fermentation
	Alcoholic fermentation	Brewing, vinification
	Use of preservatives	Addition of preservatives • inorganic (e.g. nitrite sulphite) • organic (e.g. propionate, sorbate, benzoate) • antibiotic (e.g. nisin)
Inactivation of microorganisms	Heating Ionizing radiation Decontamination	Pasteurization Sterilization Fumigation
Restriction of access of microorganisms to food	Packaging and storage etc.	Aseptic processing and packaging, etc.

References

1. I.R. Booth and R.G. Kroll, The preservation of foods by low pH, in *Mechanisms of Action of Food Preservation Procedures*, G.W. Gould (ed.), Elsevier Applied Science, London (1989), pp. 119–200.
2. Anon, *Biological Sciences—Theses*, Science and Engineering Research Council, Swindon, UK (1986).
3. F.A. Paine, *Package Design and Performance*, Pira International (1990), p. 6 *et seq.*
4. F.A. Paine, *Package Design and Performance*, Pira International (1990), Chap. 3.
5. R. Heiss, *Principles of Food Packaging*, P. Keppler Verlag AG by arrangement with FAO (1970).
6. D. Richard-Molard and L. Lesage, Food microbiology and water activity, in *Food Packaging and Preservation*, M. Mathlouthi (ed.), Elsevier Applied Science, London (1986), pp. 165–180.
7. G.W. Gould, *Mechanisms of Action of Food Preservation Procedures*, Elsevier Applied Science, London (1989), pp. 1–10.

8. W.C. Frazier, *Food Microbiology*, 2nd edition, Tata McGraw-Hill, New Delhi (1967), Chap. 6.
9. M.V. Jones, Modified atmospheres, in *Mechanisms of Action of Food Preservation Procedures*, G.W. Gould (ed.), Elsevier Applied Science, London (1989), pp. 247–284.
10. C.M.E. Catsberg and G.J.M. Kempen-van Dommelen, *Food Handbook*, Ellis Horwood, London (1990).
11. R.B. Duckworth (ed.), *Water Relations of Foods*, Academic Press, London (1975).
12. J.M. DeMan, *Principles of Food Chemistry*, Van Nostrand Reinhold, New York (1990), pp. 1–35.
13. J.H. Troller, Water activity and food quality, in *Water and Food Quality*, T.M. Hardman (ed.), Elsevier Applied Science, London (1989), pp. 1–32.
14. K. Eichner, The influence of water content and water activity on foods, in *Food Packaging and Preservation*, M. Mathlouthi (ed.), Elsevier Applied Science, London (1986), pp. 67–92.
15. F.A. Paine, Technological and health considerations of modern food packaging, in *Food and Health: Science and Technology*, Elsevier Applied Science, London (1980), pp. 345–368.

7 Fresh and chilled foods: meat, poultry, fish, dairy products and eggs

Meat

The quality of meat is affected by the growth of microorganisms, by enzyme activity and by oxidation. The three kinds of microorganisms involved during meat storage are bacteria, yeasts and moulds, and those normally present on meat cause spoilage which can be smelt and seen. Not all bacteria are harmful but some can cause food poisoning directly (e.g. *Salmonella*) or indirectly through the toxins they produce (e.g. *Staphylococcus*). Uncooked meat is an ideal medium for rapid bacterial growth, because it supplies the three necessary factors: moisture, nutrients and an environment which is only slightly acidic (where food is highly acidic, only specialized microorganisms can grow). Enzymes present in the meat bring about chemical changes, only some of which are beneficial. Finally, oxidation of the fat by atmospheric oxygen produces rancidity, which gives the fat an unpleasant flavour.

Enzymes are destroyed and microorganisms killed by high temperatures such as those used in cooking. However, warm temperatures encourage rapid microbial growth. At low temperatures, enzymic activity and the growth of microorganisms are retarded, and at very low temperatures, they virtually stop. Oxidation is retarded at low temperatures and can be prevented by packaging meat in materials which have low oxygen permeability, although materials used for short-term retailing for fresh prepacked cuts are often selected for their high oxygen transmission rate, which helps preserve the bright red colour of the meat.

Preparation of meat

Bacteria are present in very large numbers in the gut and on the fleece or hide of live animals. Good slaughter practice is essential, from the personal cleanliness of the operators to the regular cleaning and sterilizing of knives and other tools to minimize the transfer of bacteria from the hide and gut to the meat. The carcass may also be cleaned by spraying with hot water. Hygiene at the abattoir and in handling and storing carcass meat is essential for a good storage life.

Chilling and chilled storage

Immediately after the animal has been slaughtered, the carcass is at body temperature (38°C) at which all bacteria, including any food poisoning organisms, can grow rapidly. Below 10°C, bacteria grow only slowly, and hence the meat is quickly cooled to that temperature. Reducing the temperature still further to 1°C prevents the growth of spoilage organisms. Nevertheless, chilling of beef and lamb must not be carried out too quickly and the temperature within the meat must not fall to 10°C in less than 1 h or rigor mortis sets in, the muscles contract (cold shortening) and the meat becomes tough. However, rapid chilling has very little effect on pork [1].

Unchilled carcasses lose weight by evaporation of moisture from their surfaces. The extent of the loss can be reduced by lowering the surface temperature, and, therefore, the faster the chilling, the smaller the loss of weight. Quick chilling also reduces the quantity of juices lost when meat is cut. There are obvious economic advantages in this, and modern abattoir practice combines rapid chilling with careful temperature control to avoid cold shortening.

After chilling, fresh meat is stored until transported to the retailer's refrigerator. Ideal storage conditions for fresh meat carcasses are a temperature of about of 0°C, a relative humidity of 85–90%, adequate spacing for hanging and good air circulation at low air speeds of 15–30 cm/s. This allows the meat to age which makes it more tender.

The expected maximum storage lives for carcasses under such conditions are:

Beef: up to 21 days
Veal: up to 21 days
Lamb: up to 15 days
Pork: up to 14 days
Offal: up to 7 days

Cutting and boning

In the cutting room, at both the abattoir and the retail shop, the temperature of the meat should be kept as low as possible. Air-conditioning keeps cutting room temperatures low, and the carcasses are boned, cut, packaged and returned to the chiller in as little as 30 min. Precautions are necessary to ensure the cleanliness of the cutting room and the personal hygiene of the operators.

Deterioration of fresh and chilled meat

Fresh meat is a complex material in which many biological processes associated with living tissues are still active. The period of storage is obviously important, but the best means of packaging meat depends upon a number of other factors. These are visible appearance (colour and bloom), organoleptic properties (taste and odour), moisture relationships and bacteriological condition.

Visible appearance.   The most important single factor in the appeal of lean meat is colour; this is particularly true for prepackaged meat. The purple-red colour of a freshly cut meat surface is due to a pigment known as myoglobin, which is closely related to the haemoglobin of blood. The differences in the

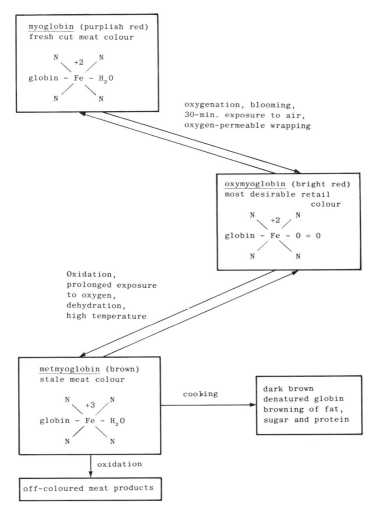

Figure 7.1   Colour changes in fresh meat. After Niven, Influence of microbes upon the colour of meat, *Am. Meat Inst. Bull.* **13** (1985).

208 HANDBOOK OF FOOD PACKAGING

colour of meat surfaces arise from the chemical changes of this pigment (Figure 7.1) [2, 3].

On exposure to air, an oxygen molecule attaches to the iron of the myoglobin, producing oxymyoglobin, which has a bright red colour. This 'bloom' is very quickly formed and is accepted as the most desirable meat colour in both unpackaged and prepackaged fresh meat. When the bright red surface is exposed to air, another reaction occurs to form a brown pigment (metmyoglobin), characteristic of stale meat. The rate of development of metmyoglobin depends upon (a) the temperature (the higher the temperature the faster the reaction), (b) the pH of the meat (higher pH gives darker meat, e.g. dark-cutting beef) and (c) the acceleration of bacterial spoilage.

Organoleptic properties. The taste and odour of meat are affected mainly by the bacterial status. The rate of oxidation producing rancidity in the intramuscular fat is higher in meat from non-ruminants (pork) than from ruminants (beef, lamb and mutton).

Moisture relationships. Loss of moisture also has a significant darkening effect on the surface colour of fresh meat. It is attributed to the migration to the surface of water-soluble pigments which become concentrated after the evaporation of the moisture. Moisture loss is an important problem associated with the storage and distribution of fresh meat. Loss of moisture can result in shrinkage and darkening of the meat surface, together with the deposition of significant amounts of drip in the package.

Bacteriological condition. Unwrapped and packaged meats are equally perishable and sensitive to bacterial attack. The rate at which bacteria grow on the meat surface will depend on the type of organism, the temperature and the nature of the storage atmosphere [4]. Under *aerobic conditions* bacteria cause:

(a) Surface slime: the temperature and available moisture influence the kind of microorganism found.
(b) Change in colour of meat pigments: the red colour may be changed to shades of green, brown or grey as a result of the production of oxidizing compounds.
(c) Changes in fats: oxidation of unsaturated fats in meats takes place in air, and bacteria may also cause breakdown and accelerate the oxidation of some fats.
(d) Phosphorescence: an uncommon defect caused by phosphorescent or luminous bacteria.
(e) Off odours and tastes: caused as a result of the growth of bacteria on the surface and the production of volatile acids. Under aerobic conditions, yeasts and moulds may also grow on the surface of meats to cause sliminess, stickiness, off odour and tastes and discolorations, whisker-

ing, etc. Spots of surface spoilage by yeasts and moulds are usually localized and can be trimmed off without harm to the rest of the meat.

Under *anaerobic conditions* spoilage is due to:

(a) Souring: giving a sour odour and sometimes taste caused by acids which can result from the actions of the enzymes in the meat during ripening or ageing, anaerobic production of fatty acids or lactic acid by bacteria, or protein breakdown without putrefaction.
(b) Putrefaction: anaerobic decomposition of protein by microorganisms with the production of foul-smelling compounds.
(c) Taint: any off-flavour or odour (bone taint is souring or putrefaction next to the bones).

Chilled transport to the retail outlet

In transport (Figure 7.2) the temperature must be maintained as near as possible to that of the chill store, and refrigerated vehicles are used for most distribution systems, and certainly for long journeys. Air circulation around the load is important and free circulation of air between the meat and walls, floor and ceiling is required. If carcasses touch the walls, heat will be transmitted from outside. Meat should never touch the floor of the van where operators walk when loading and unloading.

Wholesale packaging

Traditionally carcasses and joints have remained largely unwrapped during transport from abattoir and wholesale market to retail processors and outlets, but materials such as cotton and elasticated netting are used. Vacuum packaging is widely used for hot boned wholesale meat, especially beef.

In the past the wholesale trade was almost all in the form of whole carcasses, sides or quarters, bone-in. It is now more usual to bone-out the meat after slaughter and to supply retailers with 'primal cuts' which are intermediate in size between quarters and retail cuts. The meat is often deboned hot (i.e. before chilling) and the meat vacuum packaged in sealed heavy duty barrier bags. (e.g. Saran and Cryovac), cooled and usually supplied unfrozen to the retail trade or frozen in polyethylene lined boxes for manufacturing. Generally vacuum packaged meat has a dark purplish colour which is not attractive to customers and therefore it is not often used at retail level. Good hygiene is even more important with this method and care must be taken to avoid packaging so-called dark-cutting beef, because of its high pH which encourages more rapid spoilage under vacuum conditions. Vacuum packaging of wholesale meat is often combined with modified atmosphere packaging at the retail level [5].

A recent development from New Zealand involves an aluminium foil laminate outer bag and a 100% carbon dioxide atmosphere which acts as a

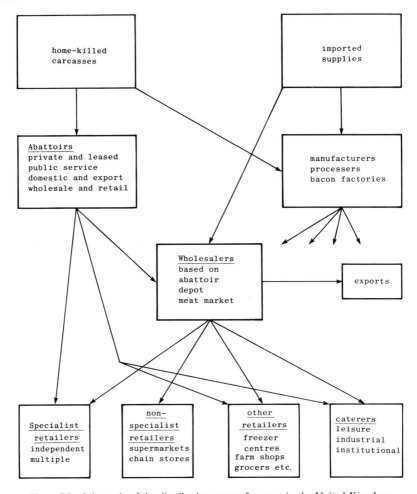

Figure 7.2   Schematic of the distributive system for meat in the United Kingdom.

flexible metal can. Beef and venison will keep packaged in this way at a temperature of $-1°C$ for up to 5 months, lamb for 4 months and chicken and pork for 3 months [6].

Retailing

Recent developments in retailing have led to considerable changes in the way fresh meat is distributed and marketed. Table 7.1 shows the decline in sales at butchers outlets and the increase in supermarket sales. The advantages of self-service retailing have been further advanced by centralized packaging of meat. Vacuum packaged boneless primal beef cuts have become an accepted

Table 7.1 Percentage household purchases of meat[a] (by volume) and place of purchase

	1980	1981	1982	1983	1984	1985	1986	1987	1988	1989	1990
Butchers	44.2	44.1	42.8	41.0	40.7	39.6	37.4	33.6	32.1	31.1	29.0
Co-ops[b]	6.9	6.9	6.7	6.8	6.8	6.9	6.7	6.4	6.0	5.9	5.9
Supermarkets	27.3	28.8	29.8	31.5	32.2	33.1	35.2	38.2	40.2	42.2	45.1
Independent grocers	4.8	4.4	4.3	3.8	3.6	3.1	2.8	3.0	3.1	3.0	3.0
Freezer centres	4.0	4.3	4.9	4.9	4.8	5.2	5.8	5.5	5.7	4.9	4.5
Other retail outlets	12.9	11.5	11.5	11.9	12.0	12.1	12.0	13.2	12.9	12.9	12.4
Total	100	100	100	100	100	100	100	100	100	100	100
(× 1000 tonnes)	1631	1637	1641	1654	1592	1568	1556	1521	1490	1395	1338

Source: AGB (table provided by the Meat and Livestock Commission).
[a] Meat includes beef, veal, mutton and lamb, pork, poultry, bacon and other meat and offal.
[b] Co-ops include the Co-op and other butchery outlets. Other retail outlets include farm shops, market stalls and meat sold in a variety of chain stores.

alternative to bone-in beef quarters in wholesale distribution [5]. These primal cuts provide retailers with partly butchered meat with a potential storage life of several weeks [7]. Supermarkets are also making increasing demands on producers to go one stage further and supply the final retail packs of meat, requiring no further in-store preparation.

Unwrapped meat should be stored under refrigeration whilst in the retail shop. At low temperatures, exposed lean meat surfaces lose water more slowly and retain colour better. Fluctuating temperatures accelerate these changes, and refrigerated displays should maintain a steady temperature which is as near $-1°C$ as is practicable. Temperatures should not exceed $+2°C$. As the meat is not packed when on display and the packaging needs are minimal, it is generally bagged for the customer in a thin high density polyethylene (HDPE) film which prevents leakage and wetting of other items in the shopper's bag.

Prepackaged meat is now common in supermarkets and is displayed in a variety of wraps. The main consideration is to retain the red colour and bloom of the meat. This means that some oxygen must reach the meat surface, therefore the packaging film used must have high permeability to oxygen as well as good appearance, sealability and low permeability to water vapour to avoid weight loss. In general expanded polystyrene or moulded pulp trays are overwrapped with stretch films such as PVC or polyethylene. It is usual to add an absorbent pad to reduce drip loss.

Meat thus packaged may keep for approximately 10 days at a steady $0°C$ before it becomes microbiologically unacceptable. However, it becomes unsaleable in less than half this time because, although still edible, it changes colour from red to an unattractive brown. Under commercial conditions in the display cabinets of a busy shop, where the air temperature can rise to 5°C or more, actual shelf life is only 1 or 2 days. Hence butchers only prepare enough meat for immediate needs. More recently microporous films with various levels of perforations, which are so small that oxygen ingress is improved considerably without great changes in its water vapour barrier, have been introduced.

As with unwrapped meat, the storage life of packaged meat is dependent on the initial level of bacterial contamination and the temperature during storage. In general, prepackaged fresh meats are refrigerated as near as possible to $-1°C$ which is the lowest temperature which can be maintained without freezing the meat. An outline of the meat packaging operation is given in Table 7.2.

Vacuum packaging of meat

The complete removal of oxygen from a pack of fresh meat ensures longer preservation against microbial deterioration than packaging in oxygen but the colour of the meat becomes darker and purplish. On opening the package, oxygen becomes available at the surface and the meat colour reverts to the desirable red colour [8].

Oxidation of meat, therefore, can be reduced by using a packaging material of low oxygen permeability (seals and closures for the packages must, of course, be 100% efficient). However, it is virtually impossible to remove all the oxygen as small quantities will be trapped within the food. In addition, anaerobic organisms which are not affected by oxygen are not reduced.

The spoilage of fresh meat in a chilled atmosphere is caused by a superficial bacterial slime. The main bacteria responsible for the production of this slime are *Pseudomonas* spp., and the accompanying odour is usually described as putrid. Spoilage of meat under these conditions normally occurs when bacterial numbers reach about $10^7/cm^2$. When fresh meat is vacuum packaged in carbon dioxide, the growth of *Pseudomonas* spp. is inhibited thus extending the shelf life of the meat.

The most common bacteria on stored vacuum-packaged beef are the lactic acid bacteria. These can reach numbers of $10^8/cm^2$ without immediately spoiling the meat. However, the sour taste which finally develops and which limits shelf life to about 8–10 weeks at 0°C is probably due to the continued action of these bacteria. The rate of development of this type of spoilage increases with increasing temperature. Spoilage also occurs more rapidly if the meat is heavily contaminated with bacteria before packaging.

Microbiological problems. If a vacuum pack is punctured, spoilage of the meat will occur. This may affect only a small surface area of the meat if the puncture is small but all the contents will be affected if the puncture takes the form of a slit. There is also a risk or spoilage with beef of high pH (> 5.9) when

Table 7.2 Meat packaging operations[a]

Operation	Packaging	Place							
		Abattoir	Depot or market	Specialized cutting plant	Independent or multiple retailer	Supermarket or chain store	Other retailers	Catering supplier	Processor or bacon manufacturer
1. Pack bone-in carcasses	S, F	×							
2. Vacuum-pack fresh boneless wholesale cuts	F, B	×		×				×	
3. Pack and freeze boneless wholesale cuts	F, B	×		×				×	
4. Pack and freeze for freezer trade caterer or retailer	F, P, B	×		×	×	×	×	×	×
5. Pack for fresh retail trade	F, P, B	×		×		×			
6. Wrapping fresh retail cuts	W, F, P			×	×	×	×		
7. Pack and freeze manufactured products	F, P, O, B							×	×
8. Vacuum-pack manufactured products	F, P, O, B							×	×
9. Can manufactured products	C								×

[a]S = cotton scrim; F = film; B = box; P = other plastics; W = paper wrapper; C = can; O = other. Original table produced by M. Ranken, Leatherhead Food Research Association.

it is stored in vacuum packs. This is caused by the growth of hydrogen sulphide-producing bacteria which are inhibited in vacuum packs of beef of normal pH. The hydrogen sulphide reacts with myoglobin to produce a green pigment, sulphmyoglobin. This type of spoilage is termed 'greening' and may occur after less than 3 weeks during storage at 1°C. Greening may be delayed by the use of films with very low oxygen permeability, but the best way to prevent it is not to vacuum pack meat with a pH above 5.8.

Pork and lamb are far less commonly vacuum-packaged than beef as there is less information on their microbiology and shelf life. It might be expected that lamb would have a shorter shelf life than beef because it has a higher pH, and that pork would have a shorter shelf life because it is more heavily contaminated by bacteria.

Suitable materials for vacuum packaging of meat must then combine a high resistance to gases and water vapour with perfect seals and good mechanical strength.

Modified atmosphere packaging

The use of controlled atmospheres to preserve food is well established [9]. In the 1930s chilled fresh beef, stored under carbon dioxide, was being shipped from Australia and New Zealand to the UK. This type of storage continued for meat and fruit until comparatively recently, but in recent years fresh food sales in supermarkets have increased considerably, resulting in changes in both distribution methods and the associated storage conditions. Most fresh and many chilled foods are now prepared and packaged in central depots from where they are distributed to retail outlets. Food must therefore remain 'fresh' throughout a longer distribution chain and still allow a reasonable shelf life in the retailer's premises.

Modified atmosphere packaging (MAP) is a process by which the shelf life of a fresh product is increased significantly by enclosing it in an atmosphere which slows down degradative processes, such as the growth of microbial organisms, whilst enhancing some beneficial processes, such as retaining the desirable red colour of meat. The term 'controlled atmosphere packaging' is also sometimes used but this is not strictly correct. The atmosphere inside any permeable package will change with time because gases diffuse into and out of the package at different rates or are absorbed and/or given off by the food. The term 'controlled atmosphere', therefore, should be reserved for truly controlled storage, such as gas storage of apples where the conditions are maintained at specific levels over the period concerned.

Table 7.3 indicates the shelf life extensions of meat possible with MAP [9]. Depending on the product and type of packaging a variety or different combinations of gases (oxygen, nitrogen and carbon dioxide) are required. The properties of these three gases are given in Table 7.4 [9]. The gases can also have a negative effect. Carbon dioxide can dissolve in the water in the

Table 7.3 Shelf life extensions for MAP meat

Product	Temperature (°C)	Normal package (days)	MAP (days)
Ground beef	2	2	4
Liver	2	2	6
Pork cuts	2	4	6–9
Beef cuts	2	4	10–12

Table 7.4 Summary of gas properties

Gas	Properties
Oxygen	Sustains basic metabolism
	Prevents anaerobic spoilage
Nitrogen	Chemically inert
	Prevents: oxidation
	rancidity
	mould growth
	insect attack
	package collapse
Carbon dioxide	Inhibits bacterial and mould activity
	Fat-soluble (hence not suitable for dairy produce)
	High concentrations can injure produce

food and the presence of oxygen can cause aerobic bacterial growth and oxidation. So it is important to balance the positive effects against the negative effects. By carefully selecting the percentage mixtures of the three gases and controlling the temperature to about $-3°C$, bacterial growth, oxidation, mould growth and enzymic action are reduced to a minimum.

The selection of gases must take into account the different spoilage processes involved. Atmospheres of 60–80% oxygen and 40–20% carbon dioxide [9,10] are commonly used for meat because the bright red colour of fresh meat is only present when oxygen is freely available to combine with muscle pigment. Mixtures of 20% oxygen, 70% carbon dioxide and 10% nitrogen slow down the rate of gas absorption.

The selected gas mixture is injected into the pack before sealing. The package itself is made of barrier materials and the internal atmosphere changes with time, so the packaging material selected is crucial to the success of MAP. Few single materials are suitable and it is more usual that a multilayer material is required. When deciding on package material, several factors should be taken into account, including type of package, barrier properties needed, mechanical strength, antifogging requirements and seal integrity. Suitable films might be a PVC/PE laminate base tray lidded with PVdC coated polyester/PE to which an antifogging coating is added. It is also

Table 7.5   Advantages and disadvantages of gas packaging of food [11]

Advantages	Disadvantages
Increased shelf life	Initial high cost of packaging equipment, films etc.
Increased market area	Discoloration of meat pigments
Lower production and storage costs	Leakage
Improved presentation: fresh appearance; clear view of product	Fermentation and swelling
Easy separation of slices	Potential growth of organisms of public health significance

usual to add an absorbent pad within the package to catch the drip. When combined with refrigeration, MAP is a highly effective way of increasing the shelf life of the product. It does have some limitations and cost implications, however, including the requirement to use very high quality raw material, strict temperature control, specialized equipment and training for operators of the gas flushing machinery. Quality control procedures and public health implications are extremely important. Table 7.5 lists some of the advantages and disadvantages of MAP [11].

Poultry

Preparation and spoilage

After slaughter the birds are defeathered and eviscerated and the meat quickly cooled with water and ice or cold air. Birds are then hung like other meat but the time is shorter. After weighing and grading whole carcasses are trussed prior to packaging in film bags sealed by clip or tape. Semi-automatic bagging machines are used.

Poultry are susceptible to spoilage similar to that of other meats but microbial spoilage, in particular *Salmonella* and *Campylobacter*, is the main problem. Packaging needs to protect the product against water vapour loss and discoloration through oxidation and subsequent loss of flavour.

Packaging

The choice of material depends upon factors such as appearance, protection required and susceptibility to damage. Similar considerations to meat apply but the different poultry shapes and sizes must be accommodated. Shrink-wrapping polystyrene foam trays with PVdC films or shrink-wrapped low density polyethylene (LDPE) is often used. Poultry is also stretch-wrapped with film laminates such as LDPE/vinyl acetate copolymer (EVA).

Modified atmosphere packaging of poultry

Under a carbon dioxide enriched (up to 25% CO_2) aerobic atmosphere, poultry has an increased shelf life and it is believed that concentrations over 25% are advantageous but care must be taken because unpleasant flavours in the cooked meat can develop [9]. MAP packaged fresh poultry in 50% nitrogen and 50% carbon dioxide can be kept for 7 days in lidded thermoformed trays [10].

Fish and shellfish

The fishing industry is made up of a number of subsidiary industries including processing, storage, retailing and farming [12,13].

Factory ships

Factory ships have been in operation since the sixteenth century, but the term has come to mean a ship which is specially designed to carry out a modern, mechanical fish processing operation, e.g. canning or freezing, entirely at sea. Trawlers have also been built for freezing whole fish, chiefly cod and haddock, in specially designed freezers. The fish are then stored at $-29°C$ on the vessel and thawed out at the ports for subsequent filleting. Factory ships also catch, freeze and can salmon, crabs, etc. Fishermen are becoming more interested in extending the shelf life of fish as a means of extending voyage times because fishing is becoming more difficult and there are advantages in staying at sea for longer periods [13].

Fish processors

Fish processing and merchandising is categorized by the size of the organization [13]. There are:

- Traditional independent merchants who employ less than 30 people and who mainly buy fresh fish or shellfish and fillet the fish or prepare shellfish and sell to wholesale markets.
- Medium sized processors who employ up to 200 people and use specific specialist processing or packaging methods such as fillet block production for portion manufacture, smoking, marinading, cook freezing, canning or MAP to produce particular product types for specific markets.
- Large processors (over 200 people) who mainly supply the frozen fillet and portion products, natural enrobed, smoked or incorporated in recipe dishes, required in high volume by the retail and catering markets.

Handling and transport

Fish is unloaded at ports, sometimes into wooden boxes holding about 50–60 kg of fish, or more usually into aluminium tubs which are normally maintained and kept clean by the market authorities. Returnable lightweight boxes are increasingly used for distribution and larger companies with their own transport facilities often use returnable aluminium boxes. Injection-moulded HDPE:polyethylene-coated paperboard or pulp containers are also available.

Usually, the fish is auctioned on the dock to fish merchants, before being packed in ice, and sent to inland markets, retailers and large-scale users. Fish is transported by road in insulated vehicles or vehicles with built-in refrigeration units. Solid carbon dioxide is also used for cooling during transit. For rail travel insulated vans with airtight doors are used.

Chilling

Fish is one of the most perishable of foods and begins to spoil immediately after catching. Short-term preservation by chilling, which should be under-taken as soon as possible, is carried out using ordinary water ice, although dry ice (solid carbon dioxide) and chilled salt water are also used. The bacteria responsible for fish spoilage are psychrophilic (cold-loving), and even when the fish is chilled at $0°C$ under the best conditions of handling, bacterial activity can result in severe losses of quality approaching inedibility after 14–16 days. On increasing the temperature to $5–6°C$ the fish spoils more than twice as quickly, and at $11°C$ spoilage occurs in less than one-fifth of the time (i.e. in 2 or 3 days the fish is inedible). The nearer the temperature approaches the freezing point of the fish, the greater the effect in reducing bacterial activity [12].

Alternatively, fish is stored in refrigerated sea-water tanks on board ship. Here the higher rate of heat transfer between the fish and circulating chilled brine gives a faster initial rate of cooling and by close temperature control the fish is maintained slightly above its freezing point. However, anaerobic conditions can arise if blood and scales are allowed to clog the filters, resulting in an adverse effect on fish quality, therefore the tanks must be well aerated. Fish such as herring break up with the movement of the water, and the soluble constitutents which are responsible for flavour are lost. Iced fish can also be maintained by close temperature control (super-chilling) at $-2°C$ in the fish hold without freezing, giving an appreciable extension to storage life.

Chill rooms and cabinets are also used to keep fish cool at a temperature of about $0°C$ in order to avoid slow freezing. Carbon dioxide gas at a concentration of at least 35% retards bacterial growth in fish if applied soon after catching, but even after 6 days, bacterial activity becomes too advanced to be suppressed by carbon dioxide. Solid carbon dioxide can be used in fish distribution, therefore, but its effectiveness is confined to its cooling effect.

Considerable research is being carried out on in-board bulk storage of gas packed fish (deep-chilling) and other methods of controlling spoilage [13].

Fish farming

Initially farmed fish were regarded with suspicion by some merchants and retailers but now the constant supply of fish of known size and quality has outweighed the possible disadvantages and differences that may arise in colour, texture and shape. Fish farming in the UK is increasing steadily but shellfish production is still quite small [12,13]. Virtually all trout and salmon consumed in the UK is farmed.

Retailing

As with meat retailing the type of retail outlet for fish has changed over the years, but unlike meat the traditional supplier, the fishmonger, still predominates. Table 7.6 shows retail sales in 1988 by type of outlet [14].

Deterioration of fresh and chilled fish

The packaging of fresh fish is beset with difficulties due to the nature of the fish itself. Of all fresh food, fish is one of the most susceptible to spoilage. It is more perishable than meat because of the more rapid autolysis by fish enzymes and because it is less acid in reaction, which favours microbial growth. Also, fish is susceptible to rancidity and odour formation due to trimethylamine production [12].

Rancidity. The fat content of fish flesh ranges from less than 1% for cod and similar species, to over 22% in herring, and is affected by marked seasonal fluctuations, particularly in the pelagic species. The fat of fish is rather oily and fluid and tends to be laid down in definite deposits, mainly beneath the skin. These fish oils are unsaturated, and as a result readily react with atmospheric oxygen, which results in the development of an unpleasant rancid flavour. The rate of oxidation is reduced at lower temperatures but the reaction can occur

Table 7.6 Retail fish (fresh and smoked) sales in 1988 by outlet (%) [14]

Fishmongers	49
Supermarkets	17
Market stalls	16
Mobiles	7
Other	11

Source: Attwoods/Seafish Industry Authority

even in the frozen state. It is catalysed by the enzyme systems in the fish, and traces of iron or copper act as pro-oxidants. The onset of rancidity can also be promoted by light.

Trimethylamine (TMA) formation.   When marine fish stale trimethylamine, $(CH_3)_3N$, accumulates as a result of bacterial reduction of trimethylamine oxide (TMO). The reaction involves the simultaneous oxidation of lactic acid to acetic acid and carbon dioxide:

TMO + lactic acid → TMA + acetic acid + carbon dioxide + water

$$2(CH_3)_3NO + CH_3CH(OH)COOH \rightarrow 2(CH_3)_3N + CH_3COOH + CO_2 + H_2O$$

This reaction is dependent on pH. Live or nearly dead fish are slightly alkaline, but after death there is a rapid fall in pH as glycogen breaks down, producing lactic acid. Fish flesh and its constituents (particularly TMO) are fairly strong buffers (the pH of fresh fish is usually around 6.6). As the fish begins to stale, the TMO is reduced by bacteria to TMA, which does not buffer at this pH and therefore the pH falls slightly. Eventually the lactic acid disappears, more alkaline material (including ammonia) is produced and the pH rises to 8 or more in the putrid fish.

Bacteriological condition.   The flesh and body fluids of live marine fish are generally sterile, but large numbers of bacteria are found in the external surfaces, gills, slime and the gut. The flora of fresh marine fish are usually psychrophilic (cold-loving) and grow well at 0°C. When fish are landed at the ports, after up to 15 or 16 days at sea in ice, bacterial counts as high as 2×10^7 per cm^2 may have developed. Coliforms, including faecal coli, can normally be detected on the fish, as a result of contamination aboard fishing vessels.

Factors affecting spoilage [3] include:

(1) The kind of fish: shape and size affect the rate of rigor mortis and fish of lower pH keep longer than fish with a higher flesh pH; also fatty fish tend to deteriorate more rapidly than other fish
(2) The condition of the fish when it was caught and the extent and type of bacterial contamination
(3) The temperature: cooling should be as rapid as possible.

Fresh fish is rarely, however, involved in food poisoning outbreaks. When this does occur it is mostly due to *Salmonella, Staphylococcus* and *Clostridium welchii* picked up in the course of human handling in distribution. Botulism from fish has occurred, though fortunately it is rare. Botulism type E has been reported in vacuum-packaged, hot-smoked, freshwater fish in some countries. Some fish, for example the mackerel family, produce a form of histamine poisoning when they spoil and some tropical species are poisonous even when perfectly fresh [12].

Prepackaging

Prepackaging of fresh fish for retail sale has been slow to develop. Early packs of expanded polystyrene trays and film overwraps, although acceptable to consumers, meant more wastage for retailers. Sales of fresh fish in the UK have declined over the last three decades but are now beginning to improve [13,14] because of expansion in retail sales and introduction of new products. Prepackaging alone does not extend the shelf life of fish but it does keep it clean and free from flies, reduces odour and makes self-service possible. Discoloration can be prevented by packaging with nitrogen and other gases.

Modified atmosphere packaging of fresh fish

The introduction of modified atmosphere packaging of fresh fish has improved the expected shelf life. The air is usually replaced with a combination of carbon dioxide, nitrogen and oxygen selected to increase product shelf life during distribution and retail display and to decrease the rate of bacteriological spoilage [15].

As with meat, carbon dioxide inhibits bacterial growth in fish but for white fish it tends to be absorbed by the tissues and therefore lowers the pH, decreasing the water-holding capacity of the fish with subsequent drip. This causes a lowering of internal package pressure and potential package collapse. A typical mixture of gases is 40% carbon dioxide, 30% nitrogen and 30% oxygen which reduces the risk of package collapse and drip development as well as extending shelf life up to 9 days at 0°C or 5 days at 2°C [9]. For fatty fish a mixture of 60% carbon dioxide and 40% nitrogen gives satisfactory results [10]. There have been concerns about the possible dangers of botulism from MAP packaged fish, especially packs excluding oxygen. In the United States [16], this type of packaging is restricted for fresh fish. However, under the right conditions and with proper management of the cold chain, MAP should be safe [49,50]. The Sea Fish Industry Authority has published guidelines for the handling of fish packed in a controlled atmosphere [17] and recommends a temperature of 0–5°C. Packaging materials used are selected according to the properties required but are similar to those used for MAP meat. Microwave-ready fish in MAP are likely to develop further in future [18].

Vacuum packaging of fresh fish

Vacuum packaging provides some increase in shelf life but care must be taken because anaerobic conditions can pose potential problems. Products should be kept at a temperature below 3°C during retailing and be of high initial quality and hygienically prepared. As with MAP, vacuum packs are leak proof and odour free and offer merchandising advantages. Packaging materials are similar to those used for vacuum packaged meat.

Milk

Milk is transported from farm to processing plant in a number of ways, depending on geographical area. In rural areas, such as some parts of India, women carry it in jars on their heads to the receiving station where it is transported in large cans to the processing plant on a lorry. However, in industrialized parts of the world the introduction of mechanization and bulk collection has provided an economic system for the collection of milk from large farms [19].

With modern milking machines the milk is drawn off in a completely enclosed system into refrigerated stainless steel tanks installed in the farm dairy, where it is cooled and held at 4.5°C or below [20]. Bacterial growth is retarded at this temperature and the milk is maintained in good hygienic condition ready for collection. Later, it is transferred by pump from the tank to a collecting tanker, instead of being filled, as it once was, into churns and transported by lorry.

Such a system is practicable only where access facilities for the tankers, which are insulated, are good, and is most economic for large farms. The main disadvantage of the system is that one load of bad milk can spoil a complete tanker load, whereas if collection is in churns an efficient inspection procedure will find any churn of bad milk and avoid mixing it into the bulk. Careful cleaning and sterilizing of churns, tanks and tankers is essential and these are usually made of tinned mild steel or aluminium. Milk is generally spoiled by bacterial growth and the production of lactic acid which causes souring or taints due to protein decomposition or fat rancidity. In cows, udder milk has a bacterial count of less than 500 per ml. When counts reach about 10^7 per ml, taints are obvious and the commercial life of the milk is over. Failure to cool milk sufficiently and failure to clean and sterilize equipment and utensils are the main factors responsible for losses due to souring.

Once it has been processed, e.g. by pasteurization, liquid milk is filled into bottles, cartons or pouches and delivered to consumers either through the normal retail sales outlets or in some countries, by 'doorstep delivery'. This was once widely practised but, because of rising costs and shortage of manpower, the UK is now almost the only country using doorstep deliveries and even here this is becoming more difficult to sustain in some parts of the country.

Quality and composition

Milk for human consumption is produced in the greatest volume by the cow and the water buffalo, although the goat is also an important source in Egypt, India, China and other Asian countries. Some goats' milk is also marketed in Europe and North America, but compared with cows' milk it is relatively unimportant. The water buffalo is an important source in India and a few

Table 7.7   Composition of various liquid milks

Milk origin	Approximate composition			
	Water (%)	Fat (g/100 g)	Protein (g/100 g)	Lactose (g/100 g)
Cow	87	3.7	3.5	4.9
Goat	87	4.3	3.5	4.3
Buffalo (Indian)	83	7.4	3.6	5.5

other countries and buffalo milk is often mixed with cows' milk during processing. The composition of these three types of milk is given in Table 7.7.

Milk is an excellent culture medium for many microorganisms; it is high in moisture, nearly netural in pH, and rich in nutrients such as milk sugar, butterfat and protein. The fat content of cows' milk in Europe reaches a minimum from April to June and a maximum from November to January, while the solids-not-fat content is at a maximum from April to June and a minimum from November to January. This is attributable to the nutrition of the cow: where pastures decline, feeds are supplemented with hay, silage, etc., to maintain both the volume and composition of the milk. Other variations in milk composition occur because of differences in feeding regime, breed, stage of lactation and climate.

The fatty materials in milk are less liable to breakdown than the protein and carbohydrate, but under certain conditions even these will deteriorate to give various taints and flavours. Milk is therefore a complex and unstable liquid food, subject to many spontaneous changes, not only at room temperature but also when chilled. The fat globules in the creamy part of the milk will also absorb odours readily.

Effect of temperature on bacterial growth

Many bacteria are most active between 10°C and 30°C. The lower limit for activity in milk is 0–1°C and the upper limit, above which growth of microorganisms is virtually nil, is around 70°C. Milk should be cooled, therefore, to below 10°C immediately after milking or processing and preferably stored at 4.5°C or below, particularly where it is destined for the liquid milk market. The freezing of milk as a method of storage and transport has not been adopted commercially, partly because of cost and partly because of the changes that take place in its taste and physical structure.

Pasteurization

Milk is a delicately flavoured, easily changed material and rigorous preservation treatments cannot be used to preserve it without changing it undesirably and making it into a different product. The preservation processes employed are therefore as mild as is consistent with achieving the desired shelf life. Pasteurization of milk kills all the pathogens, and improves the keeping quality without harming flavour, appearance and nutritional value. If the milk is required for further conversion into other products, then pasteurization will destroy all those microorganisms which would interfere with the activities of starter organisms and could cause spoilage.

The pasteurization method commonly used is high temperature, short time pasteurization (HTST) [19] in which the milk is passed through a heat exchanger and held for 15 s at not less than 72°C. It is then cooled immediately to about 3°C so that, after bottling, the temperature of the milk leaving the dairy is between 4 and 6°C.

Various tests are used to check the efficacy of the process and assess keeping quality. Pathogens should be destroyed and all yeasts, moulds and most vegetative bacteria cells eliminated.

Characteristic spoilage of pasteurized milk

Apart from oiliness and oxidative changes catalysed by light, and faults related to the inherent enzymes of milk, spoilage is mainly due to bacterial growth [12,19]. Souring or acid formation, resulting in coagulation, is usually because of lactic acid bacteria, which at temperatures between 10°C and 37°C produce large quantities of lactic acid and above 37°C, other volatiles as well. 'Stormy fermentation', which produces gases, is often due to coliforms. 'Ropy milk' results from the growth of bacteria which gain access to the milk after leaving the udder stressing the importance of scrupulously sterilizing the equipment used in processing. 'Bitty cream' or 'sweet curdling' is a common fault in warm weather where small lumps of fat appear floating on the surface. It is believed to be caused by certain organisms attacking the protein and fat constituents of milk. The characteristic stale, rancid odour of butyric acid is probably due to the enzyme lipase, either present in the milk or produced by some moulds or bacteria.

Packaging

Both bottles and cartons take into account the properties of milk and provide packaging acceptable to consumers worldwide. Glass bottles have the advantage of being easily cleaned, transparent and rigid, but have the great disadvantage of high weight and fragility.

Increasingly, milk is also packaged in gable top (PurePak, Elopak) or other cartons (TetraBrik). Even though the equipment may be expensive to install, the advantages include a lower price per unit of milk and a lower risk of contamination from the air during filling [19]. Smaller quantities have been packaged in plastic pouches, although this method is not used extensively in the UK. In Australia a cylindrical milk carton with a reclosable pouring lid has been introduced [23].

Following pasteurization of liquid milk, most UK milk is delivered to the consumer in glass bottles, the commonest size being 1 pint, although a 2 pint glass container has been test marketed [22]. The bottles are washed by machines of 'soaker' or 'jet' type. After a preliminary rinse, one or two hot detergent treatments are given, followed by a warm and final rinse. The bottles are not sterilized, but a good machine gives bottles which are virtually sterile. In bottling the milk is delivered from a filler bowl into the bottle by an automatic device and the bottle is immediately capped with aluminium foil. Contamination from the air must be carefully controlled.

While suitable for sterilized milk, glass bottles are a problem for 'long-life' milk. The question of container is therefore of vital interest to the dairy industry, as about 50% of all milk produced in the UK is sold in liquid form [24]. Sunlight can destroy riboflavin and vitamin C in milk, producing a taint by the oxidation of the fat [25] and protein. This led to some interest in the use of brown glass bottles which hold back the actinic rays responsible. However, the taint is very rare and since brown bottles are not very attractive and it has also been found that milk sours faster in brown than in colourless bottles, their use has not been great [26].

Returnable bottles

For economic reasons, the use of the returnable glass bottle has continued over many years in the UK. The advantage of a non-returnable container is that the dairies do not have to undertake the troublesome, unpleasant and space-requiring work of getting the bottles returned and cleaning them, which is only economic as long as the daily milk round continues. The glass bottle will take a long time to disappear because of its economic advantage, the traditions of the industry and the attitude of the consumer. In other countries factors such as transport costs have led to the use of non-returnable packaging although more recently 'green' considerations have led to the re-introduction of returnable bottles in some countries [27–29].

The advantages of non-returnable containers are:

(1) Elimination of returned empties
(2) Elimination of collecting, sorting and washing problems
(3) Elimination of the foreign object problem
(4) Elimination of the glass fragment problem
(5) Reduction in transport costs

The disadvantages are:

(1) Possible increases in costs of packages
(2) Consumer acceptability
(3) Delivery problems resulting in lower total sales
(4) Hygiene problems
(5) Environmental considerations [30]

Plastic containers and plastic-coated cartons are nearly sterile by virtue of their method of manufacture. No sterilizing process is necessary for pasteurized milk, but for the aseptic filling of milk, sterilization is essential and so far TetraBrik has proved most effective. Containers for this purpose must therefore be sterilized immediately before filling (see chapter 10).

Other dairy products

The dairy industry is not only one of the largest food industries but one of the largest of all industries [31]. The composition of the main fresh milk products is shown in Table 7.8 and it will be noticed that many have a high fat content. To some extent this determines the type and degree of protection they require from their packaging.

Butter

In butter-making [19], the milk is separated and the cream 'churned' until the emulsion 'breaks', i.e. the fat-in-water emulsion becomes a water-in-fat emulsion. Ripening (i.e. souring) of the cream greatly assists churning, but ripened cream butter does not possess the long keeping qualities of sweet cream butter, as the acidity accelerates metal-catalyzed oxidative faults. Variations in butter composition are attributable to differences in manufacture, but most butter is regulated by law to a limited amount of moisture. Other constituents are protein, lactose and salt. The natural colour of butter is due to carotene and similar fat-soluble pigments in the fat globules of the milk.

The flavour of butter is produced by the fermentation of bacteria in the cream. Although souring gives a fuller flavour, the use of butter cultures or starter organisms gives better control of flavour and avoids the danger of undesirable taints.

Table 7.8 Approximate composition of dairy products

Products	Approximate composition (%)			
	Water	Fat	Protein	Lactose
Butter	16	82	0.5	Tr
Whole milk	87	4	3	5
Skimmed milk	91	0.1	3	5
Semi-skimmed milk	89	2	3	5
Cream (single)	73	19	3	4
Dairy (low fat) spreads	50	40	6	0.5

Protection required

Packaging [19,32] must protect the product in relation to its flavour, body and texture, appearance, moisture and colour. In addition, butter readily absorbs odours. The nature of the emulsion makes butter especially prone to rancidity caused by oxidation of the fats, producing a 'fishy' taint. The fishy taste and smell is ascribed to the presence of metallic contamination of the butter by traces of metals which dissolve off the processing equipment.

Bulk butter is packed into 25 kg LDPE lined paperboard boxes, in competition with cans which are used for longer term storage [33]. Wrapping materials for retail sale of butter must protect the product from light, prevent oxidation and be suitably resistant to water vapour to prevent surface drying and discoloration.

Butter is normally wrapped using parchment paper or aluminium foil wrappers laminated with vegetable parchment or greaseproof paper.

Dairy spreads

Blends of butter and vegetable fats are manufactured into spreads often with a lower fat content than butter. Manufacture is similar to that of margarine (see chapter 12) and the product usually contains added milk protein, an emulsifier and a stabilizer. Beta-carotene is added for colour. Spreads are susceptible to mould growth and rancidity as with butter and cream. They are usually packaged in thermoformed polypropylene or LDPE containers with lids of aluminium foil, PVC or other film [19].

Eggs

The quality of fresh eggs deteriorates during storage as the porous shell lets moisture evaporate through it resulting in the contents shrinking. Moisture loss can be reduced by packaging in closed containers. Also moulds and bacteria can penetrate the shell. Provided the eggs are kept dry the contents can remain stable for several weeks.

Packaging requirements

The primary requirement for egg packaging is to protect against breakage and to keep the eggs clean and dry. Tremendous losses from breakages occur throughout the world from inadequate packaging. Originally they were protected by packing in layers, each egg separated from its neighbour in paperboard divisions, the layers being separated by flat sheets. Moulded pulp trays then took over and moulded pulpboxes shaped to hold six, ten or twelve eggs are also used. Trays and boxes are also made from expanded plastics. Shrink-wrapping over moulded trays is used for larger quantities. Egg boxes can also be manufactured using recycled plastics materials and paper [34].

Cream

Creams have a fat content ranging from 12 to 16% and this affects the shelf life and physical characteristics [19]. Cream is separated from milk and is then treated by HTST pasteurization. It is subject to the same spoilage factors that cause deterioration of milk but the higher fat content makes it more susceptible to oxidation, and it also absorbs odours readily.

Bulk supplies to the food industry are delivered in road tankers while bakeries and caterers may receive it in churns or bag-in-box containers. Cream for retail sale is usually packaged in plastic (polystyrene) tubs with heat sealable coated polyethylene/aluminium foil laminate covers. Aerosol creams usually contain sugar, a stabilizer and a propellant such as nitrous oxide. The aerosol is designed so that the cream is dispersed as an aerated foam.

Guidelines for packers, manufacturers, distributors and consumers

Packaging suppliers should be aware of regulations, codes of practice and guidelines to which all parts of the food chain should work. They should also be aware of the appropriate standards and other specifications for equipment and packaging materials. Many countries have their own legislation on both hygiene control and temperature control of foods.

The Institute of Food Science and Technology (IFST) has produced general guidelines on good manufacturing practice [47] and specific guidelines on the handling of chilled foods. This latter guide covers the basic rules of hygiene for food handlers and manufacturers, rules for handling chilled foods in predistribution storage, during distribution and transport and in retailing and catering, as well as advice to the consumer on how to buy, store and cook them. IFST recommended product temperatures for fresh meat, poultry, offal, fish and shellfish are -1 to $+2°C$; for milk, cream and low fat spreads 0 to $+5°C$; and for butter and other spreads 0 to $+8°C$ [48]. The EC has hygiene directives on a number of commodities [35–39] and a number of other proposals are planned. In the UK fresh and chilled raw foods are subject to hygiene regulations and temperature controls for food storage, distribution and retail outlets are laid down. All chilled foods in the UK must be kept at or below $8°C$ except foods at risk from *Listeria* bacteria (which must be kept below $5°C$); only transport vehicles carrying less than 7.5 tonnes are exempt from this requirement [40]. The Sea Fish Industry Authority guide to the gas packing of chilled fish [17] has already been mentioned but there are several other codes of practice in the UK on chilled foods. Some cover all chilled foods, including precooked ones, but they are just as relevant to fresh foods [41–46].

References

1. M.E. Ranken, Meat and meat products, *Food Industries Manual*, 2nd edition, M.E. Ranken (ed.), Blackie, Glasgow (1988).
2. H.W. Ockerman, Chemistry of meat tissue, *USDA Training Course for Processed Food Inspectors*, Ohio State University (1970).
3. C.F. Niven, Influence of microbes upon the colour of meat, *Am. Meat Inst. Bull.* **13** (1951).
4. W.C. Frazier, *Food Microbiology*, 2nd edition, Tata McGraw-Hill, New Delhi (1967), Chap. 16.
5. A. Cuthbertson, Progress in meat handling and meat quality, in *Food Technology International Europe* (1987), p. 92.
6. R. Bentley, Dressing upside down, down under, *Food Sci. Technol. Today* **5**(2) (1991), 111.
7. A. Taylor Centralised packaging of fresh meat, in *Food Technology International Europe* (1989), p. 379.
8. A.A. Brody, Controlled atmosphere packaging for chilled foods, in *Food Technology International Europe* (1990), p. 310.
9. R. Inns, Modified atmosphere packaging, in *Modern Process, Packaging and Distribution Systems for Food*, F.A. Paine (ed.), Blackie, Glasgow (1987).
10. A.A. Brody, Controlled atmosphere packaging, in *The Wiley Encyclopaedia of Packaging Technology*, Wiley, New York (1986), p. 219.
11. J.P. Smith, H.S. Ramaswamy and B.K. Simpson, Developments in food packaging technology, Part 2, *Trends Food Sci. Technol.* November (1990), 107.
12. M.J. Urch, Fish and fish products, in *Food Industries Manual*, 2nd edition, M.E. Ranken (ed.), Blackie, Glasgow (1988).
13. K.J. Whittle, Current developments in fish processing, in *Food Technology International Europe* (1988), p. 137.
14. P.R. Sheard, *A Guide to the British Food Manufacturing Industry*, 3rd edition, Nova Press, London (1991).
15. M.S. Schotland, Fish packaging and marketing in the UK, paper presented at the *4th Int. Conf. on Controlled Modified Atmosphere Vacuum Packaging*, New York, Business Research (1988), pp. 31–42.
16. E.S. Garrett, Packaged seafood and safety, paper presented at Packaging Conference Inc. on *Food Safety Packaging*, Chicago (1989), 13 pp.
17. Sea Fish Industry Authority, *Guidelines for the Handling of Fish Packed in a Controlled Atmosphere*, Sea Fishery House, Edinburgh (1985).
18. Anon, Will it swim in California? *Packaging Dig.* **26**(5) (1989), 53, 56, 60.
19. K.J. Burgess, Dairy products, in *Food Industries Manual*, 2nd edition, M.E. Ranken (ed.), Blackie, Glasgow (1988).
20. National Dairy Council, *The Quality of Milk* (1982).
21. W.C. Frazier, Food Microbiology, 2nd edition, Tata McGraw-Hill, New Delhi (1967), Chap. 3.
22. Anon, Associated's 2 pint glass bottle tackles plastic container, *Milk Ind.* **91**(12) (1989).
23. Anon, Innovation in milk packaging, *Austr. Packag.* **37**(9) (1989) 28.
24. Anon, Dairy heard, *Packaging Today* **11** (1989) 9.
25. Anon, Protecting milk from light, *Verpack Derat.* **1** (1989), 23–28.
26. G. Stehle, Trends in packaging techniques for milk products and fruit, *Neue Verpack* **41**(10) (1988), 56, 60–61.
27. Anon, Milk bottles milk bottles for the sake of the environment, *Austropack* **6** (1989), 163.
28. R.N. Hassen, A dressing down of the bottle freaks, *North Eur. Food Dairy J.* **55**(4–5) (1989), 81.
29. Anon, *North Eur. Food Dairy J.* **55** (4–5) (1989), 82.
30. Anon, The environment: How green is milk packaging? *Milk Ind.* **93**(2) (1990), 26, 27, 30–33.

31. M. Hilliam, *The European Dairy Industry*, Leatherhead Food Research Association (1990).
32. C.R. Oswin and L. Preston, *Protective Wrappings*, Cann Publications. London, (1980), p. 41.
33. Anon, *Nov 1988* Danish company uses butter in a box for competitive strategy, *Food Eng. Int.* **13**(11) (1988), 15.
34. Anon, Brand concepts for eggs—environmental aspects of egg box production, *Pack. Rep.* **12** (1989), 18.
35. EC Directive on health problems relating to minced meat and similar products: import from Third World Countries—D88 1657, *Official J. Eur. Community* L382 (1988).
36. EC Directive on Int. EC trade in fresh meat (health problems)—D72/461 and D72/462 and Modif D87/64, *Official J. Eur. Community* L34 (1987).
37. EC Directive on health problems affecting intercommunity trade in fresh meat—D88/288, *Official J. Eur. Community* L124 (1988).
38. EC Directive on Health and hygiene conditions for production and trade in game, meat products and preparations, in press.
39. EC Directive on health conditions for animal food products not covered by existing legislation: eggs—D89/437, *Official J. Eur. Community* L212 (1987).
40. Food Hygiene (Amendment) Regulations UK, SI 1431, HMSO, London (1990).
41. Chilled Food Association, *Guidelines for Good Hygiene Practice in the Manufacture of Chilled Foods*, Edinburgh (1989).
42. National Cold Storage Federation, *Guidelines for the Handling and Distribution of Chilled Foods*, London (1989).
43. The Retail Consortium, *Guidelines for Retail Operators*, London (1989).
44. Int. Inst. of Refrigeration, *Recommendations for Chilled Storage of Perishable Produce* (1979).
45. Department of Trade, *A Guide to the International Carriage of Perishable Foodstuffs*, HMSO, London (1988).
46. Department of Health, *Guidelines on Cook Chill and Cook Freeze Catering Systems*, HMSO, London (1989).
47. Institute of Food Science and Technology, *Food and Drink Manufacture Good Manufacturing Practice: A Guide to its Responsible Management*, 3rd edition, IFST, London (1991).
48. Institute of Food Science and Technology, *Guidelines for the Handling of Chilled Foods*, 2nd revision, IFST, London (1990).
49. F.P. Coyne, Effect of carbon dioxide on bacterial growth with special reference to presentation of fish. Part II. Gas storage of fresh fish, *J. Soc. Chem. Ind.* **52** (1933), 19T.
50. J.H. Hotchkiss, Microbiological hazards of CAP/MAP food packaging, in *Proc. 3rd Int. Conf. on CAP/MAP/Vacuum Packaging—CAP 87*, Ithaca IL (1987).

8 Fresh fruits and vegetables (including herbs, spices and nuts)

Fruits and vegetables

Fresh fruits and vegetables have many characteristics which make their packaging and handling (packaging cannot be divorced from handling) very specialized and often expensive. Many different types of packages are used for fruits and vegetables, and even with the most careful rationalization, a wide variety of packages would be needed to satisfy their diverse requirements.

In general, fruits and vegetables are bulky, easily damaged mechanically, consist largely of water which is readily lost, and, above all, are living and must be kept so. This means that they are sensitive to their environment, their rate of metabolism is temperature dependent, and they may be damaged by heat or cold. They are affected by the levels of oxygen and carbon dioxide, ethylene and other volatiles in the atmosphere. When fresh fruits and vegetables respire, they take in oxygen and give out carbon dioxide, heat and water vapour. They also lose moisture (wilt) rather rapidly by evaporation. While attached to the plant, the losses due to these processes are replaced by the flow of sap, but after harvest respiration and water loss continue and the plants are dependent entirely on the food reserves and moisture they contain; losses are not replaced, deterioration begins and they eventually perish.

Variability

Variability is characteristic of all biological material and is particularly important in fruits and vegetables. No two individual fruits are exactly alike; there is variability within and between trees, between seasons and between climatically different areas. There is also genetic variation, and cultural practices greatly influence the quality and behaviour of any crop.

The growing process, respiration and ripening

Fruits, and many of the vegetables which botanically are fruits, such as tomatoes and cucumbers, are not harvested until growth of the fruit is complete. Those vegetables which consist of leaves, buds, or stems are harvested at an earlier stage of growth. Achievement of desired quality, maturation and ripening of fruits is normally required before consumption, but maturation of vegetables is usually not wanted and is discouraged, although it must be recognized that the greater the maturity when harvested, the greater will be the yield.

Growth generally comprises an initial short period of cell division followed by a longer period of cell enlargement culminating in maturation. This is followed by ripening, senescence and death. These processes are controlled by hormones and are outlined in Figure 8.1. Broadly, kinins and gibberellins are concerned with cell division and differentiation, auxins with growth, and abscisins and ethylene with maturation and ripening.

A fruit can be said to be *physiologically mature* when it has reached its last, slow stage of growth and has developed the ability to ripen normally after harvesting. It may be *commercially mature* at an earlier stage, when sufficient desirable characteristics have developed to make it edible.

Practical tests to indicate maturity all have limitations; no single test is a reliable basis for forecasting the storage life or the ripening behaviour of a fruit. Respiration rate, a good guide, is not a practicable routine test. With many fruits, a change in the colour of the skin, from the 'ground colour' of the deep green of immaturity to a lighter green, or even a greenish yellow, is also a useful guide. For other fruits and vegetables, softening (a pressure test) can be useful. Thus, changes in several characteristics have to be taken into account, with calendar date based on experience being a prime consideration.

Physiological activity in harvested fruit and vegetables may be essential for the attainment of the desired ripeness or it may lead to a deterioraton in quality. The main metabolic process of all harvested produce is respiration, the breakdown of organic substrates with the release of energy and the depletion of reserves. This energy may be used for additional syntheses, as required during ripening, or it may be released as heat. The object of any storage technique is to slow down these metabolic processes and the respiration of the fruit or vegetable, and thus prolong its storage life without upsetting its normal metabolism which could result in abnormal ripening or other undesirable changes.

However, the inevitable end of all living tissues is senescence and death, and the ripening of a fruit represents the start of this process. Senescence involves the progressive disorganization of the metabolic processes of the cell. Maintenance of the integrity of the cells and their metabolism requires a constant supply of energy from respiration. In the presence of oxygen, respiration is *aerobic* and the final products are carbon dioxide, water and heat. In the absence of oxygen, respiration is *anaerobic* and much less efficient as a producer of energy—the products are compounds of intermediate molecular size such as ethyl alcohol and acetaldehyde. Aerobic respiration is much more important in harvested fruits and vegetables, but anaerobic

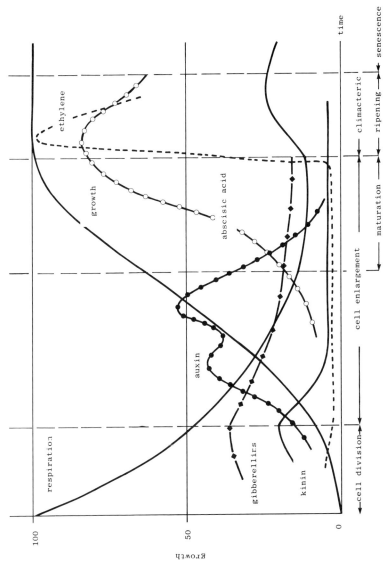

Figure 8.1 Processes controlling maturation and ripening of fruit.

respiration may be significant in senescent tissues where structural breakdown has reduced the permeability to oxygen.

In general, the rate of respiration, as measured by the production of carbon dioxide or by the consumption of oxygen, is a good measure of the rate of metabolism and of the anticipated storage life and relative perishability of the fruit or vegetable. Generally, storage life is inversely related to respiration rate. Most fleshy fruits, which are picked hard and unripe and have a characteristic and distinct ripening phase, also have a characteristic temporary, but marked, rise in the respiration rate to a climacteric maximum which normally coincides with the more obvious ripening changes. This climacteric rise is accompanied by an increased production of ethylene. Other fruits and vegetables without such a distinct ripening phase do not show this climacteric rise.

Green peas, beans and leafy vegetables have a high respiration rate, many times greater than apples, oranges or bananas, while 'hard' vegetables, like potatoes, onions and pumpkins, respire slowly. As a general rule the earlier-maturing varieties of fruits produce heat and carbon dioxide more rapidly than the later varieties and have a correspondingly shorter life. Ripening fruits respire faster than green unripe fruits.

The rate of respiration is decreased by increasing the amount of carbon dioxide present and decreasing the oxygen content of the atmosphere—these effects are the basis of controlled atmosphere (gas) storage. If carbon dioxide levels are too high or the oxygen level too low, anaerobic respiration takes over and the tissues are irreversibly damaged. Tolerance to increased carbon dioxide or reduced oxygen varies widely; tropical fruits as a class are commonly more tolerant than fruits from temperate climates, so controlled atmosphere storage is relatively more effective with them.

A low rate of respiration is desirable as it indicates a low rate of utilization of sugars, the main respiratory substrates, and other essential reserve material, and consequently longer life. The object of any storage technique is to minimize deterioration without upsetting normal life processes, and packaging can aid or hinder realization of this objective.

Ethylene is produced by all plant tissues and is the natural ageing and ripening hormone responsible for the breakdown of chlorophyll pigments, leaf fall and fruit ripening, probably by inducing ripening enzyme systems. It is physiologically active at very low concentrations (less than 0.1 ppm in the atmosphere). The production of ethylene (Table 8.1) is closely related to respiration, but the rise in its production may be before or after the climacteric rise in respiration. The amounts produced by different fruits vary; apples produce large amounts while production by the mango, pineapple and citrus fruits is very low, even though these fruits respond to external ethylene.

Control of ripening is an important part of the storage and marketing of fruit. Fruits ripen well only within a limited range of temperatures with an optimum for most fruits of about 20°C, so that ripening is controlled primarily

Table 8.1 Ethylene production rates of some common fruits and vegetables

Relative rate ($\mu l/kg\ h$)	Produce
Very low < 0.1	Cherries, dates, most citrus fruits, rhubarb, tomatoes, asparagus, most root and leafy vegetables
Low 0.1–1.0	Most berried fruits, olives, pineapple, quince, water melon, green beans, cucumbers, peppers
Moderate 1.0–10	Bananas, breadfruit, honeydew melon, lychee, mangoes, plums
High 10–100	Apricot, avocado, nectarines, papaya, pears, peaches, canteloup melon
Very high > 100	Apples, passion fruit

by control of temperature. Bananas will not ripen properly below a temperature of about 15°C, whereas some varieties of plums and pears will ripen slowly but satisfactorily at 5°C. Control of ethylene, carbon dioxide and oxygen levels is also part of the control of ripening (see below).

Bananas are specially sensitive to ethylene in concentrations as low as 1 ppm and must be stored in atmospheres free from this gas. Bananas have been kept at 20°C for 6 months in an unripe state in an atmosphere free from ethylene and containing 5% oxygen and 3% carbon dioxide. When transferred to air and given ethylene, normal ripening took place. Ripening also requires maintenance of the original water content of the tissues and therefore high humidities. The ripening of bananas is hastened by introducing a few ppm of ethylene into the ripening rooms. Citrus fruits can be rapidly degreened by exposure to a few ppm of ethylene at temperatures of around 25°C and with a high humidity. Ethylene hastens the ageing of citrus fruits and their storage is prolonged if the ethylene is removed from the atmosphere.

Temperature

The temperature at which fresh fruits and vegetables are stored and transported is very important [1]. At temperatures in the range $-3°C$ to $-0.5°C$, fruits and vegetables will freeze. The higher their water content and the lower the concentration of dissolved substances (mainly sugars) in the sap, the nearer the freezing point of the produce will be to 0°C, the freezing point of water. Once frozen, fresh fruits and vegetables are damaged, the extent of the damage depending both on the temperature and the duration of the freezing. If frozen for only a few hours at a temperature close to their freezing point, some will recover if thawed gradually in a high humidity atmosphere at a temperature not much above their freezing point. If frozen more severely, all are permanently damaged and break down rapidly after thawing. It is clear then that in storage or transport, freezing must be avoided, which means that flesh temperatures should not fall below $-1°C$ for most produce.

At the other end of the scale there are definite upper limits of temperature above which fruits and vegetables are irreversibly damaged. These limits vary rather widely for different kinds of produce. Some, such as bananas and tomatoes, are more sensitive to temperature than others and are injured and ripen abnormally when exposed to temperatures over 27°C for any length of time. Others are more tolerant of high temperatures and are not obviously injured by temperatures as high as 35°C, although quality is usually reduced. Most fruits and vegetables are rapidly damaged by exposure to temperatures of 38°C or more after harvest.

Ripening temperatures must be considered as well as storage temperatures. Most fruits ripen best at temperatures of the order of 18–22°C, nevertheless there is considerable variation in the range of temperatures at which different fruits will ripen satisfactorily. Unless ripened at suitable temperatures, quality will be poor and the fruit will not be acceptable to the consumer. Bananas are very sensitive to temperature and will ripen properly only within a narrow range from 15 to 22°C. At temperatures of 20–25°C, tomatoes ripen with the development of a full red colour and good texture and flavour. At temperatures above about 30°C the red pigment lycopene does not develop and the fruit ripens with a yellow colour, soft texture and poor flavour. At temperatures below 20°C colour development is poorer, and at about 15°C or below the fruit develops only a pale pink colour, a soft 'grainy' texture and poor flavour. Tomatoes will not ripen at all below 10°C. On the other hand, some varieties of pears and plums will ripen satisfactorily at temperatures as low as 5°C and as high as 27°C and paw-paws will ripen well enough at 30°C.

The rate of many chemical reactions is generally at least halved by a reduction in temperature of 10°C, but the effects of temperature on the rates of respiration, metabolism and storage life of fruits and vegetables are more variable and often more marked. For many perishables a reduction of about one-third is more typical, and thus a reduction in temperature of 10°C brings about a threefold increase in storage life.

Other things being equal, the lower the temperature the longer the storage life. This effect of temperature is not uniform, small changes in temperature have more effect in the range of -1 to 5°C than at higher temperatures. The life of many apple varieties, peaches and plums is about 25% greater at 0°C than at $+1°C$, and the life of William pears at $-1°C$ is almost double that at $+1°C$, but the rate of ripening of these fruits is little affected by changing the temperature from 18°C to 20°C.

Because of this greater sensitivity to temperature near freezing point, storage of many kinds of produce requires close control of temperature, and a variation in air temperature of no more than $\pm 0.5°C$ should be aimed at. The best temperature for longest storage of many fruits and vegetables is as close to their freezing point as can be safely maintained. Thus for pears and many varieties of apples it is $-1°C$, for leafy vegetables, which have a high water content, $0°C$. However, other kinds of fruits and vegetables are injured by exposure at temperatures well above the freezing point and their longest life is obtained at substantially higher temperatures. Most tropical or subtropical produce falls into this category, for example, bananas must not be exposed to temperatures below $10°C$, nor pineapples below $7°C$. Some temperate fruits and vegetables exhibit similar behaviour, e.g. potatoes and tomatoes.

The damage produced by low temperatures varies. Some fruits show external symptoms such as brown stains, spotting or pitting of oranges, pitting of mangoes, scald on apples, and grey to brown discoloration of bananas. Others are injured internally, e.g. peaches, plums and pineapples. Others, such as tomatoes and melons, show little visible signs but deterioration is hastened and rapid rotting during marketing occurs. With potatoes, the effect of low temperature is chemical, hydrolysis of starch to sugar giving undesirable sweetening. Best storage temperatures for the principal fruits and vegetables are shown in Table 8.2. The times for which different kinds can be kept vary from a few days to several months, depending largely on whether their rate of respiration is naturally fast or slow.

Composition of the atmosphere

Because the rate of respiration of produce governs its storage life, factors other than temperature which affect this rate can often be used to assist cool storage. Thus increasing the carbon dioxide content and decreasing the oxygen content of the atmosphere is the basis of 'gas' storage of apples and pears.

An adequate supply of oxygen is necessary for respiration. The 21% of oxygen normally present in air usually supports maximum respiration. The rate of respiration, and hence the rate of ripening and deterioration, can be reduced by decreasing the oxygen supply to the fruit by storage in an atmosphere with a low content of oxygen. To avoid oxygen starvation and consequent abnormal respiration leading to fermentation and breakdown, the oxygen supply must not fall below a certain critical level which is greater at higher temperatures (where respiration is faster). At cool storage temperatures this critical level is about 2% for apples and pears.

Research has also shown that increasing the carbon dioxide of the air to a level of several percent has a marked depressing effect on respiration rate. However, too high a concentration of carbon dioxide will upset respiration too much and cause breakdown. This sensitivity varies rather widely; 2–3% of carbon dioxide may improve some varieties of apples while a few are tolerant of 10%. Citrus and other fruits are very sensitive to carbon dioxide and its concentration in the storage atmosphere should not be allowed to

Table 8.2 Life of fruits and vegetables at best keeping temperature

Best storage temperature	Fruit or vegetable	Approximate life in weeks (1–12)	> 12
− 1.0°C	Grapes (most varieties)		
	Nuts etc.		up to 12 months
	Pears		up to 28 weeks
	Parsnips (topped)		up to 20 weeks
− 1.0 to 3.0°C	Apples (depending on variety)		up to 28 weeks
− 0.5°C	Apricots		
	Berries		
	Cherries		
	Figs		
	Nectarines		
	Peaches		
	Plums		
	Grapes (late varieties)		
	Quinces		up to 16 weeks
	Garden peas		
	Beetroot		up to 20 weeks
	Carrots		up to 20 weeks
	Onions[a] (later varieties)		up to 28 weeks
	Swedes		16–24 weeks
0.0°C	Asparagus		
	Broccoli		
	Brussels sprouts		
	Cauliflower		
	Chinese leaves		
	Lettuce		
	Rhubarb		
	Silver beet } Spinach		
	Mushrooms		
	Cabbage		
	Sweet peppers		
	Coconuts[a]		
	Cabbage (late varieties)		

Table 8.2 (*Continued*)

Best storage temperature	Fruit or vegetable	Approximate life in weeks (1 2 3 4 5 6 7 8 9 10 11 12)	> 12
	Celery		
	Early onions[a]		
5.0°C	Melons		
	Avocado pears		
5.0 – 7.0°C	Mandarins		
	Oranges		
7.0°C	Mangoes		
	Green beans		
	Cucumbers		
	Tomatoes		
	Avocadoes (Fuerte)		
	Passion fruit		
	Potatoes		16–24 weeks
10.0°C	Paw-paws		
	Pineapples		
10.0–12.0°C	Marrows[a]		
	Grapefruit		up to 16 weeks
	Pumpkin[a]		up to 24 weeks
12°C	Bananas		
	Sweet potatoes		16–24 weeks

[a] Require storage at r.h. below 70%.

rise above 1%. Strawberries will withstand 25% carbon dioxide for considerable periods.

Gas storage delays ripening and ageing because the rate of metabolism is largely governed by the levels of carbon dioxide and oxygen inside the fruit itself. In cool storage the carbon dioxide content of the internal atmosphere of an apple (the tiny intercellular air spaces accounting for some 30% of the volume of the fruit) is about 0.5% lower than in the air surrounding the fruit. These differences are greater at higher temperatures.

The most simple controlled atmosphere storage system is called non-scrubbed storage and involves venting with outside air. The result will be a mixture of oxygen and carbon dioxide, for example 15% oxygen and 5% carbon dioxide. Where low concentrations of oxygen and carbon dioxide are required, oxygen is controlled by air vents and carbon dioxide is controlled by circulating a proportion of the storage atmosphere over a carbon dioxide scrubber containing hydrated lime or activated carbon which absorb carbon dioxide [2]. In the gas storage of fruits such as tomatoes and hard fruits, the use of optimum conditions is a balance between temperature, carbon dioxide and oxygen. The latest generation of controlled atmosphere stores rely on microcomputer control.

Generally the gas storage of vegetables is less well developed than that of fruit, although considerable research is being carried out on many crops. However, for many root crops low temperature and high humidity $(0-1°C, < 97\%$ r.h.) appears to be effective.

Volatiles.   The behaviour of fruits and vegetables is not only affected by the levels of carbon dioxide but also by the very small amounts of various gases, such as ethylene, emitted by the produce, especially ripening fruit. Ethylene has a detrimental effect on most vegetables and therefore storing vegetables with fruit should be avoided [3]. Removal of these volatile products from the atmosphere by ventilation or by other means such as ethylene scrubbers [2] is required. Controlled oxidation with ozone and/or atomic oxygen produced by UV lamps can greatly delay ripening and ageing of a number of fruits and vegetables.

Humidity.   All produce loses water by evaporation, the rate depending on the nature of the skin, especially its waxiness, the presence of skin injuries, the shape and size of the product, and the relative humidity of the atmosphere. The rate of evaporation varies considerably between different fruits: bananas lose water rapidly, tomatoes relatively slowly. Leafy vegetables, because of their very large surface, have a very high rate of evaporation. For most fruits and vegetables, evaporation of water accounts for all the weight loss, wilting and shrivelling after harvest and during storage, transport and marketing.

Maintenance of 'freshness' in fruits and vegetables depends very much on preserving the original high water content. Therefore successful handling, storage and marketing must not only maintain a minimal respiration and prevent rotting but also keep the rate of evaporation of water as low as possible. However, high humidity to minimize weight loss and wilting encourages the growth of fungi and bacteria, therefore saturated or near saturated atmospheres cannot be used in storage. It follows that the relative humidity should be high enough to keep shrinkage to a minimum but not so high as to encourage the growth of moulds or yeasts. A relative humidity of around 90% plus for leafy vegetables and 85–90% for fruits and most other vegetables would be ideal. Onions need a lower humidity (about 70%) to discourage rotting and root growth.

Shrinkage can be usefully reduced by the use of waxed paper or plastics film liners. In long-term storage at a relative humidity of 85% and a temperature of 0–1°C, wrapped and packaged fruit may lose several percent of its weight and unwrapped fruit about twice as much. Besides loss of saleable weight, apples, pears and oranges become noticeably shrivelled when they have lost about 5% of their original weight.

Bacteriological conditions

Fruits and vegetables are subject to various diseases of physiological or microbiological origin. The latter are caused by a relatively small number of parasitic organisms, often attaching specific hosts in easily recognizable ways. As fungal and bacterial growth is temperature and moisture sensitive, it is closely related to storage conditions [4].

Handling

Fruits and vegetables are easily damaged in handling. Mechanical injuries (bruises, cuts, punctures and abrasions) are not only unsightly, but provide access for infection by rot-producing organisms, increase moisture loss and increase the rate of respiration and ageing. The intact skin is an effective natural barrier, and most rots start as infections entering through breaks in the skin. The importance of careful handling and packaging to protect against mechanical damage to the produce cannot be over-emphasized. Damaged produce should never be packaged with undamaged material as the shelf life of the whole package will be considerably reduced.

Transport

Modified atmosphere and controlled atmosphere conditions are combined with refrigeration during transport of fruit and vegetables. A number of different systems for producing and maintaining different atmospheres, including the use of gas-flushed plastic-shrouded palletized units, have been developed [2]. Usually these systems are used in sealed refrigerated freight containers which can be transported by road or rail.

Packaging

Packaging [5] may have to allow for rapid cooling of produce from high field temperatures to the desired storage temperatures, for the removal of metabolic heat during storage and transport, and, if it is a climacteric product, provide adequate ventilation capability whilst containing the product throughout the ripening process. Some commodities are very sensitive to ethylene gas (e.g. avocado) and need to avoid the build up of gas during transit to avoid premature ripening. Packaging must also protect from moisture loss and against bruising which can occur if the product is handled or packaged incorrectly.

Impact bruising, when the product is dropped onto a hard surface can be avoided by the use of cushion pads, by unitization and improved package filling procedures. Compression bruising occurs with incorrect stacking, overfilling, and inadequate package performance. Vibration and abrasion bruises generally result from movement inside the package and can be minimized by correct package sizing, and the use of internal packaging materials such as wraps, paddings and trays.

The package must also be able to perform well under all temperature and humidity conditions. Fluctuations in temperature and humidity may arise, for example, on removal from cold store which could cause water to condense on both the product and the package.

Bulk packages for fruit and vegetables include wooden boxes, crates, and corrugated fibreboard cases and these are often fitted with moulded pulp or expanded polystyrene foam trays for locating and separating individual items.

Prepacked fruit and vegetables

At retail level, fresh fruit and vegetables have traditionally been sold loose in nets or polyethylene bags. More recently stretch-wrapped goods over moulded plastic or pulp trays have been used and the demands for quality and freshness from retailers and consumers have led, in some countries, to the development of modified atmosphere packaging (MAP) systems. MAP is more advanced in Europe, particularly in the UK, than in the United States for example, because of the different system of retailing and distribution [6]. There is a slight trend towards supermarket sales of fruit and vegetables in the UK (Table 8.3) [7].

Modified atmosphere packaging

As already discussed, alteration of the atmosphere around fresh produce causes a change in respiration rate and this fact is used to retard ripening of certain produce. When the oxygen supply is normal, respiration is aerobic, but without oxygen anaerobic respiration takes place. Lowered concentrations of oxygen will give rise to a mixture of both anaerobic and aerobic respiration resulting in a balance of oxygen and carbon dioxide which is just right for

Table 8.3 Retail sales of fruits and vegetables in the UK by outlet (%) [7]

	1987	1988	1989
Multiples	34	35	39
Greengrocers	30	29	28
Market/farm outlets	23	25	23
Other grocers	13	11	10

Source: AGB.

the type of produce. This, combined with an appropriate temperature, will retard respiration. In a sealed but permeable package as respiration takes place there will be a constant change in the composition of the internal package atmosphere as oxygen is consumed. First the concentration of carbon dioxide will rise as the oxygen level falls but since the film will always be more permeable to carbon dioxide than oxygen the former will diffuse out through the walls of the package faster than the oxygen can diffuse in. The rate of permeation also depends on the partial pressure of the two gases. Bacterial growth will also be affected, not necessarily in a beneficial way, by the increase in carbon dioxide and decrease in oxygen. The suppression of respiration will also affect the production of ethylene, the principal ripening agent. Therefore MAP of fruit and vegetables needs a different approach than for meat and fish [8]. Table 8.4 shows some respiration rates of typical fresh produce [8].

Selection of an appropriate film with the correct permeability and designing a pack with an appropriate ratio of weight of product to surface area of film, allows produce to breathe and produce an *equilibrium modified atmosphere* (EMA) [9] (see Figure 8.2).

As oxygen is consumed carbon dioxide is produced in approximately equimolar amounts creating gas gradients across the film. These diffusion gradients drive carbon dioxide out of the pack and force oxygen in. At first the driving force is low but eventually an equilibrium is obtained (see Figure 8.3) and, combined with low storage temperatures, reduces the rate of respiration [9].

Film permeability is critical and choice will depend on the respiration rate of the produce. The greatest problems occur in matching respiration rates to film permeability, particularly with products with high respiration rates. Even the most permeable films will result in over-modification of the

Table 8.4 Respiration rates of some typical fresh produce (CO_2 production in mg/kg/h) [8, 15, 16]

Produce	Temperature (°C)				
	0	5	10	15	20
Broccoli	–	–	–	–	425
Calabrese	42	58	105	200	240
Sweetcorn	–	–	94	–	20
Strawberries	–	–	60	–	130
Asparagus	28	44	63	105	127
Brussels sprouts	17	30	50	75	90
Lettuce	9	11	17	26	37
Tomatoes	6	9	15	23	30
Onions	–	–	6	–	9
Potatoes	–	–	4	–	6

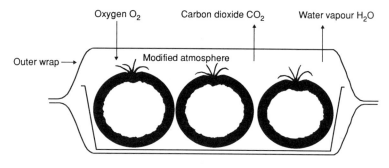

Figure 8.2 Biochemical and physical processes occurring in a retail produce pack. Reproduced by kind permission of Courtaulds Packaging Limited.

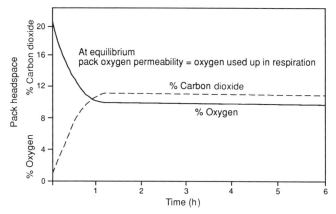

Figure 8.3 Establishment of equilibrium modified atmosphere in fresh produce packaging. Reproduced by kind permission of Courtaulds Packaging Limited.

atmosphere with subsequent deterioration of the produce. The use of controlled microperforations to modify the gas barrier properties of the films has been used effectively [10] for some PVC and PE stretch films, LDPE, EVA, OPP and LDPE/nylon laminates.

Possible suitable materials for MAP of fruit and vegetables include pillow packs of oriented polypropylene films, thermoforms of polypropylene and stretch-wrapped PET trays and punnets. Expanded polystyrene is usually used for shock absorption and anti-fogging coatings can be applied to the barrier films for lidding.

Studies on the retardation effects of MAP are continuing for both small retail packs and for larger units used in wholesale distribution.

There are also hazards in using MAP. If anaerobic respiration is allowed to progress there will be an accumulation of ethylene gas which allows potential growth of *Clostridium botulinum*. Also at low oxygen concentrations there can be an accumulation of acetaldehyde, ethanol or organic acids in fruit, producing discoloration and off-flavours [6].

The Institute of Food Science and Technology chilled food guidelines recommend [11] that fruit and vegetables should be stored between 0°C and 8°C but point out that some fruit and vegetables will suffer damage if kept at the lower end of this range.

Prepared vegetables and salads

Another group of products which have recently become popular are freshly chopped vegetables and salads which have a very restricted shelf life even under refrigeration. However, because of their very rapid deterioration as a result of discoloration of cut surfaces and microbiological spoilage, initial modification by gas flushing rather than relying on natural equilibrium is more effective [2]. Studies of commercially produced salads have shown that shredded vegetable tissue is often heavily contaminated with bacteria, and the fluid which comes out from the damaged cells may also initiate problems. The composition of the gas used in packaging these products is very important. Reduction of oxygen slows down enzyme browning but excessively high carbon dioxide can adversely affect flavour. Respiration is higher in cut vegetables and so the problems of increased loss of water vapour and wilting are greater. Antimicrobials are often used in wash water to decrease the number of microorganisms, and antioxidants are used to retard discoloration [12].

The ideal package must therefore balance atmospheric and storage conditions with package design, barrier properties and microbial load. Packaging materials used are expanded polystyrene or PVC trays covered with a stretch film, vertical form fill and seal pillow packs made from coated or uncoated oriented polypropylene and LDPE laminated bags [12]. New developments in this area include microwavable vegetables in MAP requiring little consumer preparation.

Fresh herbs and spices

Consumer demand for fresh produce has led to growers producing more varieties of fresh herbs which are becoming available in the modern supermarkets. However, fresh spices are not generally available in Europe and the United States. In the countries of origin spices are often sold with little or no packaging in markets and roadside stalls as well as prepacked.

Similar considerations exist for herbs and spices as for fruits and vegetables. Spoilage can occur from loss of moisture, bacteriological spoilage and metabolic processes. In addition there may be some loss of aroma volatiles and the oxidation of fats and oils.

Herbs are presented in film bags or flat folding punnets which protect the shape and structure of the herbs and allow easy and attractive display. MAP can also be used although it may be expensive for low volume produce. The choice of film used will depend on the permeability properties required. Bacterial contamination of herbs and spices can be very high and they therefore need 'cleaning'. Treatments used include fumigation, heat/steam treatments and irradiation which can be combined with MAP and other packaging (see chapter 10).

Nuts and seeds

Nuts and seeds keep well as long as the shell or outer coating is undamaged. They are susceptible, however, to loss of moisture, infestation and mould growth therefore storage and packaging must protect against these problems. Fumigation in storage is common at temperatures of about 4–8°C [13]. Unshelled nuts and seeds such as sunflower seeds, for the retail trade, are normally sold in plastic bags or netting.

Shelled nuts are more susceptible to moisture loss, infestation and mould growth than unshelled nuts. They are also susceptible to oxidation because of a high fat content. They are stored under similar conditions in bulk and packaging is usually in fibreboard boxes or lined sacks. For some expensive nuts and where infestation is a problem, e.g. cashews, highly protective aluminium or steel containers are used and the nuts are stored under nitrogen or carbon dioxide at 8°C [14]. In general the transport and storage of shelled nuts in bulk is very much more cost effective than that of unshelled nuts because of weight and volume considerations.

Retail packaging of nuts is more sophisticated and is designed to protect against physical damage, staling (loss of moisture), mould growth and rancidity [13]. Laminates, including foil laminates, are used to keep oxygen out and moisture in. The most expensive and the best packaging is the tinplate or aluminium container which may be nitrogen flushed before sealing. Similar packaging is used for flavoured and dry roasted nuts (see chapter 11).

References

1. J.S. Alvarez and S. Thorne, Effect of temperature on the deterioration of stored agricultural produce, in *Developments in Food Preservation*, S. Thorne (ed.), Elsevier Applied Science, London (1981), pp. 215–237.
2. J.D. Geeson, Controlled and modified atmospheres for fresh fruit and vegetables, in *Food Technology International Europe* (1987), p. 101.
3. P.W. Goodenough and R.K. Atkin (eds.), *Quality in Stored and Processed Vegetables and Fruit*, Academic Press, London (1981).

4. W.C. Frazier, *Food Microbiology*, Tata McGraw-Hill, New Delhi (1967) Chap. 15.
5. F.A. Paine, *Package Design and Performance*. Pira International (1990).
6. B. Day, A perspective of modified atmosphere packaging of fresh produce in Western Europe, *Food Sci. Technol. Today* **4**(4) (1990), 215.
7. P.R. Sheard, *A Guide to the British Food Manufacturing Industry*, Nova Press, London (1991).
8. R. Inns, Modified atmosphere packaging, in *Modern Processing Packaging and Distribution Systems for Food*, F.A. Paine (ed.), Blackie, Glasgow (1987).
9. M. Gill, Developments in films for MA packaging, paper given at *BHPA/IPR Joint Symposium on Developments in the Use of Plastics for Packaging and Handling in Agriculture and Horticulture* (1990).
10. J.D. Geeson, Microperforated films for fruit and vegetable packaging, paper given at *BHPA/IPR Joint Symposium on Developments in the Use of Plastics for Packaging and Handling in Agriculture and Horticulture* (1990).
11. Institute of Food Science and Technology, *Guidelines for Handling of Chilled Foods*, 2nd edition, IFST, London (1990).
12. T. Brocklehurst and E. Van Bentem, Packed freshness at your service, *Food Sci. Technol. Today* **4**(3) (1990), 156.
13. R.G. Booth, Snack foods, in *Food Industries Manual*, M.D. Ranken (ed.), Blackie, (1988).
14. R.G. Booth, Nuts, in *Snack Food*, R.G. Booth (ed.), Van Nostrand Reinhold/AVI, New York (1990) Chap. 12.
15. A. Ballantyne et al., MAP of broccoli florets, *Int. J. Food Sci. Technol.* **25**(1988), 353–360.
16. A. Ballantyne et al., MAP of shredded lettuce, *Int. J. Food Sci. Technol.* **23**(1988), 267–274.

9   Frozen foods

Freezing

Freezing ($-18°C$ and below) is used to slow down the growth and activity of microorganisms in food, to retard chemical reactions and prevent the action of enzymes. Related to this is the reduction in available water, as freezing not only immobilizes most of the moisture present, but also increases the concentration of dissolved substances in any 'free' water and hence reduces available water.

Whilst the numbers of viable organisms are slightly reduced by freezing, it does not completely sterilize the food. Selection for quality and preliminary preparation of foods for freezing are also important. Most vegetables are blanched (see chapter 10), and fruits may be packed in syrup, before freezing. Meats and seafood are handled in a way which minimizes enzymatic and microbial changes. Many foods are packaged before freezing, but some, if in small pieces (e.g. peas) or portions (e.g. fish fingers) may be frozen before packaging.

In freezing [1, 2] heat is removed from the product and as the temperature falls a temperature gradient is produced within the product starting at the surface layer, which is the coldest. The freezing process has three major stages:

(1) cooling the food product to the freezing temperature
(2) the formation of ice crystals and the removal of the latent heat of fusion at around $-1°$ to $-2°C$
(3) cooling down to the storage temperature.

The storage temperature depends on the type of product but is usually around $-18°C$. Ice formation is possible at higher temperatures if materials are present which seed the initial formation of ice. Similarly delays in ice formation can also occur with materials that interfere with the seeding process. Both can exist in a product or may be added, e.g. the presence of many solutes such as salts and sugars will affect the freezing point.

The second stage of freezing is the important rate determining stage. The rate of freezing of foods depends upon a number of factors, such as temperature and method of freezing, air circulation, package size and shape, kind of food, etc. The formation of ice crytals is at its greatest between 0 and $-5°C$ (zone of maximum crystallization) and ice formation occurs mainly in the liquid between the cells. Within the cells there is a depression of the freezing point by dissolved substances. Consequently, as the ice grows in between the cells, the concentration of salts in the unfrozen liquid increases to exceed that within the cell, and osmosis takes place with loss of fluid. The slower the freezing rate the greater the loss of fluid, which may cause cell distortion and a modification of enzyme behaviour. This may result in excessive drip loss in thawing and changes in texture and flavour.

In fast freezing heat is removed so quickly there is no time for the internal cell water to transfer outside the cell and ice forms within the cell. Even though tissues may have the same dimensions they will freeze differently because of the different ice-forming characteristics of the fluid in and between the cells. Also the rates at which water can cross the cell walls will affect the frozen structure. It is generally agreed, therefore, that speed in freezing is desirable. On the other hand, too fast a freezing rate can cause dimensional stresses leading to the disintegration of sensitive products.

Commercial freezing methods

The most frequently used methods of freezing [2] are blast freezing, plate freezing, and immersion and cryogenic freezing.

Blast freezers operate at air temperatures as low as − 40°C and large fans blow the air evenly over the product. Blast freezers can be used to freeze packaged or unpackaged materials, supported in trays or hanging free in either batch or continuous systems. Free-flowing products such as vegetables can also be frozen using blast freezers by passing the air upwards over a continuously moving product. The products are described as individually quick frozen (or IQF) and can be stored frozen in bulk and packaged when required.

Plate freezers normally operate at or below − 30°C and consist of a series of refrigerated plates which are pressed against both sides of the not-too-thick section of the food to be frozen. Plate freezers are normally used for freezing consumer packs of regular shape, e.g. fish blocks, which make good contact with the freezer plates through which the refrigerant circulates.

In *immersion* or *liquid contact freezing*, the food is sealed in a packaging material such as polyethylene and the package is dipped into a tank containing the cold liquid, a solution of either common salt or propylene glycol. Alternatively, liquid nitrogen can be used for rapid freezing of portion controlled foods such as burgers and frozen meals and for soft fruit tissues, shrimps and prawns where maintenance of texture and flavour are important. Other cryogenic agents, such as carbon dioxide, can be used but not all countries permit the process [2].

Storage and distribution

Quick freezing is successful mainly because of the small size of ice crystals formed during the process, However, if the storage temperature is not correct and fluctuations occur, then the ice crystals will tend to fuse together and become larger thus defeating the objective of quick freezing. Temperature control during storage is therefore as important as the freezing process itself. Immediately after quick freezing, the products should be put into cold storage where the temperature is maintained constant at about $-29°C$. Distribution is then carried out by bulk transport of the frozen food in insulated vehicles, either by road or rail. In some cases road vehicles possess their own refrigeration unit. If insulated vans are used without a refrigeration unit, the low temperature is usually maintained by the use of 'dry ice' (solid carbon dioxide). In retail cabinets the temperature should be maintained at or below $-18°C$ and the cabinet should not be overloaded with food packages. Domestic freezers and refrigerators with frozen food compartments are sold with star markings which correspond with the following temperatures and keeping times:

      * $-6°C$ maximum keeping time 1 week
     ** $-12°C$ maximum keeping time 1 month
    *** $-18°C$ maximum keeping time 3 months.

Frozen food packs must indicate the storage instructions to the consumer, especially if there are unusual circumstances pertaining to a particular product.

Protection needed by frozen foods

A good frozen food package must withstand low temperatures and sometimes high temperatures, e.g. for microwave and boil-in-the-bag products. It must be non-toxic and impart no odours or flavours to the food, must provide a barrier to the transmission of water vapour (and sometimes oxygen and/or fats), must be water resistant, and must be capable of being handled on semi- and fully automatic filling and closing equipment. It must also lend itself to graphic decoration and be tamper-resistant. The pack must not fall apart when it becomes damp on thawing, and in the retail cabinet packs must remain free from defects such as scuffing or collapsing.

To maintain frozen foods in perfect condition during storage and distribution, the packaging must provide protection against the following:

(a) *Dehydration* caused by moisture vapour escaping through the walls or seals of the package. This moisture loss dehydrates surface areas of the frozen food and causes desiccation (freezer burn). The dehydrated surface layer can be very thin, but may affect the appearance and ultimate saleability of the product.

(b) *Oxidation* promoted by enzymes which have not been eliminated by blanching can be caused if air penetrates the package.

(c) *Light* accelerates oxidation particularly in foods with a high fat content.

Heat can induce increased enzyme activity and general chemical and bacterial deterioration. Fatty foods must be prevented from transferring grease.

(d) *Flavour or odour loss* and the absorption of airborne odours are unlikely to occur whilst prepacked foods remain frozen. Special care is necessary with pre-cooked foods where evaporation of the volatile content could cause flavour loss.

(e) *Physical damage* can be caused by compression during storage and transport. Special care should be exercised in handling cases containing packs of frozen products. Further damage may occur to the bottom layers of packs if the outer containers are dropped onto a hard surface.

While microbial growth is stopped by freezing, not all microorganisms are destroyed. Spores and preformed toxins will survive and damaged bacterial cells may still be viable after a long resuscitation period. It is important therefore that raw materials are processed with good hygiene to be as free as possible of microorganisms and that no delays during processing or packaging operations occur.

Types of package

In general, frozen foods do not require a hermetically sealed container because their reliance for preservation is on the low temperature of storage, but moisture and oxygen barriers of varying degrees are required and in some instances it is also necessary to prevent leakage during thawing, e.g. in fruit products in syrup [2]. Apart from the problems of desiccation, which can cause weight loss in the pack, ice separation can also occur when packages are not completely filled. The ice separation is caused by sublimation from the product within the package if fluctuating temperatures occur in frozen storage. This may happen to packages not used in rotation but left at the bottom of display cabinets for weeks. Such dehydration of product adversely affects the quality as well as the weight.

There are many types of package in use.

Cartons made from paperboard, originally coated with wax but now more commonly with polyethylene, with locking bases and lids and coated paper overwraps are common. The coating gives the board water-resistant characteristics which are useful for wet products. The carton itself gives physical protection and protection against moisture and oxygen transfer is provided by a barrier overwrap or a sealed inner pouch or liner. A typical overwrap is coated polypropylene. Internal liners are frequently made of polyethylene.

A more modern carton is the single-component type produced from polyethylene-coated board sealed with a hot melt adhesive. Extrusion coatings of low density polyethylene are applied to both sides of the board to improve seal quality. This is an excellent container for frozen food which also lends itself to high-speed machines. Cartons can be externally printed and present an attractive appearance with convenience for use.

Direct film wraps and bags. Film materials vary from unsupported polyethylene and polypropylene films to laminates of polyethylene, polypropylene or polyester. Coating the films with polyvinylidene chloride copolymers may further increase the barrier properties.

Boil-in-bag packs. The products are packaged in bags in which they are intended to be cooked before opening and in addition they usually have an outer carton. Foods which produce a strong odour during cooking (e.g. kippers) can be cooked very conveniently in boil-in-bag packages. Examples of various materials which can be used for boil-in-bag products are high density polyethylene and polypropylene, which, although fairly permeable to gases, give a reasonable shelf life. Laminated materials are used to give a longer shelf life. More expensive materials are polyester/polyethylene and a laminate of polyethylene/polypropylene coated with vinylidene chloride copolymer.

Overwrapped trays. Moulded pulp trays are still in use for the packaging of small cuts of meat, poultry, etc., but expanded polystyrene trays which look whiter and cleaner are a more attractive alternative.

Aluminium foil, pressed into a tray form, is also used, particularly for pre-cooked fruit and meat pies. Trays may be internally coated for protection. All of these products may be overwrapped, the ends of the wrap being tucked under the tray and heat-sealed.

Thus, there is a wide range of packaging materials available for frozen foods. The choice of container has to be made carefully, bearing in mind the cost and storage performance, and the nature of the frozen product. The scope is large and embraces a wide and increasing range of foods (poultry, fruits and vegetables, fish and fish products, meat products, baked products, pizzas, confectionery, desserts and ice cream) [3].

Frozen meat and poultry

Freezing is a means of extending the storage life of meat beyond that obtained by chilling. It is effective because all microorganisms cease to multiply at low temperatures and some are actually destroyed. The spoilage of meat is reduced when the temperature falls below freezing until, at around $-8°C$, bacteria and moulds stop developing (although some still grow at $-10°C$). The physical and chemical changes in meat take place more slowly as the temperature falls, but are not completely arrested even when stored at $-30°C$. Frozen meat, therefore, will not keep indefinitely. The fat will eventually go rancid and, if exposed to light, the red pigment (myoglobin) in the lean tissue

Table 9.1 Storage times for some frozen meat products [2]

Product	Storage life (weeks)		
	$-18°C$	$-25°C$	$-30°C$
Beef			
Carcasses	12	18	24
Joints, steaks, cuts	12	18	24
Mince	10	12+	12+
Veal			
Carcasses	9	12	24
Joints, cuts	9	10+	12
Lamb			
Carcasses	9	12	24
Joints, cuts	10	12	24
Pork			
Carcasses, joints, cuts	6	12	15
Bacon	2–4	6	12
Products, sausages	6	10	–

will fade. Irreversible dehydration will also occur at the surface of the meat, unless it is wrapped in airtight, vapour-resistant material, e.g. direct film wraps in conjunction with heat-shrink films are commonly used. Special equipment is available for this operation and the film used must combine good water vapour resistance with an oxygen barrier. To maintain the quality in frozen meat stored over long periods, a low temperature is essential. This must be a minimum of $-18°C$, which is the normal running temperature of the domestic freezer, but $-25°C$ or below, the temperature of a commercial cold store, is better. Table 9.1 gives storage times of some frozen meat products.

Freezer burn

This is due to the dehydration of the surface of unwrapped or badly packaged frozen meat. Freezer burn becomes progressively worse when badly wrapped frozen meat is stored for a long time and greyish-white marks appear on the lean surfaces of the meat. Although dehydration occurs, the soluble proteins and other nutrients remain virtually intact. Freezer-burned meat is quite safe to eat but it can be dry, brittle, discoloured and unpalatable after thawing.

The freezing process

During freezing the moisture naturally present in meat changes to ice at about $-1°C$ and continues down to $-40°C$, although commercially meat would be considered frozen at about $-7°C$ when three-quarters of the water has turned to ice. Lean meat contains about 70% water, but fat contains only 10%. Meat is a poor conductor of heat and rapid freezing produces a frozen crust on the outside, with unfrozen meat in the centre. Sufficient time must therefore be allowed for the centre to freeze. Meat can be subjected to a wide range of freezing rates with little effect on quality, and consequently the speed at which meat is frozen is chosen on economic, not technological, grounds.

Commercial freezing methods

Carcass meat and joints are usually frozen in *blast freezers* in bags or boxes lined with polyethylene. For speed meat is often hung or placed on open trays. *Plate freezing* or *immersion freezing* are used for small pieces of meat such as chops and steaks. Liquid nitrogen is relatively expensive and is used for high value products.

Meat is stored at $-18°C$ and if properly packaged and handled can have a storage life of 1–2 years. Typical materials are polyethylene pouches, PVC and PVdC films. Very thick PVC films are used to protect barrier bags from puncture. Some products, such as beefburgers, are individually wrapped in film bags and then packaged in cartons [4, 5].

Frozen poultry

The frozen poultry market has grown considerably throughout the world during the past two decades and owes much to packaging, with the development of the skintight PVdC copolymer film package. The prepared poultry are inserted into bags and transferred to a rotary vacuumizing machine which packages at speeds of up to 32 packs/min with clip closures and bag neck trimming. When shrunk, the bags form a second skin around the exact contours of the birds, which are then either frozen in brine or in blast freezers, according to the preference of the particular processor.

The tough vacuumized and shrunk bags protect the birds in the brine bath and prevent freezer burn during prolonged storage. Bags are available in a variety of films and colours. Water absorption is neglible and the oxygen barrier qualities sufficient to prevent fats becoming rancid. Bags are also used for packaging frozen ducks which are usually blast frozen and vacuumizing is again carried out by machine.

Materials used include vacuum grades of PVdC copolymer film, a range of laminates, deep freezing grades of polyethylene and other deep freeze films. Considerable developments have taken place in recent years with the introduction of, for example, frozen uncooked and pre-cooked poultry portions and also prepared items such as chicken Kiev, chicken rissoles, etc.

Frozen fish

Freezing can extend the storage life of fish and fish products as it retards many of the normal reactions which occur in fresh fish [2]. Bacterial counts tend to fall; some potential pathogens are more severely affected than others. The rate of oxidation is affected by temperature, but the reaction may still occur in the frozen state.

Effect of freezing on fish

Most fish begin to freeze at about $-1°C$ and multiplication of putrefactive bacteria is stopped at $-9°C$. Although bacterial spoilage is suspended not all bacteria are destroyed: some survive and remain dormant, resuming their growth when the fish is thawed. The survival of these organisms is dependent on time and temperature of storage. In the past the freezing of fish was not very successful but it has now improved. Excessive loss of juice in the form of 'drip' during thawing and cooking resulted in a tough and dry flesh which was thought to have been due to the slow freezing process (over several days) used. The water was allowed to freeze into crystals, mostly between the cells, and slow osmosis caused the cells to collapse. Therefore although rapid freezing is essential, the freshness of the fish and temperature of storage are equally important.

Protection is also needed against evaporation in cold storage caused by the transfer of moisture. This is now usually taken care of by 'glazing' (i.e. dipping the frozen fish in water to ice coat the surface) or else by sealing the fish in a water and water vapour resistant wrapper (waxed paper or plastic film). Thus the packed weight of the product is maintained, visible surface dehydration (freezer burn) is avoided and so-called 'cold-storage' flavours (e.g. rancidity) retarded.

Table 9.2 illustrates the potential storage life of fish in cold store. The times given are the period for which the product is practically as good as the fresh product [2].

Table 9.2 Storage life of fish in cold store [2]

Type of fish	Storage life (months) at storage temperatures of:		
	$-9°C$	$-21°C$	$-29°C$
Gutted white fish	1	4	8
Ungutted herring	1	3	6
Smoke cured white fish	1	3.5	7
Kippers	0.7	2	4.5

Methods of freezing

Fish is ideally suited to the quick freezing process and the methods used are *blast freezing*, where air is chilled by circulation over cooling pipes and blown over the fish by means of fans, and *plate* or *contact freezing*, where the packaged product is placed in layers between refrigerated metal plates pressed against both large surfaces to ensure good contact.

A reduction in drip after thawing frozen fillets can be achieved if they are dipped for a few seconds in 30–50% saturated sodium chloride solution before freezing (immersion brine freezing). However, fish treated in this way can develop peculiar cold-storage flavours. In addition salt is a pro-oxidant and promotes rancidity in fatty fish. Therefore, air blast freezing and plate freezing are the preferred methods.

The thawing of fish on a large scale can also be a problem. Fish is usually left to thaw in air on running water, but quicker and more easily controlled methods are *dielectic thawing* (blocks of frozen fish pass between two metal plates charged with an alternating voltage of many thousands of volts at a frequency of about 40 million cycles per second, and energy is produced in the fish in the form of heat), or *heating by electric conduction* (current is passed directly through a block of frozen fish).

Storage

Fish requires a lower cold storage holding temperature than most comparable commodities, e.g. pork and beef. This may be caused by the fact that fish fats are more unsaturated than mammalian fats and are more easily oxidized in cold store, producing rancid flavours more readily, or possibly the delicate flavour of fish (as compared with meat) gives rise to a slight unpalatability which becomes objectionable more quickly. The temperature used for cold storage has decreased over the years and varies with expected time of storage, but $-30°C$ is fairly standard for long-term storage.

Packaging frozen fish

For packaging raw, filleted, fish portions, polyethylene films are often used either in the form of pre-made bags or for wrapping on form-fill-seal machines. Frozen fish can create additional problems as the high salt content of the fish can quickly corrode metal contact parts in the packaging process, and normally freezing can only be achieved at temperatures of $-4°C$ instead of the normal $0°C$. In consequence, stainless steel equipment is essential, especially for automatic weighing machines; ice build-up can affect weighing accuracy.

A PVdC copolymer system is also used in vacuum-packaging some fish such as whole frozen salmon. This system provides a better alternative to glazing. It eliminates moisture loss on initial freezing (and also during the first 6 months of storage), drip loss on thawing, trim loss at 4–5 months of storage, weight loss on glaze, and reduces labour and time needed for traditional glazing. The lightly vacuumized package enables the salmon to retain its freshly caught characteristics throughout the entire distribution system.

Frozen fish products

Fillets of fish or cut portions are often covered with breadcrumbs or batters. Fish fingers, for instance, are fish fillets which have been frozen into flat blocks, sawn into thin strips in the frozen state, battered, coated with breadcrumbs and finally fried for just over a minute in deep fat. This treatment colours the outside of the finger but does not thaw the centre. When the surplus fat has been drained off, the fingers are packaged and re-frozen.

Packaging materials used include $12\,\mu m$ polyester film for items such as fish cakes, in pillow pack style on horizontal form-fill-seal machines. The film is reverse printed on the treated side and laminated to $50\,\mu m$ polyethylene, which gives an excellent printed effect. This laminate possesses barrier properties and is puncture-resistant over a wide range of temperatures, giving protection in transit, stacking and in supermarket cabinets.

Cartoning systems are still used in the packaging of fish fingers and other fish products. Additional storage protection can be provided for cartoned fish fingers by wrapping 10 or 12 cartons in shrink film and then stretch-wrapping a number of these units together. These are subsequently made up into palletized loads using pallet stretch-wrap film.

Frozen fruits and vegetables

Many fruits suffer substantial damage on freezing [2]. Osmotic changes occur as a result of ice formation which destroys the cell membrane. Generally fruits do not require blanching before freezing and can be IQF, packaged in sugar, syrup or puréed before freezing. Frozen juices and concentrates are an important trading commodity and are packed in bulk.

Vegetables, however, need to be blanched before freezing to ensure enzyme inactivation, which would otherwise result in objectionable flavours and loss of nutritional value and colour. Most vegetables benefit from quick freezing which gives a crisper final product [2].

The majority of frozen fruits and vegetables are packaged in plastics films, such as deep freeze grades of polyethylene, made into pillow packs on vertical form-fill-seal machines. Some soft fruits, such as raspberries, are packaged in lidded plastic containers.

The preparation and weighing of frozen fruits and vegetables before packing can be a complex operation to accommodate, e.g. facilities for grading and inspection prior to packing are required. This is often done after freezing, the vibratory action of the conveyor units assisting the separation of the products. Inspection conveyors can be installed as a link between the cold store or freezer and weighing stations.

A controlled flow of product is necessary if optimum performance is to be achieved from the weighers. To cope with the demand for larger packs for home freezers, machines weighing up to 5 kg are available linked to suitable form-fill-seal machines. Some free-flowing products such as frozen peas can also be packed volumetrically. The line sequence would normally be: flow-freeze to volumetric cups, form-fill-seal to check-weighers, to outer packing, to frozen store. Some soft fruits such as raspberries, blackberries and loganberries are often bagged in polyethylene films before freezing. Many retail frozen vegetables are packaged in cartons, particularly in Europe. This is particularly important for packaging of frozen vegetables in liquid form, such as spinach purée, where a leakproof base is essential.

Other frozen products

Baked products account for a large part of the frozen food market. The frozen pizza market has been particularly successful. Most bakery products stale very rapidly at chill temperatures ($-2°C$) and consequently this temperature range must be avoided for storage. However, the freezing of bakery products including doughs and batters can be very successful providing the staling temperature is passed quickly. Temperature ranges for freezing baked goods vary from $-20°C$ to $-40°C$. Freezing usually takes place after packaging.

Cartons are widely used, but owing to the delicate nature of some items such as cream cakes, sponges, gateaux and cheesecakes, some form of internal packaging is often necessary in addition to the carton. This normally takes the form of wrapping in heat-sealable films on form-fill-seal machines, and sometimes inserting a carton board or plastic collar tray. Most individual cakes are then end-loaded on cartoning machines and multiple products are fed into top-loading cartons. Grease-proof and water-resistant paperboard cartons are used.

Frozen egg and egg products are supplied to the food industry in large containers of 10–15 kg capacity, e.g. cans, cartons and plastic bags. Drums with plastic bags are used for bulk shipments. However, to avoid gelation, sugar and salt may be added. Blast freezing is used to freeze egg products at -30 to $-40°C$, the time required depending upon the size and shape of the package. Free-flowing pallets of frozen eggs are also available. The shelf life for frozen eggs is approximately 6 months at $-12°C$, 1 year at $-18°C$ and up to 2.5 years at $-30°C$ [2].

The market for *frozen desserts* has escalated over past years, e.g. frozen sorbets, mousses, puddings. Cartons continue to be used as well as thermoformed high-impact polystyrene (HIPS) containers often with printed board inserts.

Ice-cream

Originally a blend of milk, cream and sugar, which is frozen to give a desirable texture, ice-cream [6] is probably one of the most popular food products in the world. The composition of modern ice-cream is shown in Table 9.3. A small proportion of the milk solids-not-fat is sometimes replaced with whey solids, and glucose syrups are often used in conjunction with the sucrose. The traditional fat sources are cream and butter, but in some countries vegetable fats or blends of vegetable fats and butter as well as butter-oil and anhydrous milk-fat are used.

In ice-cream manufacture, a pre-mix is first prepared by mixing the ingredients; this is then pasteurized, homogenized and cooled to 5°C. The ice-cream structure then changes dramatically from a simple oil-in-water emulsion to a complex food microstructure. After ageing for several hours, further processing in an ice-cream votator (revolving beater) and a lowering of the temperature to approximately -5 to $-6°C$, causes about 90% of the water in the formulation to freeze. During freezing air is incorporated so that the final product contains about 25–50% air resulting in soft ice-cream. Further blast freezing gives hard ice-cream which is stored at -25 to $-30°C$. Ice-cream is filled (by horizontal or vertical form-fill machines) into cups, cartons or plastic containers.

Table 9.4 shows the approximate proportions of the different phases of ice-cream, the main differences between hard and soft ice-creams being the ice and water phases.

The home freezer boom in sophisticated societies means that ice-cream in 2- and 4-litre packs has become a regular frozen food purchase. Plastics containers are the main packs, and one of their great attributes is their value as a household storage utensil after use.

Table 9.3 Typical composition of ice-cream [6]

	%
Milk solids-not-fat	9–12
Fat	9–12
Sucrose	14–16
Emulsifiers	0.3–0.5
Stabilizer	0.2–0.4
Colours/flavours	To taste
Water	To 100%

Table 9.4 Approximate proportions (%) of different phases in ice-cream [6]

Phase	Soft $(-5°C)$	Hard $(-20°C)$
Air	50	50
Ice	15	27
Fat	5	5
Continuous water phase	30	18

There is also a growing demand for paperboard cartons as alternative 2-litre packs. The printed carton blanks are polyethylene-coated inside, freeze-varnished on the outside and have hood cover closures to give the reclosable feature essential in this type of pack. Cold seal-coated metallized MXXTA/S cellulose film makes an attractive and effective pillow shape pack and foil and polypropylene are commonly used for choc ices and lollies although many lollies remain in coated paper wrappings [7].

Cook-freeze products

Products that are cooked and frozen at a central production point, distributed frozen and reheated at the point of consumption are known as 'cook-freeze' products. The style of package and type of material must be matched to the product and its proposed outlet (retail store, restaurant or catering establishment).

Many frozen foods are packaged so that they can be cooked straight from frozen (or after thawing) in conventional or microwave ovens. Increasingly the rise in microwave ownership has led to microwavable packs only. Packaging materials must therefore withstand low and high temperatures as well as protect the combinations of foods contained in the pack. Aluminium trays as well as ovenable cartons and plastics are commonly used (see chapter 10) [8–11].

Cooking inside a carton allows hot food to be served in clubs, railway stations, factories, schools and other bulk catering sites. The secret of closing and opening ovenable frozen food cartons lies in an adhesive which works at deep-freeze temperatures yet allows the lid flap to be peeled open after baking or reheating the ovenable carton in conventional or microwave ovens.

Some catering establishments and institutions use total cook-freeze catering systems where the food is prepared, cooked, packed and frozen on the same premises where it will be served.

Guidelines for packers, distributors, retailers and consumers

The Institute of Food Science and Technology (IFST) Guidelines on Good Manufacturing Practice includes a section on frozen foods [14] and their packaging requirements. Apart from meeting the general requirements, the IFST Guide recommends that frozen food packs should:

(1) Include a moisture vapour barrier to prevent dehydration and weight loss
(2) Carry storage instructions necessary to validate the stated 'use by date'
(3) Include production date, process details, etc. on bulk packs such as sacks or pallets for easy identification at a later date when repacking occurs
(4) Include 'use before' dates on consumer packs.

Guidance on personnel, raw materials, formulation, processing, freezing and storage are also given.

In addition there are a number of other codes of practice, including the international FAO/WHO Codes of Practice on frozen foods, which cover specific frozen foods such as frozen spinach or strawberries as well as more general guidelines on quick frozen [15] and pre-cooked frozen foods [16]; the UK Quick Frozen Foodstuffs Regulations, which control the labelling, manufacturing, processing and storage of quick frozen foods other than ice-cream; the UK Department of Health guidelines on cook-freeze and cook-chill systems used in institutional catering [13]; and the UK Association of Frozen Food Producers Guidelines for Frozen Meat and Meat Products [17] and Code of Recommended Practice for the Handling of Frozen Foods [18].

Future trends

Frozen foods rely on the low temperature at which they are kept after being rapidly frozen to preserve them in prime condition and their packaging is required to prevent dehydration (freezer burn is a typical example), oxidation of fats, etc. which is often promoted by light, flavour and aroma loss (or gain) and physical damage during handling and transport. To this end a variety of primary packagings are currently employed including plain, coated or metallized plastic films and laminates as wraps and bags, over-wrapped and cartoned trays made from coated boards or plastic, and lidded plastic trays and thermoforms often contained in paperboard sleeves or cartons.

All these developments involve changes in frozen food stores, methods of handling packaged products and their distribution in controlled temperature refrigerated transport to meet the critical requirements of retail outlets. This means future packaging must meet the newer demands of production, storage, distribution and the retailer.

To extend keeping times and quality, frozen food stores, which operated 10 years ago at -15 to $-20°C$, are now capable of better control at -25 to $-30°C$. Distribution, originally carried out in vehicles cooled by the use of dry ice, are now being maintained at lower, better controlled and monitored temperatures in specialized electrically refrigerated vehicles in which distribution temperatures will not exceed $-21°C$ for frozen foods or $-23°C$ for ice-cream. Such a temperature reduction requires the use of packaging capable of withstanding these significantly lower temperatures without embrittlement. This, plus modern merchandising methods involving the use of article numbering (ANA coding) systems, means that the use of shrink- and stretch-wrapping methods for shipping are likely to fade out and corrugated cases (containing modular unit packages) with a 'footprint' related to the international 1200×1000 mm pallet dimensions with a height of 1650 mm will become the standard.

Storage and distribution costs are clearly critical for frozen foods because of the cost of the frozen systems (cold stores and refrigerated transport). Consequently most frozen food manufacturers use the smallest size packaging possible to minimize costs.

In respect of merchandising it is interesting to compare frozen and chilled foods particularly in the ready meal field. Transparent packaging gives a merchandising advantage to chilled foods in display while frozen foods, because of the below zero temperatures, will always spoil transparency by deposits of frost and recognition must rely on opaque packaging with good graphics. Graphic design these days must also share the available space on the packaging with the bar code, any instructions for opening the pack and preparing the food for consumption, a 'best before' or 'use by' date, an ingredients list and nutritional information as well as other legal requirements.

The demand from consumers for rapid meal preparation both in fast food outlets and the home has led to the development of microwave cooking and in turn to packaging which can be inserted without difficulty directly into the microwave oven, cooked or simply reheated in minutes straight from the freezer without defrosting.

The development of ready meals for microwave is not so easy as might be supposed because different meal components require different times for heating unless their processing has taken account of this. The use of susceptors to provide 'browning' and 'crisping' of pizza cases, for example, has received much attention and is a field that has still to develop fully.

The combination of microwaving and packaging can also make a considerable contribution to hygiene in the provision of hot snacks for serving in railway buffets, etc. where, once packaged, the product is not handled by anyone before the consumer receives it in a heated condition over the buffet counter.

With products where the food is in intimate contact with its packaging, such as boil-in-the-bag, compliance with recent legislation covering plastics in contact with food must be ensured. These requirements oblige the food manufacturer to carry out all the necessary testing to ensure that migration from the plastics into the food is well within the limits of the proposed legislation. The technical and administrative problems to comply with the regulations are considerable in terms of expertise, time and cost.

Finally, now and in the next decade much emphasis will be placed on environmental matters. Questions must be asked (and answered) during the design and development of packaging for frozen foods as to the best packaging material to meet not only the protective needs of the food and its packaging line and selling requirements, but also the impact on the environment, in terms of material resources, energy conservation and most importantly, recyclability, which will result in minimizing the growing landfill problem.

We must, however, remember that no activity, whether initiated by nature or by man, is neutral to the environment: there are always pluses and minuses. Paper and board are derived from renewable resources and can be recycled but have minuses in respect of energy requirements and pollution of the air and water during initial manufacture and in reclamation. But there are exceptions to the rule. For example, virgin board from Scandinavia made using hydroelectric power as the energy source may well be more environmentally friendly overall than recycled board produced in other parts of Europe using fossil fuels as the energy source.

Another important environmental point is the ability of the packaging material to be recycled in one way or another. Plastics probably provide the use of less material per unit of food than any other material but they are derived at present from non-renewable fossil fuels and recycling, except possibly for energy recovery, is currently not much practised. The monolayer barrier materials used in the packaging of frozen foods are more easily recycled than the multilayer barriers necessary to protect foods stored at ambient or even chilled conditions but this will not always be so in the future because plastics can be made from renewable vegetable sources. An agreed method of balancing such positive and negative aspects is required which measures the degree of 'greenness or environmental responsibility' and produces an 'eco-balance'.

References

1. D.S. Reid, Basic aspects of freezing, in *Food Technology International Europe* (1990), p. 89.
2. S.D. Holdsworth, Freezing and refrigeration, in *Food Industries Manual*, M.D. Ranken (ed.), Blackie, Glasgow (1988).

3. Euromonitor, *The European Food Report*, Vol 1, Euromonitor Publications Ltd, London (1989).
4. A.L. Brody, Food packaging, in *The Wiley Encyclopedia of Packaging Technology*, M. Bakker (ed.), Wiley, New York (1986), p. 361.
5. D.G. Jones, Films, flexible PVC, in *The Wiley Encyclopedia of Packaging Technology*, M. Bakker (ed.), Wiley, New York (1986), p. 309.
6. S. Jones and M. O'Sullivan, *The Control and Manipulation of Ice Cream Quality.*
7. T. Cumming, Cool customers, *Packaging Weekly* **5**(11) Suppl. (1989), p. 18.
8. J.D.T. Faithful, Ovenable thermoplastic packaging, *Food Technology International Europe* (1988), p. 357.
9. L.D. Brugh, Ovenable paperboard packaging, *Food Technology International Europe* (1989), p. 365.
10. P. Aggatt, The heat is on, *Packaging Weekly* **5**(7) (1989) 18, 19.
11. Anon, Fresh from the freezer, *Packaging Report* **4** (1989), 48, 51, 52–56.
12. *UK Quick Frozen Foodstuffs Regulations*, HMSO, London (1990).
13. Department of Health, *Guidelines on Cook-Chill and Cook-Freeze Catering Systems*, HMSO, London (1989).
14. Institute of Food Science and Technology, *Food and Drink Manufacture—Good Manufacturing Practice: A Guide to its Responsible Management*, 3rd edition, IFST, London (1991).
15. FAO, *Recommended International Code of Practice for the Processing and Handling of Quick Frozen Foods*, CAC/RCP 8–1976.
16. FAO/WHO, *Recommended International Code of Practice. General Principles of Food Hygiene* CC/RCP 1–1969 FAO/WHO.
17. UK Association for Frozen Food Producers, *Guidelines for Manufacturing Practice for Frozen Meat and Meat Products*, London (1987).
18. UK Association for Frozen Food Producers, *Code of Recommended Practice for the Handling of Frozen Foods*, London (1977).

10 Heat-processed foods (including irradiated foods, etc.)

Heat processing

The destruction of microorganisms by heat is due to the inactivation of the enzymes required for metabolism. The heat treatment selected will depend upon the kind of microorganism, other preservative methods to be employed and the effect on the food [1].

Commercial sterilization

Temperatures of 110–130°C are used to destroy or render inactive all microorganisms that could cause toxicity or spoilage. In fact, total sterility is unnecessary, and absolute destruction of all microorganisms is seldom obtained.

Factors affecting resistance of microorganisms to high temperatures

Cells and spores of microorganisms differ greatly in their resistance to high temperatures, as a result of several factors [2]:

(1) Temperature–time relationship: generally the killing time for spores or cells decreases as the temperature increases.
(2) Initial concentration of spores: the more spores or cells, the greater the heat resistance.
(3) Previous history: the conditions under which cells or spores have grown will affect their resistance to heat. In general, the better the culture medium, the more resistant the spores.
(4) Incubation temperature: resistance increases as temperature of incubation increases.
(5) Phase of growth: heat resistance varies with phase of growth. Resistance is greatest during the lag phase and the least resistance occurs during the logarithmic phase. Very young spores are less resistant than mature ones.
(6) Desiccation: dried spores are harder to kill than moist ones.
(7) Constituents of the substrate: constituents such as water, sugar, salt and colloidal materials affect resistance.

(8) Hydrogen ion concentration (pH): this is the most important factor influencing heat processing. Cells or spores are generally most heat-resistant in a substrate that is at or near neutrality. Any increase in acidity or alkalinity hastens killing by heat, but an increase in acidity is much more effective than a corresponding increase in alkalinity. The processing time required in the heat processing of foods will therefore increase as the pH moves from acid through neutral to alkaline conditions. Most of the heat-resistant pathogenic organisms (e.g. *Clostridium botulinum*) occur in less acid foods at a pH above 4.5, and therefore the minimum heat processing applied to foods with a pH greater than 4.5 must eliminate all risk of botulism. Satisfactory processing of acid foods (pH less than 4.5) is possible in boiling water or steam at atmospheric pressure for relatively short periods. Sterilization is followed by immediate cooling.

Factors affecting the rate of heat penetration

In order to establish a process for any particular product, the rate of heat penetration at the slowest heating point of the container must be measured. This is usually the geometric centre of the package for products which heat up by conduction. Where heating is by convection (i.e. for non-solid products), it is usually a little below the geometric centre.

In any event, the rate of heat penetration will depend upon several factors [3] including the *container material* (heat penetrates glass more slowly than the tinned iron walls of a can) and the *size* and *shape* of the container. Larger cans will take longer than smaller cans to reach a given temperature at their centre. Although the initial temperature of the food in the can makes practically no difference to the time needed for the centre of the can to reach the required temperature (as food at a lower initial temperature heats faster than the same food at a higher initial temperature), food with a higher starting temperature is in the lethal range for microorganisms for a longer time and the average temperature during heating therefore is higher. This is important in the processing of those foods which heat slowly, e.g. canned meat products.

The *size* and *shape* of container contents also affect the heating process. Foods that do not break up, such as peas, heat in almost the same way as small food pieces in brine. Heating is delayed when the food pieces are large, because heat must penetrate to the centre of each piece of food before reaching the retort temperature. Foods which disintegrate on cooking heat slowly, because heat penetration is mostly by conduction rather than convection. In foods which are in layers, e.g. asparagus and spinach, the layer formation interferes with the convection currents which travel mostly up and down. The composition of sauces also affects the consistency of the container contents and starch concentrations of up to 6% can also interfere with heat processing, but salt (sodium chloride) has little effect. Increasing the sugar

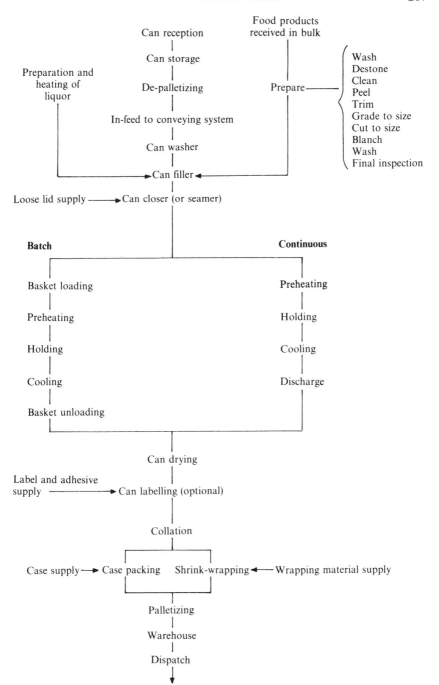

Figure 10.1 Food canning operations.

concentration decreases the rate of heat penetration, but a decrease in viscosity also occurs with increase in temperature. The higher the *retort temperature*, the quicker the food reaches the lethal temperature. *Rotation* and *agitation* will accelerate heat penetration when the food is fluid, but may also cause undesirable physical changes. Agitation can be very helpful with foods that layer.

Other factors which affect the rate of heat penetration are the amount of fill, the residual headspace, and the position and stacking of the containers in the retort.

Once the heat penetration rate is obtained, it is used to determine the lethal effect with respect to the thermal death time of any particular microorganism [3, 4]. The heat resistance of a microorganism is expressed in terms of a thermal death time (D value) which is defined as the time taken, at a specific temperature, to kill a stated number of organisms or spores under specified conditions [5–8]. The F_0 value is the number of minutes it takes at 121°C for the heated food to reach the desired sterility. Heat penetration data can be mathematically converted from one can to another and from one retort to another, but it is always advisable to check the result experimentally.

The processing needed is usually determined by the requirements of microbiological stability and is the minimum necessary to achieve this in order to minimize the cooking effect. There is one basic rule: all canned foods with a pH greater than 4.5 must be given a safe process or 'botulinum cook'. However, this will not ensure freedom from spoilage organisms more resistant than *Cl. botulinum*. In order to decide what is necessary, further information regarding the bacteriological quality of the raw materials and their previous history is required. The most traditional methods of heat sterilization are canning and bottling. An outline of the canning process is given in Figure 10.1.

Pasteurization

Pasteurization is the term applied to a milder heat treatment (60–75°C) given to certain products to kill most, but not all, of the microorganisms present. Normally non-sporing pathogens, such as *Salmonella*, are killed and the shelf life of the product is thereby prolonged. Temperatures below 100°C are employed and the heating is by means of steam, hot water, dry heat or electric current. The products are rapidly cooled immediately after heat treatment. Pasteurization is used when:

(1) A more severe heat treatment would harm the quality of the product
(2) The pathogens must be 'inactivated', e.g. in milk
(3) The main spoilage organisms are not very heat-resistant, e.g. yeasts in fruit juices
(4) Any surviving spoilage organisms can be controlled by other preservative methods, e.g. chilling

(5) Competing organisms are to be killed to allow a desired fermentation from an added starter organism, e.g. in cheese-making.

Several other preservative methods are used to supplement pasteurization, e.g. refrigeration, packaging the product in a sealed container (to keep out microorganisms), packaging in a sealed container which has been evacuated to maintain anaerobic conditions, addition of high concentrations of sugar or the presence or addition of chemical preservatives. The times and temperatures used in pasteurization depend upon the method of heating and the product.

Blanching

Raw food for processing should be freshly harvested, properly prepared and inspected. Many vegetables and some fruit are blanched or scalded briefly by hot water or steam (at 100°C) before canning, freezing or dehydration. The purpose of blanching is:

(1) To clean the raw materials
(2) To destroy enzymes which could otherwise produce undesirable flavours and/or discoloration, or cause the loss of vitamin C
(3) To bring about shrinkage in order to pack sufficiently high filling weights
(4) To prevent an increase in pressure developing in the cans during processing
(5) To prevent loss of colour and flavour.

Immediate filling of cans after blanching reduces blanching times, since the closing temperature of the can is higher. If immediate filling is not carried out, water-cooling should follow blanching to reduce heat degradation of the product and to prevent bacterial multiplication, which can take place rapidly in a warm product. Blanching can result in a loss of nutritive value and therefore blanching times should be kept as short as possible (2–10 min). Microwaves have also been used for blanching.

General spoilage problems with canned foods

Spoilage in canned foods [1, 2] is usually as a result of *corrosion* or *bacterial* spoilage. Corrosion is most common in canned fruit and some non-acid foods and headspace control is very important in the avoidance of corrosion. Processed foods should be closed with a partial vacuum to reduce the strain on the can ends during the sterilization process. This also reduces the amount of oxygen which can cause discoloration, vitamin losses and oxidation in some foods, but it also reduces corrosion and rusting. For some acid foods, such as canned fruit, hydrogen is produced in the can by acid attacking the steel base of the tin plate in the can. Once sufficient hydrogen has been produced the can becomes 'blown' so the initial vacuum level often controls the shelf life of the product. In agitating cookers, the movement of the headspace gas assists heat transference in the product. Bacterial spoilage of canned foods can also, but not always, result in a 'blown' can. Bacteriological examination is necessary to determine the organism responsible.

Organisms which survive the heat process will be those which form spores. In contrast non-spore formers will be present if contamination has arisen from leakage (leaker spoilage) or from inadequate processing. Moulds are usually destroyed at high temperatures but some can survive the milder processes used for products such as bottled fruit and produce enzymes which cause the product to disintegrate.

Good hygiene during processing is just as important as with other forms of preservation and cans should be stored in dry non-corrosive atmospheres to keep them in good condition.

Canned meat products.   The process time used in canning meat products varies considerably depending on the meat product to be preserved. Most meat products are low-acid foods and are good culture media for any surviving bacteria. Rates of heat penetration range from fairly rapid in meat soups to slow in tightly packed meats or pastes. Permitted additives in meats, e.g. spices, salt, nitrates or nitrites, also affect the heat processing, usually making it more effective. Examples [1] of typical heat processing times and temperatures for various products are:

- *Stewed steak* (16 oz/454 g) requires 110 min at 115°C or 80 min at 121°C followed by water cooling.
- *Minced beef*, containing flavour, tomato purée, salt, hydrolysed protein, caramel, monosodium glutamate and spices requires 90 min at 115.5°C followed by water cooling.
- *Steak and kidney puddings and pies* need careful evaluation as there will be differences in ingredients, pastry formulation and pie structure and also heat penetration through the pastry is slow.
- *Ham* if pasteurized needs to reach an internal centre temperature of 66°C as a minimum, but for a commercially sterile product 2–3 h at 110–115°C will be required according to the can size.

Some problems with canned meats.   The release of sulphur from protein during heat processing and the susceptibility to discoloration of certain products give rise to a number of problems in specifying internal can finishes. For uncured or slightly cured semi-fluid products, e.g. stewed steaks and meat-and-pastry products, it is usually sufficient to specify a lacquer with reasonable resistance to sulphur staining. However, in the case of products such as ham, pork and luncheon meats, which retain their shape when removed from the can, metal sulphide deposits may transfer on to the meat at points of unavoidable metal exposure during can manufacture, particularly the body side seam. Such corrosion is aggravated by the presence of curing salts, particularly nitrates and polyphosphates. Some meat canners prefer cans with plain bodies and lacquered ends, while others prefer fully lacquered cans and, because of side-seam discoloration, use cans with an internal side stripe of lacquer applied after the side seam has been formed.

For products containing gelatin, it is important that there is no residual sulphur dioxide because this will inevitably lead to a black discoloration, which, although harmless, is unpleasant in appearance. Because of the difficulty in removing some solid meat products from cans, lacquers containing an approved quick-release or slip agent are used. It is important with all products to minimize residual air in the cans at the time of closing by avoiding air pockets during filling.

Canned fish and fish products. Fish such as herrings, sprats and pilchards can be successfully canned. Like meats, fish are low-acid foods, have a slow rate of heat penetration and hence are difficult to process. Some seafoods soften and many even fall apart when the can is retorted and hence are packaged into cans which are not heat processed at all but are preserved by refrigeration, e.g. oysters. A few (e.g. crabmeat) are canned and pasteurized and then refrigerated until used.

Most canned fish products, however, are heat processed to be commercially 'sterile'. The process varies with product and size and shape of container. Many fish products are canned in sauces or oils. Examples of fish canning times for herring are $115°C$ for 60 min for 14 oz oval cans or 50 min for 7 oz oval cans.

Although fish may be packed in conventional round cans, many flat oval cans or flat rectangular cans are employed. They are usually internally lacquered with lacquers specially formulated to withstand the can-forming operation and to resist sulphur staining and corrosion. The ends of oval cans are usually coated with a sulphur-absorbent lacquer in order to further reduce sulphur staining. Easy open ring pull ends have also been introduced for small round or flat cans [9].

Fish (and meat) spreads are normally packaged in glass jars closed with tin plate non-venting caps which are vacuum sealed. This protects the product from oxidation which darkens the product surface. In addition small drawn aluminium pots fitted with aluminium easy-open ends are used for patés and similar chilled products as well as ceramic and plastics containers.

Canned fruit and fruit products. The pH of most canned fruits lies between 2.7 and 4.3. Bacterial spores do not germinate and grow at acid pH values (i.e. below 4.5) and even if they are not destroyed they are unlikely to develop during storage. Therefore milder temperatures than those required for less acid products can be used in fruit canning [1, 10].

Fruit shrinks on cooling and is therefore packed tightly in the can. Syrup is usually used in fruit canning because it keeps fruits firm and prevents loss of colour as well as sweetening the product. Different fruits require different syrup strengths. In addition permitted colours are often used.

Because of their acidity, spoilage of canned fruit is less likely to be due to microorganisms but 'flat sours' and 'hydrogen swells' may occur. Flat sour spoilage is usually caused by underprocessing, which allows thermophilic bacteria to survive. No gas is produced but lactic acid develops in the food giving it a sour taste.

Corrosion is always a potential problem in canned fruit. This can cause etching of the internal can surface, internal rusting, tin dissolution, production of hydrogen with swelling of the can (hydrogen swell), staining of the tinplate and discoloration of can contents. Special lacquers are used on the internal can surface for those products most at risk. For unlacquered cans control of the headspace is essential.

Fruit is also packaged in glass. Similar considerations to cans apply but because the package is transparent, grading of the fruit and freedom from blemishes are even more important. Underprocessing must be avoided because bottled fruit is susceptible to mould growth.

Fruit pie fillings are also canned but the added syrup and starch will decrease the speed of heat penetration and therefore longer processing times are required.

Canned vegetables.   Most canned vegetables are packed in brine at a pH of between 4.7 and 6.3 and are therefore more susceptible to bacterial spoilage from surviving spores than canned fruit. All harmful organisms must be destroyed during processing [1, 10]. Spoilage may therefore arise from flat sour organisms (see fruit) or thermophilic anaerobes (TA) depending on the pH. TA spoilage is caused by a spore-forming organism which metabolizes sugar and produces both acid and gas. The gas is a mixture of hydrogen and carbon dioxide and swells the can, which may eventually burst. The food usually has a sour odour. This type of spoilage can usually be avoided by proper blanching of the vegetables before processing. Sulphur staining may also occur with some vegetables, but laquering the internal surface of the can will avoid this.

Bottled vegetables are less common than bottled fruit but some vegetables are packaged in glass, where visual appearance is important.

Other canned products.   Evaporated and condensed milk and cream are also sold in cans [1]. The main type used is the double seamed-end conventional can which has now replaced the vent-hole can previously used. Condensed milk receives no post filling heat treatment because preservation is mainly because of its increased solids content. Empty cans may, however, contain mould spores and must be pre-sterilized. Headspace is not so important here but is kept to a minimum to avoid mould growth, oxidation or rusting. Aluminium collapsible tubes are used as an alternative to double seamed-end cans.

Baked products such as sponge puddings and cakes are also canned successfully in open top cans. Cake cans may have plastic reclosable lids attached [1] and are usually cooked in the can which is removed from the oven before fully baked. The can is then closed and returned to the oven to destroy any moulds or yeasts. Alternatively the cake may be placed in a can which is hermetically sealed under vacuum or gas packaged. The cans are usually lacquered and the external decoration must withstand the baking procedure.

Cans and jars are also used for a number of combination products such as *soups, ready meals, desserts* and *babyfoods.* Other ingredients will obviously affect the heat processing times required and for infant foods especially, selection of good quality raw materials and good hygiene are particularly important.

The retort pouch. A more recent development is the heat-sterilizable flexible package [3], usually in the form of a retort pouch produced from a 2-, 3- or 4-ply flexible laminate which, when sealed, is as hermetic as a can or jar, is product-resistant and can be sterilized by retorting. Typical materials used are:

- 2-ply laminate: $12\,\mu$m nylon or polyester/$70\,\mu$m polyolefin
- 3-ply laminate: $12\,\mu$m polyester/9–$12\,\mu$m aluminium foil/$70\,\mu$m polyolefin
- 4-ply laminate: $12\,\mu$m polyester/9–$12\,\mu$m aluminium foil/$12\,\mu$m polyester/$70\,\mu$m polyolefin

All these constructions are capable of producing microbiologically stable packs. However, the 2-ply specification without aluminium foil offers a relatively poorer barrier to gases, moisture and light, all of which can lead to earlier deterioration of the product. The materials used therefore depend upon the barrier required, the process temperature which will be applied, product compatibility and handling needs.

The retort pouch has five main advantages over other packs. Because it is thinner in cross section, the thermal process time for a retort pouch is significantly shorter than for a standard metal or glass container of the same capacity, and in many instances this leads to improved product quality (less overcooking). The shelf life of a pouched product is generally as long as, if not longer than, the equivalent frozen product and does not require the frozen storage and distribution chain and the marketing restrictions that this imposes. Product/container interaction is less of a problem than with cans. *Vacuum sealing* in retort pouches without brine, syrup or sauce permits up to 40% saving in weight and volume and allows some products to be packaged which are difficult to can successfully. Materials are supplied in reels or as pouches and offer significant savings in storage space and weight compared to rigid containers. The retort pouch is easy to open with scissors and it can be provided with a tear notch. For boil-in-the-bag applications the preparation time is less than for the equivalent frozen food.

However, there are a number of significant differences which require strict control procedures to ensure satisfactory filling and sealing. Although the retort pouch comes second to canning in terms of capital investment, it allows for more automation and therefore direct labour costs are lower.

Apart from Japan, however, the retort pouch has never really made the grade in the western world, probably because of its much slower filling and closing speeds and the fact that the pouch itself must be protected by a carton during distribution.

High-barrier plastics packaging

The major area of technological innovation for heat-processed packaging development and investment has been in barrier plastics forming and manufacturing processes. The production of a marketable pack containing a heat-processed, low-acid foodstuff requires a satisfactory combination of a number of factors

(1) Fundamental container construction:
   (a) barrier/shelf-life performance;
   (b) extraction/taint and odour.
(2) Closing method:
   (a) type of closure;
   (b) pack integrity.
(3) Product/processing regime:
   (a) ingoing product quality;
   (b) process method;
   (c) post-process handling/distribution.

The majority of the new plastic packages used for heat-processed foods are produced from multilayer laminate structures based around polypropylene as the structural polymer and contain an oxygen barrier of either EVOH or PVdC. Adhesive layers hold the structure together, and in many a reclaim layer (produced from excess material generated in the manufacturing process) is incorporated. The development of high-integrity heat sealing and of CMBs 'Tor' closing process opened up the opportunity for commercial testing of products in alternative shapes of 'Lamipac' barrier containers. (The 'Tor' closing process is a patented vacuum closing process, which deforms the lidding material after sealing, to produce a virtually hydraulically solid 'stress-free' pack.)

In 1985 and 1986 market activity in the United States and in Germany started what was to become Europe's largest market for 'ready meals'.

In the United States, the Omni can, a double-seamed container produced by American National Can was introduced. This is co-injection moulded with an EVOH barrier material and desiccant to counteract the loss of barrier in the EVOH as a result of moisture absorption. The products in Germany were generally full meals in two or three compartment trays, which were closed by a foil structure that required cutting with a knife to remove it from the container.

Through 1989 and 1990 in both the UK and the rest of Europe, a whole range of new products in this form of packaging, was introduced, the market segmenting into three key areas:

(1) Premium petfoods: high-barrier packages have now been used for a range of 'Whiskas' and 'High Society' products
(2) Ready meals: Most of the major food retailers now carry a range of 'own label' ready meals in microwavable high-barrier trays.
(3) Snacks: A variety of snack products in heat-sealed and double-seamed bowls, generally of lower fill weights than the ready meal products and at a lower price, e.g. Campbell's take-away and Batchelor's Micro-Chef ranges, Heinz Lunchbowls, Wilson's Micro-Quick in the UK, and in New Zealand a range of Hotshot products packed by J. Wattie.

The current market for retortable barrier containers in the UK is around 100 million units per annum (approximately half of the total market in Europe) and is growing.

A number of other forms of retortable packaging have been exploited on a limited scale and represent interesting technical pack concepts, e.g. Stepcan and Letpak.

Stepcan

This combines a plastic body (high molecular weight homocopolymer PET) with conventional metal ends and is filled, closed and processed on standard canning lines (STEP = stretch tube extrusion process). An extruded plastic tube is blown and stretched into a long thin-walled 'bubble' which is then cut into suitable lengths for a heat treatment, and these in turn are cut for individual container bodies. One metal end is then seamed on by the container maker and after filling the packer seams on the final end to close the package.

Letpak

Introduced in early 1978 by a Swedish company, the 'Letpak' (see Figure 10.2) is a three-piece can with an extruded body constructed from a laminate of polypropylene on both sides of aluminium foil, the foil providing the necessary water vapour and gas barrier [3, 11]. The body has smoothly curved corners and provides a good printing surface for multicolour designs over the entire

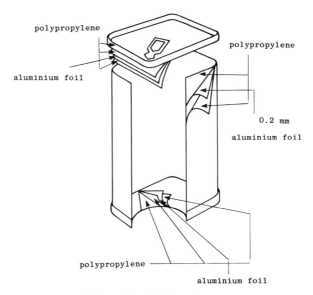

polypropylene

polypropylene

aluminium foil

0.2 mm

aluminium foil

polypropylene

aluminium foil

Figure 10.2 The Letpak 'can'.

surface. The ends are injection moulded and lined with the same type of laminate used for the body. They are welded into place by high-frequency sealing and may be produced in different colours. No tools are required to open the lid. The sizes and dimensions are standardized to suit the distribution system and consumer needs. The material used is heat-resistant and retortable and suitable for food contact. The container requires bottom-sealing equipment in the packer's works, and speeds of up to 50–75 packs per minute are possible. Processing can be carried out in a similar manner to cans, with a maximum recommended temperature of 130°C. However, the Letpak was not well accepted as early examples tended to split when dropped.

More recently a cylindrical seamless 'can' (Omni can) has been introduced with sturdy flanges that will accept a seamed-on can lid and which incoporates a ring tab pull. It is claimed to have substantial economies over metal cans as well as attractiveness and easy opening [3].

New systems are being developed but their success will depend upon the economics and improving technology of competing aseptic packaging.

Aseptic processing

Asepsis preserves food by preventing microbial contamination from the raw state to the finished product [12]. Aseptically packaged products require:

(1) Sterilization of the starting material

(2) Sterilization of the packaging material

(3) Maintenance of sterile surroundings while forming and filling the pack

(4) The production of packaged units which are sealed efficiently to prevent any reinfection.

Aseptic processing, therefore, allows better use of packaging materials and systems because, unlike conventional canning, aseptic processing causes less thermal damage to the product and less stress on the packaging. Besides improving product quality, it allows the use of materials other than the traditional metal can or glass jar. Although cans and jars are used in aseptic processing, laminated paperboard or plastic containers of various shapes may be used instead and this almost invariably reduces the cost of the packaging materials.

Producing a sterile product in a continuous process involves three steps:

● Heating it to raise the temperature to the desired level

● Passing the product through a temperature holding section for a predetermined time

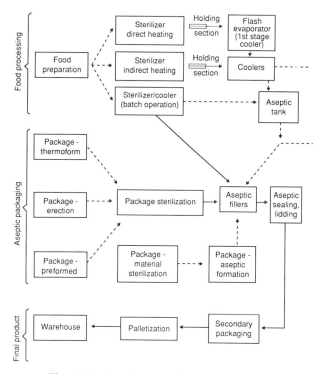

Figure 10.3   Aseptic process flow diagram [13].

● Cooling the product as rapidly as possible to a temperature of 35°C or less prior to filling.

Figure 10.3 shows the conventional canning process and the aseptic packaging process in diagrammatic form [13].

The type of heat exhanger [3, 13] used for aseptic sterilization will be determined by the nature of the product, i.e. whether it is a mobile or a viscous liquid, and the amount and size of any particulate matter. In any continuous process the size of any particles is limited to 8–10 mm diameter; larger particles (20–35 mm) need special equipment. Ideally the temperature used should be high enough to obviate the need for a holding section, but this is not practical for several reasons. First viscous products are difficult to heat uniformly; second, the food may contain heat-resistant enzymes which are more likely to survive the process; third the presence of particles makes it important to have a holding section to allow time for them to reach heat equilibration. Also the shorter the holding section the more difficult it is to exercise control.

Whilst the hazards connected with aseptically packaged foods are not considered to be different from in-pack sterilization, the risks are different. Consequently low-acid foods (pH < 4.5) should receive a heat process which is equal in effect to the 'botulinum cook'. Usually the UHT range is 130–150°C. For acid products, a pasteurization temperature of 80–105°C is used because the spoilage organisms, mainly yeasts and moulds, are relatively more heat sensitive [13].

Aseptic canning

In aseptic canning [1, 12] the product is heated in a continuous flow to a temperature of around 149°C at which temperature sterility is achieved in a few seconds. High-temperature steam is the usual heating method, and it may be injected directly into the product or indirectly through a heat exchanger. The rate of flow of product through the sterilizer is adjusted so that a small quantity only is in the heating and cooling sections at any time, and thus the heat may be introduced into and removed from the product very rapidly. Filling and closing are carried out under aseptic conditions in a steam-filled chamber, the empty can and the can ends having previously received a dry heat treatment in a special magazine attached to the seamer.

Aseptic packaging using flexible materials

Cartonboard and plastics laminates have achieved significant growth over the past few years for the aseptic packaging of liquids [14–16]. The containers used are either preformed or formed from a web of material by a form-fill-seal technique. Where extended shelf life is required, the barrier properties are increased by incorporating a layer of aluminium foil or by coextruding a good oxygen barrier plastics layer within the container material. Container and lid sterilization is normally carried out by heat and the use of hydrogen peroxide, UV or gamma irradiation [12, 13].

Packaging materials that have been developed for aseptic processing are:

- Preformed containers, principally tubs and cups which are sterilized, filled and closed with a heat-sealed lidding, usually coated foil
- Vertical pouch forming systems
- Lidded thermoform/fill/seal systems fed from a reel
- Systems starting from plastic beads which are extruded into a parison which is blown into a container and subsequently filled and closed
- Bag-in-box systems.

The most important aseptic systems using flexible materials are the 'Tetra Standard/TetraBrik' range, the gable top 'PurePak' or Elopak systems and the PKL Combibloc system [17]. The laminate construction of two common materials for such cartons is shown in Figure 10.4.

In the Tetra range the packaging material is fed into the filling machine from a reel (Figure 10.5) and then passed through a bath containing hydrogen peroxide (6 s). A certain amount of hydrogen peroxide is carried along with the web but most is removed by a pair of squeeze rollers to leave a thin hydrogen peroxide film on the surface. After passing the bending rollers, the web travels downwards and is formed into a tube which is sealed longitudinally. Small amounts of sterile air are constantly emitted into the area above the product level to prevent re-infection, while transverse seams in the tube are produced by a pair of jaws simultaneously pulling down the

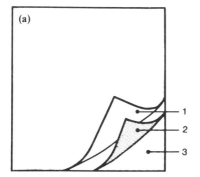
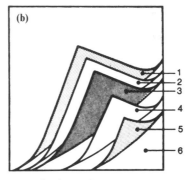

Figure 10.4 The sandwich construction of two common laminates for carton containers is shown. (a) Typical laminate for short-life products like fresh milk consists of (1) exterior PE, (2) paper and (3) interior PE. (b) Typical laminate for long-life products consists of (1) exterior PE, (2) paper, (3) Surlyn, (4) Al-Foil, (5) Surlyn and (6) interior PE.

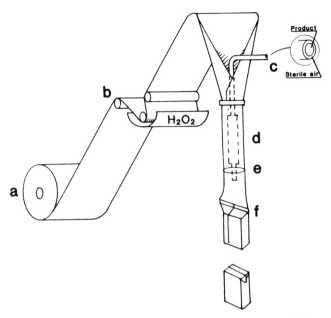

Figure 10.5   The TetraBrik aseptic concept is shown. (a) Packaging material on a reel is fed to the machine passing through a hydrogen peroxide bath (b) and formed into a tube by a longitudinal seam. The tube heater zone is indicated by (d). The liquid is admitted through (c) and filling and sealing is performed below surface level (e). Finally the package is given its final shape (f).

packaging material. Following longitudinal sealing, the tube passes a heater by means of which the inside of the tube is heated (110–120°C). Hydrogen peroxide evaporates to give oxygen and water, and sterilization of the packaging material is completed.

The product is filled into the sterile tube and transverse sealing is carried out below the level of the liquid. Sealing is by means of heating induced in the aluminium foil, which in turn melts the plastic in the vicinity of the area sealed. Pressure is applied and sealing completed. The package is then cut free from its neighbour and passes through a final folding unit to be formed into its final brick shape. The preformed gable top style carton is shown in Figure 10.6 and the filling and sealing procedure in Figure 10.7.

Aseptic processing of milk and dairy products

Plastic materials used in the aseptic packaging of milk products are polyethylene, polypropylene, polystyrene as tubs or bottles or plastic film laminates with paperboard or aluminium in the form of cartons, e.g. TetraBrik, etc. High pressure steam is used to sterilize product lines and

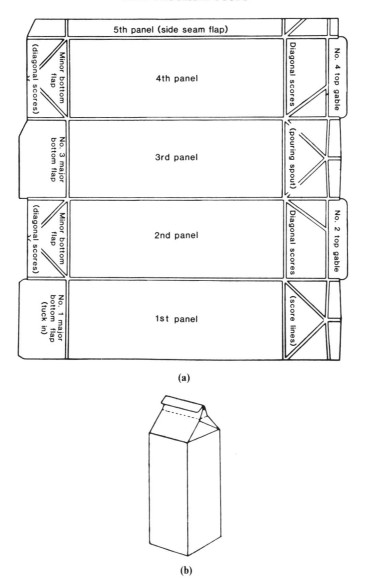

Figure 10.6   (a) An unfolded gable-top blank is shown. Scorelines necessary for folding are indicated. (b) Purepak carton.

hydrogen peroxide with heat or UV radiation for containers and lidding materials.

Plastic tubs are either supplied preformed or thermoformed as part of the process in the machine. Closure is by a plastic-coated aluminium foil lid which is heat-sealed to the top of the cup. Plastic bottles are blow moulded

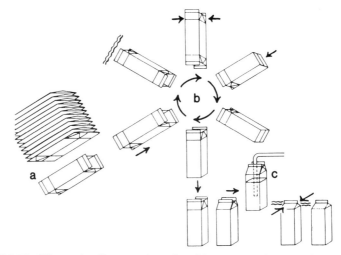

Figure 10.7 The filling and sealing procedure of a gable-top carton from a prefabricated blank is shown. (a) The blanks are fed to the machine from a magazine, unfolded and introduced into a mandrel (b) where the bottom is heated, folded according to scores, and heat-sealed. Now an open box, it is removed from the mandrel, filled (c), and top-sealed by means of heat and pressure.

and the heat generated in forming is sufficient to sterilize the container. Laminated carton systems such as TetraBrik and Combibloc are widely used as aseptic containers for milks and other drinks, while plastic pots and tubs are more widely used for creams and desserts.

Recontamination of the UHT sterilized product is the most common cause of failure. This can arise from the plant, atmosphere or packaging system. Microscopic holes in packages or faulty seals are other sources of contamination.

Milk has also been bulk packaged (30–250 litres) aseptically using 'bag-in-box' systems. The bags are presterilized by UV radiation and can be clear plastic or precoated metallized polyester for longer shelf life [18].

Aseptic processing of fruit and vegetable juices

Fruit and vegetable juices, syrups and ready to drink beverages have also been successfully packed aseptically in flexible packagings. Since these products are particularly susceptible to microbial spoilage, aseptic processing can provide a longer ambient shelf life without the use of preservatives. Flash pasteurization of the juice or drink is carried out at 85–95°C for 15–60 s and then cooled and filled into sterilized containers. Bottles, laminated cartons and plastic or aluminium foil bags in boxes are used. Sterilization of containers is usually carried out with hydrogen. peroxide and heat treatment. Bag-in-box and bag-in-drum containers have also been used for bulk

Table 10.1 Classification of aseptic filling and packaging systems

Principle	Examples
Metal and rigid containers sterilized by heat	
Steam/metal containers	Dole canning systems
	Drum fillers, e.g. Scholle, Fran Rica
Hot air/composite can	Dole hot air system
Webfed paperboard sterilized by H_2O_2	TetraPak (BrikPak)
Preformed paperboard containers	Combibloc
	LiquiPak
	Elopak
Preformed, rigid plastic containers	Metalbox Freshfill
	Gasti
Thermoform-fill-seal plastic containers	Benco Asepack
	Bosch Servac
	Conoffast
Flexible barriers	
Bag-in-box type	Scholle
	LiquiBox
Pouches	Asepak
	Prepac
	Prodo Pak
	Inpaco
Blow moulded	Bottlepack
	Serac

packaging up to 1000 litres [19, 20]. Bulk aseptically filled containers, e.g. plastic or plastic-lined metal drums and large tanks are also used for 100 kg to 1 tonne packs for foods such as tomato paste. Table 10.1 lists some examples of aseptic filling and packing systems.

Problems with particulates

Many foods, including juices, pastes, soups and sauces, lend themselves to aseptic processing. In recent years the aseptic packaging of products containing particles has become of great interest. However, there are problems with larger particulates which must be overcome.

The first problem relates to the mechanical behaviour of particulates. Special designs for valves, pumps and heat exchangers, etc. are required to avoid clogging and damage to the particles. The second problem concerns the time taken for the centre of the particles to reach the required sterilization temperature. This is essentially a heat transfer problem. In a liquid, heat is transferred mainly by convection, while in solids the transfer is by conduction, which is much slower. The heating and holding sections must be designed therefore to deal with the delay between liquid reaching the temperature and the transfer of heat to the centre of the larger solid particles. Size and uniformity of the solids will be of importance. In addition enzyme reactivation may occur. Ohmic heating (electrical resistance heating) overcomes the heat transfer problem by permitting the direct passage of electric current through the continuous flow of food product. Heat penetration is more rapid and even.

A great deal of research has been carried out into how particles are transported in fluids, and we know that particle velocity is affected by a number of parameters, including pipe diameter, particle concentration, mass flow rate and fluid viscosity [21]. Semi-continuous systems are being investigated to overcome some of the problems.

The Institute of Food Science and Technology guidelines on good manufacturing practice [22] cover heat-processed foods, including aseptic technology, and further information on the processing of low-acid foods can be found in the Campden Food and Drink Research Association guidelines for the processing and aseptic packaging of low-acid foods [23].

Sous vide

The term 'sous vide' means 'under vacuum'. Foods are cooked in sealed, evacuated heat-stable pouches or thermoformed trays so that the natural flavour, aroma, and nutrient quality are retained. This was introduced to extend the shelf life and keeping quality of fresh food in order to provide the consumer with ready-to-eat, microwavable, convenience foods.

The various steps in *sous vide* processing are [24]:

(a) Basic preparation: addition of fresh, best quality ingredients and sauces under strict hygiene and quality control conditions
(b) Packaging: ingredients weighed, placed in a plastic dish and covered with a plastic seal that is impermeable to air and contaminants
(c) Air extraction and hermetic sealing: air is removed with a vacuum packaging machine and the product is hermetically sealed
(d) Pasteurization: slow heating in an autoclave with a precise electronic regulator
(e) Quick-chilling
(f) Cold storage (0–3°C)
(g) Reheating: for 10–15 min in a boiling water bath or 4–5 min in a microwave oven
(h) Service

The effectiveness of the process depends on high quality raw materials, pre-cooking if necessary, packaging in heat-stable barrier bags under vacuum, sealing and pasteurizing at a particular temperature for a specific time. The pasteurized product is cooled to 4°C within 2–3 h and stored and distributed under refrigerated conditions (−4°C). A shelf life of 21–30 days is possible.

One major concern is that *sous vide* processed products are not shelf stable and could be a potential public health risk if subjected to temperature abuse during production, storage, distribution and marketing. Also, since many of these products are available in packs which the consumer traditionally views as shelf stable there is a high risk of temperature abuse in the home, mishandling and over-extending the product's shelf life.

The main microbiological hazards are from *Clostridium botulinum* which can survive the heat treatment and grow at temperatures as low as 3.5°C. Spores can germinate at 10–12°C. Non-spore forming bacteria should be destroyed by *sous vide* processing but may be important if the processing is inadequate or raw materials of poor microbiological quality are used.

Packaging materials used in *sous vide* processing must withstand pasteurization and chill temperatures as well as protecting and maintaining product quality.

Packaging for microwavable foods

With the increase in popularity of the microwave oven, convenience foods which can be heated in the microwave in their original packaging have increased. Microwave heating of food products is caused by the absorption of electromagnetic energy. Two frequencies have been set aside for microwave application namely 915 MHz (896 MHz in the UK) and 2450 MHz [24–26].

The heating effect of microwaves is achieved by transfer of energy to a dipole within the food. The most common molecule which exhibits dipolar behaviour is water which is present in all foods at levels as high as 90%. When food is subjected to microwave energy, the dipoles interact with the electrical signal and try to rotate in order to line up with the alternating field, but as this changes rapidly from positive to negative the dipoles are constantly moving and therefore generate heat. The rapid rise in the temperature of the water molecules allows the other parts of the product to heat by conduction and/or convection.

The rate of heating depends on the electrical and physical properties of the food and the shape and type of container in which the food is placed. Choice of packaging materials will also depend on the overall performance required, i.e. a frozen pack may be subsequently microwaved [27].

Heating uniformity is of major importance not only in terms of microbiological safety of the products ensuring that pathogenic bacteria do not survive, but also in producing an evenly heated product without hot and cold spots.

While microwaves heat food principally through the direct absorption of energy by the water molecules, food components also influence microwave absorption [26, 28], e.g. carbohydrates and proteins tend to bind to some of the water and thereby reduce the heating effect. Fats will heat more quickly in a microwave. Dissolved ions (usually sodium and chloride ions from salt) also contribute to the microwave heating mechanism. Both water and the salt content of the food are therefore important. Pretreated flavour components are sometimes added to foods at a later stage because the short heating time does not always give enough time for normal flavours to develop.

Frozen foods will also be affected differently and packaging must be able to withstand not only microwaving but also freezing and frozen storage and subsequent thawing. Many ingredients, such as some starches or wheat flour, are not stable after freezing and thawing and special starches have been developed for use to avoid water separation in the product or lumpy textures when the product is defrosted and then reheated in the microwave.

When packaging foods for microwaving, sharp edges in the pack should be avoided as well as very thin layers of food which may burn easily. Packages with rounded edges and corners are best because the sharp curve of an edge protruding out into the microwave field will affect energy distribution and concentrate energy to the corner or edge [29]. Packages should also be flat and not too deep (less than 5 cm). The effect of pack shape, size and material is of paramount importance and the combined properties, in relation to each component, must be carefully evaluated when designing microwavable foods and their packaging.

Packaging materials

Packaging materials used for microwavable foods [26] are either microwave passive, microwave active or microwave reflective.

Microwave passive materials include china, glass, paper and plastics and are transparent to microwaves, i.e. they do not heat directly in a microwave field and allow maximum energy to be absorbed by the food. The most common type of package for microwavable foods are plastic thermoformed trays. Choice of material depends upon the temperature to which the food will be subjected. The temperatures within a high water-containing foodstuff will rarely exceed 100°C but high impact polystyrene plastics which melt between 90 and 100°C will be unsuitable for many products. High density polyethylene (HDPE) which has a melting point of 125–130°C is used. HDPE cannot be used, however, for high fat/sugar content foodstuffs where temperatures exceeding 100°C may be reached. The most commonly used packaging materials are polypropylene, where a good barrier to water vapour is important, and crystalline polyethylene terephthalate (CPET). Both have melting points above 210°C and are consequently suitable for most foodstuffs. CPET has the additional advantage of being suitable for both microwave and conventional oven use.

Paperboard containers coated with PET offer a lower cost alternative [30] but are likely to distort and char at higher temperatures. HDPE and some high temperature polystyrene-based polymers can also be used for some applications.

Developments in this area are moving rapidly particularly for cook-freeze and cook-chill products [31]. However, packaging solely for the microwave is gaining ground and developing into its own sector. Container structure has changed with a view to attaining greater product homogeneity, a greater surface/volume ratio and in some cases hollow areas are created within the food to allow energy waves to move more rapidly.

Microwave active materials (susceptors and receptors)

A common defect of microwave cooking is its inability to produce browning and crisping of the food. To overcome this microwave susceptor films have been developed which are designed to absorb and convert the electrical component of microwave energy into infrared or radiant energy which is then transferred to the food [32, 33]. Susceptor performance is controlled by several factors [26]:

(1) Distance of product from susceptor: the best results are obtained when the distance between the susceptor and the food is less than 3 mm and the use of raised edges is avoided.

(2) Initial product temperature: frozen temperatures tend to produce crust browning and crisping.

(3) Ratio of food to susceptor energy absorbance: control of this ratio is important.

(4) Food and susceptor size: balance between the two sizes is necessary.

(5) Susceptor heating: susceptor heating is known to be dependent upon its vertical position in the microwave oven.

(6) Oven wattage.

Examples of susceptors are aluminium coated polyester films laminated onto a rigid support such as paper or board which prevents shrinkage at high temperatures. The key factor is the amount of aluminium deposition which absorbs the microwaves and acts as a secondary heat source, causing the food to heat locally, thus crisping and browning it.

There are several drawbacks to current suspectors [24] but considerable research is being carried out in this area using other metals such as nickel, cobalt, iron and stainless steel. Combinations can be used to produce customized susceptors so that foods can be selectively shielded from or allowed to absorb or transmit microwave energy within one pack.

Microwave reflective materials. Most metals reflect microwaves and do not heat. Reflection helps the microwaves to travel directly to the food and this is the reason why the oven walls are made of metal. Materials which reflect microwaves should not be used alone in microwave ovens as under certain conditions they can result in arcing. However, metal containers and aluminium foils can be used successfully in microwave ovens under the right circumstances but should be used with caution.

In summary, suitable materials for packaging microwavable foods include polyester coated and bleached paperboard, moulded unsaturated polyester with high heat resistance, thermoformed fractionated CPET, high density polyethylene and polypropylene [26]. All are suitable for conventional oven use as well as microwave oven use. Aluminium trays can also be used under certain conditions.

Irradiation

Ionizing radiation [34–36] was first discovered just before the start of the twentieth century. Although there was considerable interest in the biological effects, the first patent for the use of radiation to sterilize food did not appear until 1930, and real research on the preservative effects did not start until the 1940s. In 1943 it was demonstrated that X-rays could extend the shelf life of a hamburger. Since then research into the effect of irradiation has continued around the world, but the technology has been slow to gain acceptance.

In 1981 the FAO/IAEA and WHO joint committee on the wholesomeness of irradiated food concluded [36] that the 'irradiation of food up to an overall dosage of 10 kg presents no toxicological hazard and introduces no special nutritional or microbiological problems'.

Currently irradiation is permitted in 36 countries of the world [37, 38] and the UK Government [39] has recently permitted it subject to certain controls. In many countries packaging must be made identifiable by the use of an internationally agreed logo (see Figure 10.8).

Electrons or photons, whether they come from an accelerator or a radioactive source, such as cobalt-60, constitute a form of energy, just like

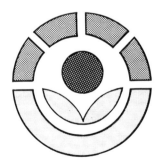

Figure 10.8 The internationally recognized logo for irradiated foods.

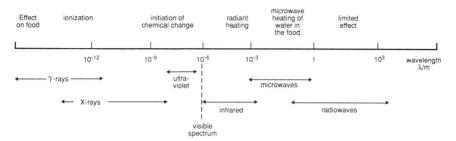

Figure 10.9   The electromagnetic spectrum.

visible light and microwaves. In fact they form part of the spectrum of energy which extends from radio waves through the visible to high-energy electrons (Figure 10.9). This radiation energy is capable of inactivating micro- and other organisms but the wavelength is much shorter than that of visible light and because of this and the higher energy, such radiation has a higher penetrative power. Hence one of the advantages of irradiation is that the product can be treated in its container, so that recontamination after processing is prevented. The product itself does not become radioactive and there are no residues left behind; the products of radiolysis differ very little from those produced by conventional heat treatment. Once treated, foods are ready for use or consumption, whether they are fresh or processed before packaging and irradiation.

The inactivation of living food spoilage microorganisms is brought about by changes in their DNA molecules. Beacuse of its size and other properties, the DNA molecule is far more sensitive to radiation than the molecules of the food undergoing processing. As a result, bacteria, moulds and yeasts are killed long before any undue changes can take place in the food itself or in its flavour. Changes in the food product are much less than in traditional heat processing (canning, bottling and pasteurization), since the temperature is not raised significantly and no cooking occurs. This can be a major advantage where heat-sensitive products are involved.

The effect of irradiation depends on the exposure dose, which in turn is determined by the residence time in the irradiation chamber. Since the power of any one source is constant, this is the only variable in processing so the process is reliable and reproducible.

Methods of irradiation

There are two possible methods of irradiation for foods at present: low-dose gamma irradiation and low-dose irradiation by electron beam [37, 40].

Low-dose gamma-irradiation uses a radiation source with protective shielding and conveyor systems both inside and outside the radiation chamber, together with control and safety equipment. The cobalt unit forms the main part of the system which uses cobalt-60 assembled in the form of tiny cylinders to form rods 450 mm long which are fully screened by hermetically welded double walls of stainless steel. A vessel filled with water, as a shield against radiation, holds the cobalt unit.

The 1.8 m thick concrete walls of the irradiation chamber and spaces between them ensure a safe shield for the surrounding areas, which include the control room and the production hall. Safety equipment is also housed in the control room. Optimum utilization of the plant is obtained using a system of conveyors allowing continuous flow of products 24 h a day for 7 days a week. The process is supervised by a telemetric monitoring system, and bulk goods can be handled in pallet loads.

Low-dose electron beam irradiation employs an accelerator consisting of an evacuated tube to which an electric potential difference is applied. An ion source at one end injects charged particles into the tube which arrive at the other end with higher energy. These then impinge on the food and their energy is transferred to it as it passes by on the conveyor. Since the penetration of the beam is low, the treatment must be repeated from the other side. Exposure time depends upon the nature and quantity of the food but is of the order of minutes.

Applications

Ionizing radiation can be used to preserve food [37, 40, 41] as follows:

(1) Inactivation of pathogens: low doses of irradiation are sufficient to inactivate pathogens, such as *Salmonella*, *Campylobacter* and *Listeria*, most of which are more sensitive to irradiation than the normal spoilage flora. Food can be irradiated after being hermetically packaged, so that re-infection is avoided. However, in meat and meat products, for example, a balance is needed between achieving the desired effect and keeping organoleptic changes to a minimum.

(2) Disinfection of herbs and spices: many spices are highly contaminated with bacteria and fungi and irradiation can reduce microbial contamination to below 1000/g using a radiation dose of 0.5–0.8 Mrad.

(3) Extension of shelf life of fruits and vegetables: fruits and vegetables vary enormously in their response to radiation and some are not suitable for such treatment. Time of harvesting is important, e.g. if strawberries are irradiated before they are ripe the red colour does not develop. The combination of irradiation and modified atmosphere packaging may have a synergistic effect and hence the desired result can be achieved with lower doses of radiation.

(4) Insect disinfestation of grain, grain products, tropical fruits: useful for export products and has the advantage that it avoids the use of fungicides and pesticides.

(5) Inhibition of sprouting in potatoes, onion and garlic.
(6) Sterilizing food for prolonged storage, e.g. foods for space travel and emergency stores. A combination of short heat treatment sufficient to inactivate enzymes and irradiation can give a long-life product without the use of preservatives or refrigerated storage.
(7) Decontamination of soils: the undesired growth of fungi and other organisms which lead to decay can be prevented.

The quality of the raw material, the dose of irradiation applied, temperature, type of packaging and storage conditions before and after irradiation all affect the success of irradiation. The penetration of gamma radiation depends upon the energy of the rays and the density of the food.

Irradiation can also be used to sterilize packaging materials before aseptic processing and potentially can be used in combination with other methods of preservation, e.g. MAP [37] although every food requires careful evaluation to determine optimum combination strategies. Also foods can be treated after the final packaging operation has been carried out, thus minimizing the possibility of post-irradiation contamination. However, the packaging materials used must not be adversely affected by irradiation and new packaging systems need to be thoroughly evaluated [41]. The FDA in the United States has concluded that the following packaging materials may be safely subjected to irradiation at the prescribed levels:

Up to 1 Mrad	Up to Mrad
Coated cellulose film	Vegetable parchment
Glassine paper	PE films
Waxed paper board	PET
Kraft paper	Nylon 6
PET	PVdC-PVAc copolymer
Polystyrene	Acrylonitrile copolymer
PVdC film	
Nylon 11	

Currently the main type of packaging used for irradiated food products is polymer films. Table 10.2 shows the main polymer types and their use or potential use for specific irradiated foods [38].

To some extent all plastic films are affected by ionizing radiation with the degree of chemical breakdown depending not only upon the type of polymer and its physical state, but also the environment and the dose rate. However, analysis usually reveals that the majority of the polymer is unaffected. Some volatile aldehydes and ketones and polymer additives may migrate into some foods. Many polymers, such as polyethylene, rapidly undergo oxidative degradation in visible and UV light and this can lead to loss of mechanical strength and optical transmission, due to cracking and discoloration. PVC

Table 10.2 Main polymer types for irradiated food containers [27]

Polymer	Form	Food
Polyethylene Polypropylene	Bulk containers Bags	Potatoes, onions
Polyethylene Polypropylene laminates	Pouches	Shellfish, spices

Polyethylene
  terephthalate
Polystyrene
  PVC
	Containers and lids	Spices, shellfish, dried
		  fruit

Expanded polystyrene
  PVC
Polyamide laminates
	Trays and overwraps
	  (MAP)
		Poultry, meat, fish
		  vegetables

can also be discoloured by irradiation. Therefore not only must the package survive irradiation and subsequent storage without losing its particular properties, such as mechanical strength, barrier properties (where required) and optical transmission, but hazardous materials must not migrate into the food and seal integrity must remain.

Detection methods

More research has been carried out on food irradiation than on any other process and while many governments are in favour of its introduction consumer resistance is high in some countries. Objections to the process are mainly concerned with safety and control and, in particular, with methods of detection; currently there is no test to determine whether or not foods have been irradiated.

One of the main difficulties in developing a method for the detection of irradiated food is that the changes induced by the process are similar to those produced by conventional processing. Several methods are currently under investigation but each is specific to a range of foods [42]. Electron spin resonance (ESR) spectroscopy has been applied to foods with bone or shell and some fruit such as strawberries; thermoluminescence has been used for dry ingredients; and the formation of volatile compounds from lipids (including long chain hydrocarbons and 2-alkyl cyclobutanones) has been used for meat but has potential use for a wide range of foods.

The development of these methods looks promising. However, while consumer suspicion of the irradiation remains, is reflected by the retail and manufacturing trade, the uptake of food irradiation is likely to be small.

Ultraviolet light

The UV wavelengths from 136 to 390 nm are germicidal; those between 250 and 280 nm are especially effective. Radiations in this range are part of the electromagnetic spectrum just below the wavelength of visible light and are absorbed by the purine and pyrimidine groups of nucleic acids in microbial cells, causing mutation and death. Examples of applications for UV light are:

(1) Treatment of water in the production of beverages
(2) Ageing of meats
(3) Treatment of knives for slicing bread
(4) Treatment of bread and cakes
(5) Packaging of sliced bacon
(6) Prevention of growth of film yeasts on pickle, vinegar or sauerkraut vats
(7) Killing of spores on sugar crystals
(8) Packaging and storage of cheeses
(9) Treatment of air for or in storage and processing rooms
(10) Packaging for aseptic processing.

Ultrasonics

Ultrasonic frequencies (over 20 000 Hz) can be used to preserve foods; most of the killing effect of sonic vibrations on microorganisms is believed to be as a result of the heat produced, although disruption of cells and membranes may also take place. Physical damage to foods, destruction of vitamins and failure to affect microbial enzymes are the disadvantages of such treatments. Sound waves have not as yet found commercial use in food preservation.

High pressure techniques

High pressure technology is another experimental technique which could be used to replace some high temperature processes. Protein can be denatured and microorganisms deactivated by the use of high hydrostatic pressure $(1000-10\,000\,kg/cm^2)$. High pressure processing has the advantage of reducing flavour and nutrient deterioration and unique food properties are formed as a result of protein denaturation.

Many new technologies, such as high intensity light, infrared heating, induction heating and electromagnetic heating, are emerging for new applications. When commercially accepted these will also have an impact on food packaging.

References

1. J.A. Rees and J. Bettison, Heat preservation, in *Food Industries Manual*, 2nd edition M.D. Ranken (ed.), Blackie, Glasgow (1988).
2. W.C. Frazier, *Food Microbiology*, 2nd edition, Tata McGraw-Hill, New Delhi (1967), Chap. 6.
3. R.C. Griffin, Retortable plastic packaging, in *Modern Processing, Packaging and Distribution Systems for Food*, F.A. Paine (ed.), Blackie, Glasgow (1987).
4. J.H. Hotchkiss, Canning, food, in *The Wiley Encyclopedia of Packaging Technology*, Wiley, New York (1986), p. 86.
5. C.R. Stumbo, *Thermobacteriology in Food Processing,* 2nd edition, Academic Press, London (1973).
6. J.J. Nickerson and A. Sinsky, *Microbiology of Foods and Food Processing*, Elsevier Applied Science, London (1972).
7. F.S. Thatcher and D.S. Clarke, *Microorganisms in Foods—Their Significance and Methods of Enumeration*, University of Toronto Press (1973).
8. F.S. Thatcher and D.S. Clarke, *Microorganisms in Food 2—Sampling for Microbiological Analysis. Principles of Speciation and Application.* University of Toronto Press.
9. M.J. Urch, Fish and fish products, in *Food Industries Manual*, 2nd edition, M.D. Ranken (ed.), Blackie, Glasgow (1988).
10. W.C. Frazier, *Food Microbiology*, 2nd edition, Tata McGraw-Hill, New Delhi (1967), Chap. 22.
11. Letpak: the autoclave plastic box, *Emballering* (1980), 34.
12. F.A. Paine, Aseptic packaging, in *Modern Processing, Packaging and Distrinution Systems for Food*, F.A. Paine (ed.), Blackie, Glasgow (1987).
13. D. Rose, Aspects of aseptic technology, *Food Technology International Europe* (1988), p. 61.
14. F. Freeman, Ascending aseptics, *Packaging Rev.* **102**(3) 29–35.
15. E.J. Mann, *Aseptic Filling, Dairy Industry International* (1980), p. 37.
16. A. Iverson, Cartons for liquids, in *Modern Processing, Packaging and Distribution Systems for Food*, F.A. Paine (ed.), Blackie, Glasgow (1987).
17. O. Schraut, Aseptic packaging of food into carton packs, *Food Technology International Europe* (1989), p. 367.
18. K.J. Burgess, Dairy products, in *Food Industries Manual*, 2nd edition, M.D. Ranken (ed.), Blackie, Glasgow (1988).
19. K. Veal, Key stages in fruit juice technology, *Food Technology International Europe* (1987), p. 136.
20. A.J. Francis and P.W. Harmer, Fruit juices and soft drinks, in *Food Industries Manual*, 2nd edition, M.D. Ranken (ed.), Blackie, Glasgow (1988).
21. L. Alkskog and L.-G. Mejvik, Sterilization of food containing particulates, *Food Technology International Europe* (1990), p. 29.
22. Institute of Food Science and Technology, *Food and Drink Manufacture—Good Manufacturing Practice: A Guide to its Responsible Management*, 3rd edition, IFST, London (1991).
23. Guidelines for the Processing and Aseptic Packaging of Low Acid Foods CFDRA. Part 1 *Tech. Man.* **11** (1986).
24. J.P. Smith, H.S. Ramaswamy and B.K. Simpson, Developments in food packaging technology: Part 2, *Trends in Food Science and Technology* (1990), p. 116.
25. P.S. Richardson, Microwave heating—the food technologist's tool kit, *Food Technology International Europe* (1990), p. 65.
26. R.M. George and S.-A. Burnett, General guidelines for microwavable products, *Food Control* **2**(1) (1991), 35–43.
27. G. Pre, Packaging materials available in Europe for microwave ovens, paper presented at *4th International Conference on Microwave Food Packaging* (1989).
28. T. Ohlson, Controlling heating uniformity—the key to successful microwave products, *Eur. Food Drink Rev.* Summer (1990), 7.
29. H. Wilden, *Pack Marknaden,* **10** suppl. Livsmedel (1989), 3–4.
30. L.D. Brugh, Ovenable paperboard packaging, *Food Technology International Europe* (1989), p. 365.

31. P. Aggett, The heat is on, *Packaging Week* **5**(7) (1989), 18–19.
32. T. Faithfull, Ovenable thermoplastic packaging, *Food Technology International Europe* (1988), p. 357.
33. Anon, Fresh from the freezer, *Packag. Report* (1989), 48, 51, 52–56.
34. F.A. Paine, Food packaging for the 1980s, paper presented to *Packaging Index Workshops* (1979).
35. M.A. Jimenz and R.C. Griffin, The case of the retort pouch, paper presented at the *4th Annual IFT*, Atlanta, Georgia (1981).
36. WHO/FAO, *Food Irradiation: A Technique for Preserving and Improving the Safety of Food*, WHO/FAO (1988).
37. D. Robbins, *The Preservation of Food by Irradiation*, IBC Technical Services Ltd, London (1991).
38. K. Goodburn, Food irradiation legislation and consumer acceptability, *Food Sci. Technol. Today* **4**(2) (1990), 83.
39. *Irradiation in Food Regulations*, HMSO, London (1991).
40. K. Nielsen, Use of irradiation techniques in food packaging, in *Modern Processing, Packaging and Distribution Systems for Food*, F.A. Paine (ed.), Blackie, Glasgow (1987).
41. H. Stevenson, The practicalities of food irradiation, *Food Technology International* (1990), 73.
42. H. Stevenson, Detection of irradiated food, *Food Sci. Technol. Today* **5**(2) (1991) 69.

11 Dried and moisture sensitive foods

Reduction of available water

The essential feature of this method of preservation is that the moisture content, or water activity, of a foodstuff is reduced to a level below that at which microorganisms can grow. This reduction is accompanied by a corresponding reduction in the weight and volume of the food, rendering dehydrated foods particularly valuable when it is important to have the maximum amount of food available at the smallest weight and storage space. The bulk of some dehydrated foods can be further reduced by compression. By drying, foods are stabilized against attack by microorganisms, and enzyme and biochemical degradation is limited. Additional advantageous properties conferred on foods by drying are lowered transport costs, and the ability to keep longer without refrigeration.

The disadvantages introduced by drying are a slow deterioration by non-enzymic browning, and the susceptibility of certain foods to oxidative changes. These changes are initiated by the bringing together of complex biochemical components when normal cellular structures break down. The addition of sulphur dioxide can limit these chemical changes and oxidation can be reduced by the use of antioxidants. Scalding or blanching is essential for most vegetables to inactivate the enzymes which can cause deterioration in colour and flavour during drying and subsequent storage. Blanching also improves the texture of the final product and assists in the retention of certain vitamins. Although the prevention of water vapour transfer is an important aspect of packaging dried foods, the transmission of oxygen and sulphur dioxide must also be considered if quality characteristics are to be maintained.

The demand for dehydrated foods is increasing slightly. Sales of dehydrated foods in Europe are shown in Table 11.1 and consumption by country in Table 11.2 [1, 2].

Methods of drying

Sun or air drying. This traditional method of drying is limited to countries with a hot climate and/or a dry atmosphere, and to certain products, such as raisins, figs or prunes, which are spread out on trays and turned during drying. Considerable research into more sophisticated solar drying methods

Table 11.1 Sales of dehydrated foods (US$m) in Europe 1982 and 1987 [2]

	1982	1987[†]	% change
Dehydrated/convenience foods	1896	2140	+ 2.8
Dry beverages	4015	4456	+ 1.9
Dehydrated sauces and mixes	1167	1266	+ 1.5
Total	7078	7862	+ 2.1

Source: Frost and Sullivan, *Dehydrated and Powdered Foods in Europe* (1982).
[†]estimate

Table 11.2 Consumption of dehydrated foods [2]

	kg/annum	% total food market
UK	5.5	29
Germany (West)	4.8	39
Netherlands	4.7	5
Belgium	4.0	3
France	3.3	14
Italy	1.3	10

have been carried out, however, and will be useful in countries with adequate amounts of sunshine [3].

Mechanical drying. Methods of mechanical drying vary according to the materials being handled, but usually involve heated air (with controlled low relative humidity) or steam being passed over the food to be dried. Radiant heat or other forms of heat can also be applied to dry the product. There are numerous types of drier, and many variations on each type; the simplest is the evaporator or kiln, where the natural draught from rising air brings about the drying of the food. More common industrial methods include spray and roller drying, fluidized bed drying and extrusion or explosion puffing [4].

In *spray dryers*, a stream of finely dispersed droplets are injected into a stream of hot air. In *roller drying* a hollow rotating drum is heated internally by steam and the wet or liquid food is fed to the outside surface of the drum from which a doctor blade scrapes it off when dry. Typical conditions are temperatures of up to 135°C at 15 rpm. Newer methods under investigation include dielectric and microwave heating [3]. For some products two or more dehydration methods are used in combination.

Increasing the solids content. Drying here is accomplished by a reduction in the amount of available moisture in the food.

(a) *Increased sugar content.*   If the solids content of a food (e.g. jam and condensed milk) is increased by the addition of sugar, the amount of available water is reduced. The high sugar content and the low pH preserve the product after the pack has been opened and during its period of use. The effect of the high sugar content is twofold:

(1) It reduces the food's water activity (a_w) to below the level at which most yeasts or moulds grow
(2) The high osmotic pressure of the concentrated sugar solution results in the movement of moisture from the cells of microorganisms which are thus virtually dehydrated.

(b) *Increased salt content.*   Certain foods such as fish may be heavily salted, so that moisture is drawn from the flesh by osmosis and bound by the solute. In this form it is unavailable to microorganisms.

Vacuum drying.   Using hot water, steam or radiant heat the food is dried under vacuum. As the process requires lower temperatures than hot air drying, heat damage is minimized and oxidation during drying is eliminated. Vacuum drying can also be combined with freeze-drying [4].

Freeze-drying.   This method dehydrates the food by freezing and then subliming the ice formed. The ice is removed directly as water vapour under carefully controlled conditions of temperature and pressure, to leave a structure restorable to its previous state by the addition of water. The process enables perishable materials to be preserved for long periods at ambient temperatures when suitably protected from moisture ingress, light and oxygen.

There are several other methods of dehydration under investigation e.g. acoustic dehydration [3] but these have yet to find commercial use. More traditional processing methods such as baking also produce moisture sensitive foods.

Moisture levels of dried and moisture sensitive foods

Dried foods fall readily into categories based upon moisture content and these tend to define the packaging requirements of each group.

Group 1.   Instant beverages such as tea and coffee, freeze-dried and vacuum-dried foods contain very low levels of moisture in the range 1–3%. Although potato crisps, chocolates and some sugar confectionery such as boiled sweets contain very low levels of moisture in this range, for convenience their packaging requirements are described in the snackfoods (Group 2) and sugar confectionery (Group 4) categories respectively.

Group 2. Most dehydrated vegetables, herbs and spices, meat, some fish, milk, eggs, breakfast cereals, soup powders, some biscuits, snackfoods (e.g. most nuts) contain moisture in the 2–8% range. Tea and roast and ground coffee are also in this range.

Group 3. Dried fruits, cereals, cereal products, pulses, oilseeds, flours, some nuts and some baked products have moisture contents in the range 6–30%.

Group 4. Jams, jellies, some sugar confectionery and high moisture dried fruits usually have moisture levels of between 25 and 40%. Some dried and fermented meat (e.g. salami) and salted fish products also fall in this range.

Specifications for dried foods invariably state the residual moisture content present. Moisture level *per se* is only of limited use to the packaging technologist. Data which are more important to them include equilibrium relative humidity (e.r.h.) or water activity (a_w) and the sorption isotherm (see chapter 6).

General spoilage considerations of dehydrated foods

Dehydrated foods are susceptible to changes in their environment and if left in the open will first become stale and then as they absorb water will deteriorate microbiologically. Enzyme activity and oxidative rancidity will occur as the water activity increases [4].

Ideally dried foods should be barrier packaged in a vacuum or inert gas to keep out moisture vapour, air and light. However even in this ideal system browning will eventually occur, depending on the temperature, moisture level and reducing sugar content of the food. Bitter flavours and a toughening of the product texture will accompany these changes. Packaging and storage therefore are designed to delay these changes.

Under temperate conditions and when properly packaged and stored, dried foods have a shelf life of up to 2 years but this will be reduced to 3–6 months at 37°C. The lower the moisture content the longer the shelf life.

Oxygen scavengers

Dried foods are susceptible to deterioration caused by minor changes in moisture or oxygen level. Sometimes desiccants, or oxygen absorbents or scavengers, are sealed in with the foods. The packages for these scavengers must transmit gases and vapours but be sift-proof to avoid contaminating the food.

Several types of oxygen scavenger are commercially available [5]. For example, finely divided iron powder contained in a sachet which is highly permeable to oxygen and water vapour under appropriate humidity conditions will use up residual and incoming oxygen to form non-toxic iron oxide. Some scavengers also contain calcium hydroxide which absorbs carbon dioxide. This system has been used for ground coffee where carbon dioxide removal reduces the chance of a flexible package bursting.

Considerable research into other scavengers is being carried out and recently oxygen scavengers have also been used to protect baked products. For example the shelf life of white bread in polypropylene film can be extended from 5 to 45 days at room temperature by incorporating an oxygen absorbent into the package [5, 6]. Crusty rolls can also have their shelf life extended [5, 7].

Oxygen absorbents have also been used to prevent oxidative rancidity in high fat foods, such as nuts, crisps and beef jerky, and to prevent flavour and colour defects and control insect infestation in cereals.

These scavengers have some disadvantages. Apart from requiring the free flow of air around the sachet for maximum efficiency, they may also contribute to the growth of potentially harmful anaerobic bacteria.

Active packaging systems

Oxygen scavengers are one example of what has been termed 'active packaging systems'. Table 11.3 outlines the main areas where oxygen scavenging and the other active packaging techniques are employed.

Table 11.3

Process	Materials used	Products involved
Oxygen scavenging	Powdered iron Ferrous carbonate Iron/sulphur Metal (Pt) catalysts Glucose/oxidase enzyme Alcohol oxidase	Cookies Cured meats Pizza crusts Breads Rice cakes
Carbon dioxide scavenging	Iron oxide powder/lime Ferrous carbonate/metal halide	Coffee, fresh meat, fish
Release of preservative	BHA/BHT, sorbitol, Hg compounds, zeolites	Produce, meat, fish, bread, cereals, cheese
Ethanol emission	Alcohol spray, encapsulated ethanol	Cakes, buns, breads, tarts

Packaging requirements for different moisture levels

Group 1 (1–3% moisture). Dried foods which contain only 1–3% moisture have e.r.h. values which are below 20% and may even be below 10%. As the humidity of ambient air is rarely in this low range, such foods absorb water vapour freely from the surrounding air. These foods often have a porous nature and a high surface-to-weight ratio, thereby increasing their hygroscopicity. Uptake of water vapour may be so rapid that the packaging operation must be carried out in an air-conditioned room in which a low relative humidity is maintained. Some very low moisture dried foods are highly susceptible to oxidative deterioration, and must be packaged in containers which are impermeable to water vapour and to oxygen. Glass jars hermetically sealed with aluminium foil under the cap, lever lid cans and heat-sealable multiwall laminates with an aluminium foil layer are generally used. They are often packaged initially under a vacuum or in an inert gas. These products are relatively expensive and can tolerate a high packaging cost. If inadequately packaged, they will not reconstitute and lose much of their volatile flavouring components.

Group 2 (2–8% moisture). Dried foods in the 2–8% moisture range have e.r.h. values which fall in the 10–30% region. Their packaging requirements are not quite as exacting as those for foods included in Group 1, and although deterioration is similar it is usually at a faster rate.

In bulk, these dried foods are frequently stored and transported in lever lid containers of 20–25 litre capacity or in open-top drums containing up to 200 litres which can be hermetically sealed. For smaller packs triple laminates consisting of polyethylene, aluminium foil and paper are used as well as flexible films with very low water vapour transmission rates (WVTR).

Most dried foods in this category are better packaged in an inert gas but this is not common commercially. Dehydrated meats exposed to the air develop rancid flavours although this can be minimized by the presence of antioxidants. Oxidation also leads to a bleaching of meat pigments. Dehydrated fish, on the other hand, when exposed to air, gradually develops a strong 'fishy' odour which ultimately becomes ammoniacal and affects the flavour.

Dehydrated vegetables are frequently packaged in laminated sacks or polyethylene bags. Although these packs give adequate protection against moisture intake for relatively long periods, they do not protect against oxidation and additional protection is required for oxygen sensitive products. Packaging is often done in air-conditioned premises. Dried vegetables generally contain between 500 and 2000 mg/kg of sulphur dioxide, and the packaging materials used should prevent serious loss of this preservative.

Dried meat and vegetables for catering are often nitrogen flushed in large cans and at retail level, tear top aluminium cans, foil-sealed glass jars or composite containers such as lined paperboard drums with metal or plastic ends are common, but more generally hard plastic tubs with sealed foil tops or heat-sealed flexible pouches are used.

Dried herbs and spices and a few dried foods such as onion, garlic and other seasoning mixes have additional requirements because of the volatile components which may interact with plastic packaging materials. In finely divided powdered form these products are very hygroscopic and must be hermetically sealed, in a total barrier against water vapour and oxygen. In bulk they are packed in plastic buckets, multiwall sacks or drums. At retail level, glass or plastic shaker dispensers are common but once opened the dispenser cannot be perfectly sealed again, e.g. polyethylene terephthalate containers offer clarity and barrier properties when sealed with a polypropylene lid [8].

Dehydrated soups require thorough protection from oxygen and moisture as well as from loss of volatile flavourings. As these products are often mixtures of ingredients, the requirements of the most sensitive component are of primary importance. Products with relatively high fat content must also be protected from light and oxygen and from interacting with the packaging material. Most of the small retail packs for soups are therefore made from heat-sealable laminates containing an aluminium foil barrier. In-package desiccation is not practical but inert gas packing is used. For catering packs, cartons containing paper or laminated bags or large cans with resealable lids are common.

Dried milk powders are produced by roller drying, spray drying or vacuum roller drying but the first two are more common. Spray drying produces a better quality product. The creamy colour, granular and almost crystalline form and tallowy or 'oxidized fat' odour are a characteristic of roller dried powder. A milder heat treatment is used to produce spray dried powder which is a fine white powder with a smaller particle size and which reconstitutes more easily. However, roller powder has the better bacteriological quality due to the more rigorous heat treatment. The packaging of dried milks therefore must take account of their sensitivity to moisture, oxidation and bacteria. Even though bacteria such as *Streptococcus thermophilus* may be completely destroyed by the heat treatment, the products of their metabolism, such as lactic acid, remain undestroyed. If the moisture content is below 3%, bacterial growth may be ignored and the only serious deterioration is due to slow oxidation of the fat, which can be delayed by holding in an oxygen-free atmosphere, by coating with an impermeable material by gas packaging or by the addition of antioxidants. Drying milk therefore produces a material of small bulk and long shelf life. Materials used in packaging dried milks (including skimmed milk powders and other more specialized powders such as caseinates, coffee whiteners, infant formulae and yoghurt powders) are varied and depend upon the degree of protection required as well as the presence of other ingredients.

Bulk supplies of dried milks can be packed in 25 kg paper sacks with 50–200 μm polyethylene liners which may be closed with ties or better heat-sealed. Retail packs are generally composite paperboard containers, aluminium cans with resealable plastic lids or heat-sealed aluminium foil/paper laminates. Cartons and plastic bottles are also common. For products which require additional protection, such as infant formula, aluminium cans or laminated aluminium sachets contained in cartons overwrapped in a clear film are used.

Eggs are usually pasteurized before spray drying. Moisture and micro-biological contamination are therefore both important. *Dried egg* is supplied to the food industry in bulk in polyethylene laminated paper sacks and is sold at retail level in paper sachets contained in cartons or paperboard drums with reclosable lids.

The keeping quality of *breakfast cereals* [9] depends largely upon the quality and quantity of the fat. Products made from cereals which have a low oil content (wheat, barley, rice, maize grits: oil content 1.5–2.0%) have an advantage in keeping quality over products made from oats (oil content 2–11%). The stability of the fat depends upon its degree of unsaturation, the presence or absence of antioxidants or pro-oxidants, time and temperature of the treatment, the moisture content of the material when treated and the conditions of storage.

As breakfast cereals are often quite brittle, protection from mechanical damage is necessary. Common packs for breakfast cereals, such as cornflakes, are lined paper or pearlized polypropylene or other plastic film bags, selected for their moisture and grease resistance, contained in an outer printed carton. Resistance to flavour loss and grease can also be achieved using vegetable parchment or glassine. Wet strength was once and still is occasionally provided by waxing. Plastic bags and lined cartons are also used to give barrier to moisture for oats and mueslis.

The relative humidity of freshly baked *biscuits* [9, 10] is very low, and hence to prevent the rapid uptake of moisture from the atmosphere, packaging with good water vapour resistance is required. The textural characteristics of biscuits are highly sensitive to moisture changes. Packaging must therefore take place as soon as possible after baking, and the sealing of the package is a key factor in determining its efficiency. Since biscuits also contain fat, which is sensitive to oxygen and odours, a non-tainting material with good oxygen- and grease-resistant properties is necessary. Biscuits are brittle and the packaging must provide mechanical protection. This may be achieved by wrapping the biscuits together with the right degree of tightness, which provides a mutual reinforcing effect. There are various kinds of packs [11] (see Figure 11.1): small nibble size products are often packed in sealed bags. Larger biscuits are piled up in a column (stack pack) or in smaller piles side by side (pile pack). Flexible films are now the most common wrapping material with coatings on the surface of the film for moisture barrier and sealability including metallized foil wrappings with tear strip openers [12, 13]. Outer casings are often made of corrugated paper to give protection during distribution.

Chocolate coated biscuits require less moisture protection and packaging

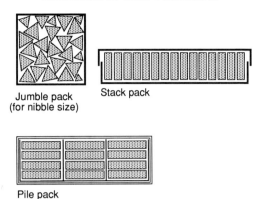

Jumble pack
(for nibble size)

Stack pack

Pile pack

Figure 11.1   Package styles [11].

Table 11.4   Snackfood sales (US$m)[a]

	Potato crisps		Savoury snacks		Baked		Nuts	
	1982	1992[a]	1982	1992[a]	1982	1992[a]	1982	1992[a]
Benelux	36	120	18	75	15	50	22	54
France	64	130	39	115	118	245	105	260
Germany	186	280	78	190	97	220	104	230
Italy	50	150	43	110	71	120	50	85
Netherlands	55	95	67	120	24	48	57	125
UK	681	1290	236	485	84	200	158	250
Total	1072	2065	482	1095	409	883	496	1004

[a] Forecast.
Source: Frost and Sullivan Report E 1005.

is often in aluminium foil laminates. Tins and rigid plastics boxes are used for quality assortments.

The *snackfood* market in Europe is expanding as shown in Table 11.4 [14]. The term 'snackfoods' means the more conventional products such as potato crisps, corn tacos, savoury and extruded snacks (made from maize or a maize/potato base fried, or dried and flavoured), nuts (uncooked, salted, flavoured, dry roasted, coated and mixed with dried fruits) and other salted, flavoured savoury nibbles. Many of these foods have a moisture content of about 5–6% and are often high in fat and therefore susceptible to both moisture uptake and oxidation.

Packaging requirements [15] will obviously depend upon the specific requirements of the product and generally packaging takes place as soon as possible after the final stage of processing. Moisture changes affect the texture of the product and will eventually result in enzymic and microbial activity. Packaging is usually required to give a 12 week shelf life. By then the product will have spoiled and would not be edible. A particular problem for fried or baked snacks is oxidation which occurs as a result of residual oxygen in the pack reacting with the fatty constituents of the food which goes rancid. Oxidation can also occur as a result of free radicals from the effects of UV light. Antioxidants can be used but increasingly both effects are controlled by careful selection of packaging materials and the use of gas packing.

Snackfoods are often fragile and therefore susceptible to damage during transport. The use of flavour materials such as salt and spices also means that the packaging must control aroma and flavour loss. Package seals are therefore important.

Crisps and savoury snacks require packs which will prevent the intake of both oxygen and moisture. Laminated bags of plastic, foil or paper are used. Larger products can be stacked like biscuits or packaged in paperboard drums with resealable plastic or metal lids. Metallized films provide light and gas blocking properties and give good visual impact.

The roasting of nuts extends shelf life by destroying lipolytic enzymes and preventing the production of disagreeable flavours. The reduction of water content also slows down bacterial growth, but nuts should not be allowed to dry out. They are particularly prone to oxidation (because of their fat content) and to mould growth in high moisture conditions. Nitrogen flushed tin plate or aluminium containers are often used for high quality products. Alternatively foil laminated bags provide a good barrier to keep moisture and oxygen out (see also chapter 3).

Coffee is principally traded in the form of green beans (12% moisture) and is commonly shipped in 60 kg bags [16]. *Roasted beans* and *ground coffee* have lower moisture content (4–5%) and are more susceptible to staling after exposure to air, losing flavour and aroma after only 1 day. Both beans and ground coffee are usually sold in vacuum or gas packaged cans or flexible pouches which give satisfactory oxygen and moisture barriers. Carbon dioxide is evolved after grinding and therefore any pack must avoid the build-up of gas or else the pack will burst. Vented bags or carbon dioxide scavengers must be used or packaging must be delayed until the carbon dioxide has been removed.

For bulk *tea*, the traditional foil-lined plywood tea chest remains a common form of packaging [16]. The aluminium foil is used as a moisture barrier. Attempts to use corrugated fibreboard cases have met with some success for bulk tea but are not in common use. Multiwall paper sacks laminated with foil and polyethylene are becoming more common, providing a superior moisture barrier and minimizing physical damage. Loose tea for retail sale is packaged in a multitude of different shapes, sizes and materials, including metal boxes with snap on lids, cartons with foil or paper laminated linings and sometimes overwrapped with films. Recently a 'fresh' tea has been launched in the UK. This is packaged in foil pouches which are hermetically sealed following nitrogen flushing to remove oxygen. Research has found that tea volatiles react most strongly with oxygen in the first 10 weeks after picking and therefore the tea is vacuum packed at source.

Group 3 (6–30% moisture). Cereals, pulses, some nuts, oilseeds and dried fruits are often held for long periods without packaging. This is possible because their e.r.h. values correspond roughly with the relative humidity of ambient air in the storage areas. During prolonged humid spells, however, unpackaged cereals and nuts, for example, may absorb sufficient moisture to support microbial growth. Moulds which produce aflatoxins are capable of developing on peanuts under humid conditions.

Cereals are the fruits of cultivated grasses, members of the family Gramineae. The principal crops are wheat, maize, rice, barley, oats, rye and millet. Bread, biscuits and cakes are all derived from these grains. Cereals are harvested, transported and stored in the form of grain. The ripe grain of the common cereals consists of nitrogenous compounds (mainly proteins), fat, mineral salts and water, small quantities of vitamins, enzymes and other substances, some of which are important in the human diet, and carbo-hydrates, which are quantitatively the most important constituents forming about 83% of the total dry matter. The situation for stored cereals, nuts and oilseeds is essentially the same as that for wheat although their critical moisture content is different.

Combine-harvested *wheat* is usually stored in bulk silos. The moisture content of the grain going into immediate storage should not exceed 15%; above 15% moisture, there is a risk of mould growth. Moreover, moisture is produced by the living wheat grains, which are continuously respiring. Respiration also generates heat which is not quickly dissipated because wheat is a poor conductor of heat. Respiration is relatively slow at 14% moisture content and 20°C, but accelerates as the temperature and moisture content increase. Enzyme activity and significant levels of fatty acid production have also been observed in wheat grain stored at high relative humidity or moisture content at ambient temperatures and this has been coupled with respiration [17]. Damp wheat (15.6–30% water) is an ideal substrate for fungi, but above 30% moisture content, bacterial growth also occurs which leads to rapid spoilage and the production of more heat, even charring the grain: the bacteria themselves are killed off at about 60°C. Insect life also becomes more active as the temperature rises, and because they respire, insects present in grain also raise the grain temperature. Damp grain can, however, be safely stored in airtight bins as the intergranular air is soon used up and respiration ceases. Grain so stored, however, is only fit for animal feed.

Whatever its initial moisture content, the grain will slowly come to equilibrium with the atmosphere, gaining moisture under conditions of high r.h. and losing it in a dry atmosphere. Regard to the prevailing atmospheric conditions in fixing safe moisture levels must be taken, remembering that ventilation in bulk bins is less than for sacked grain. When the grain arrives at the mill, it is conditioned to bring it to optimum moisture levels before milling into flour.

In the US *rice* has been packaged at retail level in resealable (Zip-Pak) packs which protect the product and which are convenient to use [18].

Flour made from wheat (and other grains) is stored in bulk bins or bags made of jute, cotton or paper. Heavy jute twill bags are returnable and must be cleaned before reuse; non-returnable bags made from paper (plain coated or laminated) are more expensive per trip, but their use reduces considerably any infestation hazard. However, storage of flour in bulk bins and delivery in bulk containers has many advantages over delivery and storage in bags. Although constructional costs of bulk storage facilities are high, the running costs are low because handling is very much reduced and warehouse space is more efficiently utilized.

The hazards to flour in storage are similar to those of grains, i.e. mould, bacterial attack and insect infestation, together with possible oxidative rancidity and tainting. Optimum moisture content for the storage of white flour from wheat is 10–13%. At moisture contents above 13% mustiness due to mould growth may develop, even if the flour does not become visibly mouldy. At moisture contents below 13% there is an increased risk of fat oxidation and development of rancidity catalysed by heavy metal ions such as Cu^{2+}.

Freedom from insect infestation during storage can only be assured if the flour is free from insect life when it is put into store and if the store itself is free from infestation. Good housekeeping in the mill and the milling of clean grain should ensure that the milled flour contains no live insect larvae or eggs. As a precautionary measure, flour is sometimes passed through a machine which consists of a rapidly rotating disc within a fixed housing. The flour is fed in centrally and is thrown with considerable force against the casing. At normal speeds of operation the machine is extremely effective for the destruction of all forms of insect life and of mites, including any eggs. The insect fragments, however, must be subsequently removed.

Self-raising flour is used for making puddings, cakes and pastry. The moisture content of the flour should not exceed 13.5% in order to avoid premature reaction of the leavening agents. Bulk flours are packed in fibre drums or multiwall sacks. Retail flours are packaged in paper bags, paper/plastic laminates, paperboard drums with reclosable lids or paper bags in cartons. Other flours such as *corn flour*, *soya flour* and *rice flour* have similar requirements.

Dried extruded products, such as *pasta* and *textured vegetable proteins*, are also susceptible to moisture pick-up. Packaging for pasta can be in as many forms as there are pasta shapes but plastic bags and paperboard boxes are usual. Textured vegetable products are packaged in paper or plastic bags. Flavoured products require additional protection from loss of flavour and are often packaged in paper, laminated bags or sachets.

The deterioration of *bread, cakes, pies* and other bakery products is called 'staling' and is due mainly to physical changes during storage which are not yet fully understood, although it appears that the mechanism for bread is slightly different to that for cake.

Bread leaves the oven with the crumb at a temperature slightly below 100°C and with a moisture content of about 45% at the centre. The crust is hotter and much drier (1–2% moisture) and cools rapidly. During cooling, moisture moves from the interior of the loaf outwards towards the crust and then into the atmosphere. If the moisture content of the crust rises considerably during cooling, its texture becomes leathery and tough and the attractive crispness of freshly baked bread is lost. The art of cooling is to lower the temperature without causing too great a loss in texture.

On ageing, the bread crumb becomes firmer and more opaque and the water-absorbing capacity is reduced. It is believed that this is associated with physical changes in the amylopectin fraction of the starch. The maximum staling rate occurs at 4°C, and reheating a stale loaf (to over 60°C) will reverse the process and improve the freshness of the loaf. Bread can be stored successfully for months in a frozen condition below $-15°C$, because at this temperature any change is very slow.

The chief types of microbial spoilage of baked bread are mouldiness and ropiness. The temperatures attained in the baking process are usually high enough to kill mould spores on the crust, so moulds must reach the outer surface or penetrate it after baking. They can reach the surface from the air during cooling, or thereafter from handling or packaging materials. Mould growth usually commences in the 'creases' of the loaf or between slices of sliced bread. The objective of bread wrapping is to keep the bread in a fresh condition for several days: it is carried out as a hygienic measure, to conserve moisture, and to prevent rapid drying out and staling. Barrier films, e.g. polyethylene bags, often perforated, are used to prevent moisture loss and staling.

Very similar considerations apply to *cakes* [11, 19, 20]. Most cakes have a fairly high e.r.h. and consequently tend to dry out fairly rapidly under normal conditions of storage. The staling of cake appears to have two causes: firstly, there is a movement of moisture through the cake, the dry cake crust attracts moisture and the moisture content decreases from 28% to 20%. At the same time, there is a 'bread-like' staling process ascribed to changes in the amylopectin fraction of the starch. The maximum staling rate of cake occurs between 20 and 25°C. Cakes come in a variety of shapes and sizes and packaging must protect the cake from damage and deterioration and present the product attractively.

There are many other types of *baked product*, e.g. pastries, pies, pizza, etc. with various proportions of the main ingredients, namely starch protein, fat, sugar and water. The ingredients, other than starch, will also affect packaging requirements but in general the primary function of packaging these products is to prevent drying out. However, the degree to which the drying out can be retarded must be limited by the need to prevent the growth of micro-organisms such as moulds and yeasts which will have contaminated the product between baking and packaging. The higher the e.r.h., the greater will be the tendency to dry out, and the greater the potential for mould growth. A controlled rate of moisture loss is therefore required. This is usually achieved by applying a wrapper which is sufficiently permeable to water vapour to allow the surface of the product to dry to a level below that required for mould growth within a day or two. Carbon dioxide flushed packages have also been used to retard the rate of microbial spoilage of baked goods. However, each product must be considered on its merits to determine the best form of package.

Also in food manufacturing, edible packaging films based on, for example, starch, have been developed to prevent fruit toppings seeping into pastry bases [21].

The principles of gas packaging of bakery products [5, 22] are very different from meat or fish. The gas composition required is different (90% carbon dioxide and less than 2% oxygen), less gas is required because the package tends to conform more closely to the product dimensions and the product is stored at ambient temperatures. A laminate of polyester, polyethylene and polyvinylidene chloride (PVdC) has been successful with the polyester providing strength and an aroma/flavour retention barrier and the PVdC and polyethylene providing gas and water vapour barriers and a heat-sealing medium. Metallized or partly metallized films have also been used to increase barrier performance. Oxygen scavengers can also be used in combination with gas packaging of baked products.

Some cakes, because of their high sugar/water ratio, are much less susceptible to mould growth than bread, but plain cakes with a low sugar/water ratio need some drying of the surface of the cake to avoid mould growth.

Dried fruits contain 17–20% moisture [4] and their e.r.h. ranges from 50–70%. Their high sugar content and, in the case of dried tree fruits added sulphur dioxide, inhibit microbial growth. Sulphur dioxide is used mainly to preserve colour by inhibiting non-enzymic browning which results when amino acids and reducing sugars react. Loss of this preservative by gaseous diffusion will shorten the storage life of the products. Thus dried fruits should be protected both during bulk storage and during retail distribution by packaging in material which has a low transmission to sulphur dioxide. Flexible film pouches, waxed paperboard drums and boxes are often used.

Group 4 (25–40% moisture). This category includes dried fruits such as prunes which are sold in flexible film pouches as *high moisture* packs. They contain 35–40% moisture and have e.r.h. values in excess of 80%. This renders them susceptible to attack by a range of yeasts and moulds. These microorganisms are controlled by hot filling the prunes into pouches at a temperature of 80–85°C. Hence, the package must be able to withstand this temperature, and it must also have a low water vapour transmission rate so that the packaged fruit does not dry out too rapidly.

The preparation of *preserves, jams* and *jellies* [23] goes back many centuries as a means of preserving fruit. The high sugar content and the low pH preserves the product after the package has been opened. Nowadays preservatives such as sulphur dioxide are also used but there are restrictions on their use in certain products. The most popular container for retail preserves is the glass jar, although cans, pottery and plastics containers are also used. To ensure protection from mould growth a hermetically sealed pack is obtained by using a lacquered metal closure to protect against corrosion. Jams for the food industry are usually hot-filled into large polyethylene lined paperboard cartons.

Jelly tablets which are susceptible to crystallization and mould growth are individually wrapped in film and may then be packed in individual cartons.

Throughout the world there are many different types of *dried meat* or *fish products*, e.g. beef jerky, sausages, rackling, spelding, etc. with varying moisture contents from 5 to 40%. These products are usually smoked or salted as part of the drying process or fermented. Packaging requirements will differ depending on the characteristics of the product, i.e. its moisture content, salt content, other compositional characteristics and will range from very simple wrappings to glass jars, cans, barrier films and vacuum packs (see also chapter 12).

Sugar confectionery [24–27] has four major ingredients: sucrose, glucose, invert sugar and water. Alone these make confections such as fondants and boiled sweets; add milk solids, fat, protein and modified starches and the

Table 11.5 Types of confectionery

Sugar confectionery	Possible ingredients
Boiled sweets	Sucrose; invert sugar, fructose, glucose syrup

Caramels, fudge, butterscotch	Sugars plus:
	   Sweetened semi-skimmed or condensed milk
	   Spray dried skimmed milk
	   Butter oil, vegetable fats and oils
	   Lactose, whey powder

Fruit pastilles, gums, jellies	Sugars, etc. plus: Hydro colloids, e.g. edible gums, gelatin and starch, etc.
Marshmallows, nougat	Incorporate a whipping agent such as egg albumen, soya protein or hydrolysed casein to give a light texture

Chocolate confectionery	The fat phase is mainly cocoa butter through which the
	   other ingredients (sugar, cocoa and milk solids) are
	   dispersed as fine particles

collection enlarges to give products such as jellies, pastilles, toffees and caramels. Chocolates are essentially finely ground sucrose suspended in solid cocoa butter. A wide variety of confections may be obtained by combining these materials (Table 11.5).

Confectionery as usually prepared may have moisture content from about 23% to almost moisture-free (jellies to chocolates). The 'mouth feel' of the confection increases in hardness (in the broad sense of the word) as the moisture content decreases. This rule, however, is relatively easily altered by manipulating the ingredients.

How confections deteriorate. The sugar component of most confections is in a soluble form, and it is important that it stays in this form since crystallization of the sugar is one of the major defects of confectionery. A confection that has a sugar/water ratio that under normal temperatures and conditions promotes sugar crystallization must be kept in the same condition as at manufacture. For example, toffees and boiled sweets, which can best be described as supersaturated, supercooled sugar syrups, rely on their viscosity to prevent sucrose crystallization. If moisture is picked up from the atmosphere, however, the surface viscosity will break down quickly and crystallization will take place.

Confectionery deteriorates when exposed to heat, moisture change, light, moulds, yeasts, odours and mechanical damage [24]. The principal factor which requires control is moisture, and Table 11.6 gives some indication of the type of deterioration and the equilibrium relative humidity above or below which this may occur.

Table 11.6 Effect of humidity on various types of sugar confectionery

Type of confection	Deterioration caused	e.r.h.
Boiled sweets	} Graining and stickiness	< 30
Toffees		< 30
Gums and pastilles	} Stickiness, growth of moulds and yeasts	50–65
Liquorice paste goods	}	55–65
Turkish delight		60–70
Fruit jelly goods	} Fairly stable at ambient conditions	60–70
Creampaste goods		65–70
Marshmallow		65–75
Marizpan	} May dry out or grow mould	70–85
Fondant cream		75–85
Jam		75–85
Milk chocolate	} Syrup forms on surface, sugar bloom	~ 80
Plain chocolate		~ 85

If the confectionery contains fat, then rancidity can occur as a result of oxidation. Flavour reversion occurs when oxidative off-flavours are thought to be produced by lipid hydroperoxides or because there is no natural antioxidant.

Lipolytic or hydrolytic rancidity from enzymes in ingredients such as milk, cocoa powder and nuts, and microbiological rancidity can be a problem, especially in high milk containing products, although it is not very common. Fat-containing confectionery will also react with moisture to produce fatty acids and their degradation products.

Odour absorption by oils and fats can also be a problem. Odorous compounds are often very soluble in oils and fats and can be readily absorbed from materials such as paints, printing inks, petroleum oils and disinfectants. When the product is eaten the odours are released in the mouth producing objectionable flavours.

Determining the protection required. The type of packaging which will protect the confection is determined from the equilibrium relative humidity at which the product neither gains nor loses moisture. Below this value, moisture is liable to be lost, and above it the product will pick up moisture. From the ingredients in mixed confections the e.r.h. can be calculated. Table 11.7 gives a comparison between some experimental and calculated values. To complicate the issue even further, in multicomponent confections moisture is transferred from one component to another, as well as the possible overall loss or gain of moisture from the atmosphere. Table 11.8 illustrates the point.

The different types of wrapping available allow varying amounts of water

Table 11.7   Equilibrium relative humidity of various kinds of confectionery

Type of product	Average values	e.r.h. of typical specimens	Usual limits
Jam	77 (78)[a]		75–82
Fruit jellies	76 (75)	59, 67, 72	59–76
Turkish delight	66 (64)	65, 70, 61, 64	60–70
Marshmallow	72 (72)	69, 68, 71	
		69, 68, 71, 66, 82, 63	63–73
Liquorice	64 (65)	66	
		65, 55, 53, 56, 62	53–66
Gums and pastilles	60 (55)	61, 64	
		60, 62, 61, 53, 53, 57	51–64
Toffee	47		Below 48
Boiled sweets	28	30, 39	Below 30
Fondant cream	–		75 to 84
Marzipan	68 (69)	83, 77[b]	Approx. 70
Cream paste	65 (65)	67, 68, 68	65–70

[a] Values in parentheses are calculated values.
[b] For use by bakers.

Table 11.8 Moisture migration tendencies in mixed confections

Components	e.r.h. (%)	Direction of migration of H_2O	Moisture content (%)
Jelly	77		20.8
Fondant	83	↑From fondant to jelly	12.0
Marshmallow	80		22.5
42 De glucose syrup		↓ From marshmallow to jelly	
Jelly	77		20.8
Marshmallow	76		22.5
63 De glucose syrup		↑↓ May move either way	
Jelly	77		20.8
Marshmallow	77		22.5
63 De glucose syrup		↓ From marshmallow to caramel	
Caramel	73		13.0
Nougat	59	↑↓ May move either way	9.5
Toffee	59		9.4

vapour transmission. Nylon and cellulose films show significant water vapour transmission. Hermetically sealed containers may also be ideal because an equilibrium between free water in the confectionery and the water in the low volume of air will be rapidly established giving a stable atmosphere [27].

The ideal film wrapping may not be one that completely protects the confectionery from the atmosphere. Preventing total movement of moisture also means condensation will occur if the product is not sufficiently cooled before packing or if it is subjected to differences in temperature during storage or distribution. Water will therefore be trapped in the wrapped product which will affect its quality. Fluctuations in temperature are often unavoidable and so the quality of confectionery is best served by wrapping in a film that allows some water movement.

Chocolate can be affected by condensation giving sugar bloom, in which a fine layer of sugar crystals forms on the surface of the product. This renders it unsaleable and if left unchecked will lead to mould growth.

Many types of packaging are employed for sweets; the amount of protection necessary varies according to the type of goods. Boiled sweets, at one end of the moisture scale, are protected by packaging with individual wrappers in sealed containers (metal, glass or foil laminates), toffees are wrapped in waxed paper or film, and so on. Individual sweets may be packaged in twist wraps in film, foil/film bunch wraps, or wrappers for bars in papers, films and foils. Multiple packages may consist of roll wraps, bags, pouches, sachets, strip packing, cartons (with bags inside and with overwraps as well as on their own), rigid boxes, glass jars, metal cans and plastics containers. This assortment of packaging can provide the properties required for most types of sweets and also permit an attractive surface and appearance.

References

1. Frost and Sullivan, *Dehydrated and Powdered Food in Europe* (1982).
2. J. McKeon, Market trends in dehydrated foods, in *Concentration and Drying of Foods*, D. MacCarthy (ed.), Elsevier Applied Science, London (1986).
3. J.G. Brennan, Dehyration of foodstuffs, in *Water and Food Quality*, T.M. Hardman (ed.), Elsevier Applied Science, London (1989).
4. J.N. McN. Dalgleish, Dehydration and dried products, in *Food Industries Manual*, M.D. Ranken (ed.), Blackie, Glasgow (1988).
5. J.P. Smith, H.S. Ramaswamy and B.K. Simpson, Developments in food packaging technology Part II: storage aspects, *Trends Food Sci. Technol.* **1**(5) (1990), 114–116.
6. Mitsubishi Technical Information, *Techniques for the Preservation of Food by Employment of an Oxygen Absorber*, Mitsubishi Gas Chemical Co. Ltd., Japan (1983).
7. J.P. Smith, B. Ooraikul, W.J. Koersen, E.D. Jackson and R.A. Lawrence, *Food Microbiol.* **3** (1986), 315–320.
8. Anon, PET spice jar has tamper evident seal, *Packaging News*, July (1989), 38.
9. C.R. Oswin and L. Preston, *Protective Wrappings*, Cann Publications, London (1980), p. 60.
10. F.A. Paine, Problems with the flexible packaging of biscuits, PIRA Report PK8(R)(1977).
11. G.M. Townsend, Cookies, crackers and other flour confectionery, in *Snackfood*, G.R. Booth (ed.), Van Nostrand Reinhold, New York (1990), p. 41.
12. Anon, Foil with tear strip is biscuit first, *Packaging News*, March (1989), 100.
13. D. Manley, *Technology of Biscuits, Crackers and Cookies*, Ellis Horword, London (1991).
14. D. Blenford, Satisfying a growing appetite for snacks, *Food Technology International Europe* (1990), p. 145.
15. E.M.A. Willhoft, Packaging for preservation of snackfood, in *Snackfood*, G.R. Booth (ed.), Van Nostrand Reinhold, New York (1990) p. 349.
16. D.J. Millin and D. Cruikshank, Hot beverages, coffee, tea, cocoa and others, in *Food Industries Manual*, M.D. Ranken (ed.), Blackie, Glasgow (1988).
17. J.E. McKay, The behaviour of enzymes in systems of low water content, in *Dehydration of Foodstuffs in Water and Food Quality*, M. Hardman (ed.), Elsevier Applied Science, London (1989), p. 169.
18. Anon, Rice in resealable zip-pak, *Prep. Foods* **158** (1989), 173.
19. J.A. Cairns, C.R. Oswin and F.A. Paine, *Packaging for Climatic Protection*, Newnes Butterworth, London (1975), p. 83.
20. R.C.F. Guy, 1980, Recent research into cake making, paper presented at FMBRA Open Day.
21. Anon, Edible and protective, *RIA* **417** (1983), 35–36.
22. R.A. Inns, Modified atmosphere packaging, in *Modern Processing, Packaging and Distribution Systems for Food*, F.A. Paine (ed.), Blackie, Glasgow (1987).
23. R.W. Broomfield, Preserves, in *Food Industries Manual*, M.D. Ranken (ed.), Blackie, Glasgow (1988).
24. C.R. Oswin and L. Preston, *Protective Wrappings*, Cann Publications, London (1980), pp. 34–38.
25. S.H. Cakebread, *Sugar and Chocolate Confectionery*, Oxford University Press (1975).
26. B.W. Minifie, *Chocolate, Cocoa and Confectionery: Science and Technology*, 2nd edition, AVI, New York (1980).
27. B. Brockway, Applications to confectionery products, in *Dehydration of Foodstuffs in Water and Food Quality*, T.M. Hardman (ed.), Elsevier Science, London (1989).

12 Other processed foods

Salting, drying (see chapter 11), smoking, curing and fermentation, either singly or in combination are traditional methods of preservation. While the shelf-life is prolonged by these methods the food itself is changed resulting in characteristic flavours, aromas and textures.

Temperature, water activity (a_w) and pH are important criteria in determining the types of microorganisms that can grow on food and these can be altered artificially to prevent or delay spoilage. The pH can be artifically lowered by the addition of acids or through fermentation, which produces lactic or acetic acids. Other deteriorative changes such as oxidative rancidity can be altered by the addition of antioxidants.

Preservation by chemical means

Classes of chemical preservative

Chemical preservatives are used to inhibit spoilage agents and to complement other food preservation techniques. Their effectiveness depends on the concentration of the chemical, the nature, number, age and history of the spoilage organisms, the processing temperature and time, and the chemical and physical characteristics of the food. There are several different classes of preservative and their use in food is controlled by law [1].

Inorganic chemicals

These include inorganic acids and their salts, alkalis and alkaline salts, halogen peroxides and gases. Some examples of the most commonly used are (i) sodium chloride in brines and curing solutions, (ii) hypochlorites, usually calcium or sodium, used in the treatment of drinking water, for processing, etc. and (iii) nitrites and nitrates used in the curing of meats, e.g. bacon.

Organic acids and their salts

These include acids such as lactic, acetic, propionic and citric acids and their salts. Also of importance is benzoic acid which can be added to a variety of foods, but in particular fruit drinks: sodium benzoate exerts its effect by increasing the hydrogen ion concentration in foods, but the concentration of the undissociated acid is probably of greater importance. An alternative is sorbic acid which has a similar effect and is used for products such as wrapped hard cheeses, delicatessen products, margarine and fat spreads. Sorbic acid has also been used to extend the shelf life of cakes, jellies and pickles. Potassium sorbate is also used for meat products, fruit products, wine, pickles, etc.

Antioxidants

Oxidative rancidity occurs in fatty foods and causes objectionable odours and flavours and possibly deleterious nutritional effects. Antioxidants retard the oxidative deterioration of fats by capturing peroxy free radicals and inhibiting auto-oxidation. Tocopherols are examples of antioxidants which occur naturally in fat but some fats are deficient in such natural protection and need an added antioxidant. However, the addition of antioxidants is strictly limited by law [2] and they must be toxicologically acceptable. Sometimes two or more substances act together in such a way that the combined effect of the two is greater than the sum of the individual effects (synergism), e.g. citric acid acts synergistically with the phenolic compounds in oils due to its ability to inactivate metals such as iron.

Antibiotics

In general, the use of antibiotics in food processing is prohibited by legislation in practically every country. The risk of unintentional sensitivity, the reduction in incentive towards food hygiene and the possibility of the development of resistant organisms in the human body are the reasons for this prohibition. However, one antibiotic which is acceptable as a food preservative in a number of countries is nisin. Produced by a number of cheese starter organisms, e.g. *Streptococcus lactis*, nisin is also naturally present in certain cheeses. However, not all pathogenic organisms which are found in foods are sensitive to nisin (e.g. *Clostridium botulinum*) and therefore the indiscriminate use of nisin to reduce the times and temperatures of processing of certain canned foodstuffs may increase the risk of botulism. In the UK nisin is permitted in cheese, clotted cream and canned food [1]. The latter is defined as 'food in a hermetically sealed container which has...been sufficiently heat processed to destroy all *Cl. botulinum* in that food or container and which must have pH of less than 4.5'. Cured pig meats require less thermal processing and therefore may not contain nisin, as the process would be below the minimum required.

Cured and smoked foods

Curing and smoking can be considered as a combination of drying and chemical preservation. They are often used in conjunction with refrigerated distribution.

Nitrite and nitrate curing

Nitrites and nitrates are used in the curing of meats, primarily to fix a desirable red colour. However, nitrites also have a bacteriostatic effect in acid solutions. Pickling of pork is carried out with a solution of common salt and sodium nitrate, and occasionally sugar, at a pH of 6.5. After treatment with the pickle, maturing takes place. During maturation, the salt, nitrate and nitrite produced and the sugar, if used, gradually diffuse through the meat, so that the concentration tends to become more uniform. During this process the appearance of the meat alters; the colour of the flesh changes from the dull pink of pork to a reddish tint of bacon, and the formation of nitrite continues. At the same time the characteristic flavour of bacon is developed.

Colour and the curing process

Colour is an important objective in curing meats: the intensity and stability of the colour are very important in prepackaged cured sliced meat. The colour of the cured product is governed by:

(a) The quality of the meat
(b) Ratio of fat to lean
(c) Temperature of curing
(d) Curing ingredients and the formula used
(e) Curing techniques employed.

The stable red or pink colour is produced by a series of chemical changes which depend upon the reaction of nitric oxide (NO) with myoglobin to produce nitrosomyoglobin, which is a pink-red pigment. To obtain nitric oxide, sodium or potassium nitrate and/or sodium or potassium nitrite are added to the curing mixture. To become effective, the nitrate must first be reduced to nitrite and this can be brought about by bacterial reactions:

$$\text{Sodium nitrate } (NaNO_3) \xrightarrow[\substack{\text{bacterial} \\ \text{reduction}}]{} \text{sodium nitrite } (NaNO_2)$$

As curing schedules become faster, the direct addition of nitrite without nitrate is becoming more popular. The nitrite is converted to nitrous acid (HNO_2) and finally to nitric oxide (NO). Low pH, ascorbic acid and other reducing conditions accelerate these reactions.

$$\text{Sodium nitrite} \xrightarrow[\substack{\text{low pH} \\ \text{ascorbic acid}}]{} \text{nitrous acid} \xrightarrow[\substack{\text{low pH} \\ \text{ascorbic acid}}]{} \text{nitric oxide}$$

$$(\text{NaNO}_2) \qquad\qquad\qquad (\text{HNO}_2) \qquad\qquad\qquad (\text{NO})$$

The nitric oxide then reacts with myoglobin to produce nitrosomyoglobin (also called nitric oxide myoglobin):

$$\text{Myoglobin} + \text{nitric oxide} \longrightarrow \text{nitrosomyoglobin}$$

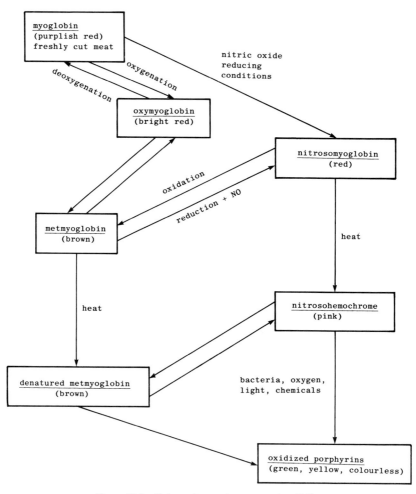

Figure 12.1   Colour changes in meat curing [33].

The nitrosomyoglobin may be oxidized to an undesirable brown pigment, metmyoglobin. If the curing process is halted at the nitrosomyoglobin stage, then oxygen-impermeable packaging material is required to prevent the formation of metmyoglobin. With heating, the nitrosomyoglobin is converted into a pinkish-red pigment, nitrosohemochrome [3] (also called nitric oxide denatured myoglobin) (see Figure 12.1).

The oxidation of nitrosomyglobin and nitrosohemochrome is facilitated by exposure to light, which presents a major problem for the display of cured meats in illuminated cabinets. An alternative to the exclusion of light for preserving the colour is to exclude oxygen, and this forms the main basis for modern practice in the packaging of cooked cured meats. Dry curing, using a dry cure rather than brine, is also used for some meat products, such as parma ham.

Bacterial spoilage

Spoilage of cured meats is related to the curing process. Nitrates inhibit growth and spore formation of anaerobes, but there is also a tendency for lactic acid bacteria to contaminate the product causing discoloration.

Packaging of cured meat products can have a significant effect upon the keeping quality of the product. Shelf life can be improved because the packaging material may prevent the growth of moulds and provide a barrier to prevent continued contamination when the product is handled (where the product is cooked in an impermeable casing, shelf life is greatly increased as the product cannot be contaminated after cooking). The development of a microclimate around the product can also have a selective and/or inhibitory effect on microorganism growth.

As with fresh meats, the organoleptic properties and the moisture relationship of the product can be affected by the type of packaging material and/or system used with cured meat products. Plastic barrier bags are often used but it is important to select the correct packaging material because for some products permeability of the casing is required as a requisite part of the process, for example, in the drying of salami.

Bacon is vacuum-packed mainly to preserve its colour. However, the growth of spoilage bacteria on bacon is inhibited by vacuum packaging resulting in a long shelf life which may finally be limited by the growth of lactic acid bacteria which cause souring. Other types of spoilage, such as the production of cabbage-like odours by *Proteus inconstans*, may occur in bacon from time to time. The spoilage rate is mainly dependent on the storage temperature, but is also affected by the levels of curing salts in the bacon.

Meat spoilage due to bacterial growth is generally a surface phenomenon. When a joint of meat is cooked virtually all the bacteria on the surface of the joint are killed. However, the meat may be recontaminated with bacteria during slicing and vacuum packing. Vacuum-packaged cooked meat cannot therefore be treated as a shelf stable product, and must be stored under refrigeration. In fact refrigerated storage of vacuum packed meats is imperative to prevent the growth of food poisoning bacteria should these contaminate the meat prior to packaging. Lactic acid bacteria will usually grow on vacuum packaged cooked meats and will limit their shelf life by souring. Good hygienic practice in slicing and packaging the meat, allied to low temperatures of storage, is essential.

Cured fish

Only a small amount of fish is now salt-cured in the UK, e.g. dried salt cod and pickled herrings, but in other countries salted fish is more popular. Fatty fish cannot be dry-salted because the fat easily becomes rancid during exposure to air, but they are immersed in a saturated solution of brine pickle where they are protected by a liquid barrier. These products are generally sold in glass jars or cans which are not sterilized and therefore require refrigeration. Bacterial spoilage can occur from halophilic (salt-tolerant) bacteria.

Smoking

The smoking of foods usually has three main purposes; to add desired flavours, to inhibit microbial growth and to retard fat oxidation. Other desirable effects may also result, such as an improvement in colour and appearance, and a tenderizing action. Apart from the preservative action, the characteristic taste and aroma of smoked foods are imparted by the creosote vapours distilled from the burning sawdust. The most important features of smoking are:

(1) The removal of moisture
(2) Penetration of the acidic vapours from the smoke into the meat
(3) The attainment and control of the critical temperature to avoid scorching and fat loss.

Traditionally, the smoke is obtained by burning wood chips and/or sawdust—preferably from hard woods, like hickory and oak—and it contains a large number of volatile compounds that differ in their bacteriostatic and bacteriocidal effects. Smoking can be carried out hot or cold. In cold smoking the temperature of the smoke is no greater than 30°C and the food remains uncooked. In hot smoking the temperature is raised to 70°C or above and the food is cooked. Wood smoke, usually at a temperature of 43–71°C is more effective against living bacteria than against bacterial spores.

An alternative to smoking is the use of smoke concentrates and essences which are deposited onto the product surface when it is electrostatically charged.

Smoked meat.   Smoking of meat [4, 5] is often combined with drying, curing, fermenting or salting, such as in sausages and hams. Products like biltong, jerky or pemmican are dried and smoked and because of their reduced weight have been an established means of storing and transporting meat in rugged conditions. Smoking provides some preservative action for meats, but refrigeration is often needed. The meat is often first treated with a cure, by injection, and then enclosed within a waterproof casing and pasteurized before cooling and slicing. The slices are then smoked and dried simultaneously and vacuum packed, which protects against microbiological contamination and fat oxidation.

Smoked fish.   The keeping qualities of smoked fish [6, 7] depend on the degree of drying and smoking and the temperature at which it is stored subsequently. Refrigeration is required or spoilage will occur in a few days. The first signs of visible spoilage are mould growths introduced with the sawdust. The flesh begins to taste sour or the smell of ammonia becomes noticeable. With fatty fish the oil becomes rancid.

Fish to be preserved by smoking are split, cleaned and brined, and hung in the smoke from smouldering fires of wood shavings and sawdust. The preserving effect is due to a partial drying resulting from evaporation of moisture from the surface of the fish and thermal breakdown of the cellulose and lignin fraction of the wood giving rise to simultaneous precipitation of vaporous aliphatic and aromatic chemicals. During the process the fish is toughened, coloured and acquires its characteristic taste and smell. Products such as smoked salmon require a high salt content, which is incorporated by dry salting. In mildly smoked fish the 2–3% salt content required is obtained by immersion in a nearly saturated salt brine for varying periods of 1–30 min depending on the size of the fish.

The addition of permitted colouring and flavouring materials to smoked fish has now facilitated milder smoking and drying procedures so that smoked fish are almost as perishable as fresh fish. Packaging requirements are therefore very similar to fresh fish or frozen fish. Vacuum packaging of smoked salmon and kippers is common and kippers are also often marketed in the frozen state in consumer packs and 'boil-in-bag' packs. Here products are intended to be cooked before opening the bag with the advantage of confining the strong odours during cooking.

Fermented foods

Food fermentation is carried out to produce new and desired flavours and to change the physical characteristics of the food, hence producing a different food product, and/or to inhibit spoilage factors by the production of alcohol or acids, usually in combination with other methods of food preservation such as low temperature, heat, anaerobic conditions, or added salt, sugar or acid [8].

Acidity development plays an important part in the preservation of foods such as sauerkraut, pickles, green olives, fermented milks, cheeses, certain sausages and various fermented foods of plant origin. Full acidity development from the available sugar is required for foods such as pickles and green olives, or fermentation can be stopped before maximum acidity is attained by chilling or canning, as in fermented milks or sauerkraut. In drinks such as beer, ale, wine and fermented fruit juices, the growth of microorganisms is prevented by the low pH, the carbon dioxide and alcohol content, the presence of hop extracts (which have antiseptic qualities) and the low storage temperature.

However, spoilage of fermented foods can occur; the most likely causes are the wrong conditions during fermentation or the subsequent oxidation of lactic and other acids in the fermented product by film yeasts and moulds which then permit the growth of microorganisms that cause undesirable qualities.

Cheese

Cheesemaking was originally a seasonal industry; a proportion of the summer milk surplus in many dairy regions was converted to cheese. The basic processes in cheesemaking are coagulation, cutting and heating to produce a curd, which is then pressed and ripened to separate the whey. The curd is then manipulated to produce the desired characteristic of the cheese. The process is a microbiological one [9], in which enzymes produced by bacteria develop the desired flavour and texture. The commercial manufacture of cheese is changing rapidly due to new methods of production, and different starter organisms and variations on the manufacturing method are used for the many different varieties of cheese. The basic process for a hard cheese, such as cheddar, is as follows.

The milk is first 'flash' pasteurized (70°C for a few seconds or 66°C for 15 s) and soured by the addition of the starter organism (e.g. *Streptococcus lactis* or *Streptococcus cremoris*) which begins the development of lactic acid from the milk sugar, lactose. When the correct acidity has been reached, rennet is added which denatures the milk protein to produce a 'coagulum'. This coagulum is cut into pieces (cutting) and dried out by 'scalding'. Steam is used for this purpose and the temperature of the steam depends on the type of cheese to be produced. Scalding has the effect of drawing the cheese particles together and squeezing out the curd. Acidity, temperature and freeness of cutting accelerate curd formation.

The curd is then allowed to fall to the bottom of the vat and is heaped together (pitching) where it begins to mat. The whey then runs out or is pumped out of the vat. The combined effect of rennet, acid and heat changes the casein so that matting takes place more rapidly and the curd changes from a rubbery consistency to a dough-like state which is cut into small pieces (nulled). Salt is added to dissolve some of the protein and assist matting, as well as to control bacterial activity during ripening, and the pieces are packed into hoops or moulds where they are pressed into cheeses. Finally these cheeses are turned, washed, greased and bandaged and placed into store to ripen. The type and flavour of cheese can be attributed to the method of manufacture and type of starter bacteria used.

Manufacturing methods for mould ripened soft cheeses, such as Brie and Camembert, usually depart from this method at the pressing stage. Instead of pressing the curd is transferred to cylindrical perforated moulds where the whey is allowed to drain away overnight. This is followed by dry salting or brining. Ripening takes place over a period of 14 days and during this time moulds develop and grow over the surface of the cheese giving it its characteristic flavour.

Cream cheeses and cottage cheese are soft acid-coagulated cheeses prepared from single cream and skimmed milk (or partly skimmed milk) respectively. In cream cheese manufacture, the cream is pasteurized, cooled and inoculated with starter culture, rennet, and incubated overnight. The coagulated cream thus produced is cooled, drained and salted. Cottage cheese differs from this in that the curd particles are kept separate through controlled washing. After inoculation with starter culture and rennet, the cheese is incubated. When the pH falls the curd is cut into cubes and scalded to 49–53°C over 2 h to control the growth of spoilage organisms. Whey is then drained off the curd which is washed three times with water at progressively lower temperatures. After drying the curd is usually mixed with a cream dressing which is homogenized, pasteurized and cooled before blending with the curd.

Packaging requirements of cheese. Storage and packaging requirements of cheese [9, 10] are particularly important for the shelf life. Storage conditions are dependent upon the type of cheese, temperature and the preferences of the market. However, there are some basic packaging requirements which are the same for all types of cheese.

Firstly, oxygen must be excluded to prevent mould growth and rancidity, and secondly, moisture must be retained to preserve the texture and avoid weight loss. When small cheese portions are prepared, considerable wastage occurs because the rind has to be rejected and because the traditional shapes and sizes of cheeses do not lend themselves to economic cutting of identical portions. Processing which produces rindless cheeses has therefore been developed by allowing rectangular cheeses to ripen whilst enclosed in rectangular bags of plastic film.

Particular problems can arise when the cheese contains too much moisture before wrapping or is crumbly in texture; therefore there is a requirement for such a cheese to be cut and packed semi-automatically. Processed cheese, which requires the same protection from moisture loss and oxidation, is much more stable than natural cheese and is usually packed (without vacuumizing) in aluminium foil with a heat seal coating on the inner surface.

In general, the higher the moisture content of the cheese the more likely it is to spoil on storage. Moulds such as *Penicillium* are the most common cause of spoilage of hard cheeses and occur on the cheese surface in cracks and holes. The rind of the cheese is not normally dry enough to prevent mould growth. Storage at low temperature and some acidity in the cheese are no deterrent to mould growths.

Soft cheeses, cottage cheese and *cream cheese* with high moisture content are very susceptible to spoilage from yeasts and moulds [11]. This is un-desirable in unripened varieties but these organisms form an essential part of the development of ripened varieties such as Camembert. When bacterial spoilage occurs, it is usually due to the growth of psychrotrophic bacteria. Storage temperatures should be between 0°C and 5°C and under these conditions microbial growth will be retarded as will enzyme activity and other biochemical reactions which spoil the flavour. However, even under such conditions soft cheese should be consumed within a week. Preservatives such as sorbic acid are sometimes added to cheeses to increase shelf life but their use is restricted [1]. Fruit and other ingredients are also added to cottage cheeses and cream cheese. A recent development is the growth of the fromage frais market [12, 13], a liquid cream cheese-like product often flavoured with fruit which, for packaging purposes, has similarities to yoghurt.

Soft cheeses are usually packaged into paperboard or plastic cartons, plastic films or tubs, vegetable parchment, waxed paper or aluminium foil. Fromage frais is packaged in plastic containers with heat sealable covers. As the spoilage organisms are aerobic, these products should be packaged so that oxygen is eliminated from the surface. Packaging in hermetically sealed containers in a vacuum or in inert gases has been successful.

General requirements for cheese film wrappings

(1) Mechanical strength: this must be assessed in relation to the method of sealing and bulk packaging.
(2) Absence of organoleptic and toxic effects: must be compatible with the product and allow no significant migration.
(3) Moisture control: must have sufficient water vapour resistance to control loss of moisture during the period for which it is in store (e.g. cheddar cheese should lose about 4% moisture in the first month and 1% per month thereafter during ripening).
(4) Oxygen permeability: it is essential to keep the oxygen tension inside the wrap at a low level if mould growth is to be prevented. In practice efficiency of sealing is more important than the oxygen permeability of the film.

(5) Carbon dioxide permeability: carbon dioxide is produced by all living organisms, but the microflora of a well-made cheese produce so little that it normally dissolves in the moisture present or diffuses away. Sometimes certain starter organisms or bacteria produce carbon dioxide so quickly that it cannot diffuse away, and then holes are produced. A film absolutely impermeable to carbon dioxide could balloon in such a case. In practice, however, an atmosphere of carbon dioxide tends to suppress most microorganisms and may assist by preventing mould growth.

Materials. The main types of material used for film wrapping of cheese are polyethylene, polyvinyl chloride, copolymers of polyvinyl and polyvinylidene chloride, polyamide or nylon, cellulose with various coatings, aluminium and paper with various coatings. The bag wrapping of a cheese involves the shaping of the film with the product by shrinking which is usually accompanied by vacuum and heat. This system has the advantage that any shape can be accommodated and oxygen is removed. The short heat application melts the fat on the cheese surface and so forms a tight bond. After wrapping, cheese blocks are stored in rigid containers such as plastic or wooden boxes to prevent damage during storage and distribution.

Yoghurt

Yoghurt is a cultured milk product which owes its characteristic flavour and texture to the growth of a mixture of bacterial cultures of *Streptococcus thermophilus* and *Lactobacillus bulgaricus*. The raw material may be cows' milk or goat or buffalo milk. Soya milk is also used these days to produce soya yoghurt. There are many different types of yoghurt consumed all over the world but it is generally classified into two categories on the basis of its viscosity, stirred yoghurt and set yoghurt. Stirred yoghurt may be subdivided into those which pour slowly and those which are fluid and pour easily. The main difference between the two types is the method of manufacture. Stirred yoghurt is inoculated with the starter and incubated in bulk prior to packaging, while set yoghurt is inoculated and incubated in the retail container. Typical composition of yoghurts is given in Table 12.1 and stages in the production of yoghurt are shown in Figure 12.2 [14]. The manufacture

Table 12.1 Typical composition of yoghurt (from McCance and Widdowson)

	Water	Fat	Protein	Sugars
Plain/whole milk	81.9	3.0	5.7	7.8
Low fat	84.9	0.8	5.1	7.5
Soya	82.0	4.2	5.0	3.9

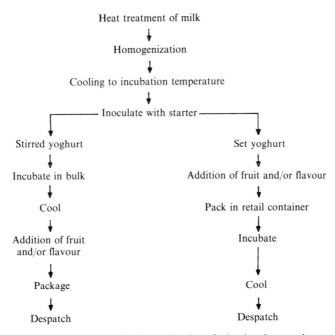

Figure 12.2   The stages in the production of stirred and set yoghurt.

of cows' milk based yoghurt is covered by a code of practice in the UK which gives composition standards and details of permitted ingredients [15].

A good quality, clean (free from antibiotics) milk (raw or pasteurized) should be used to make yoghurt. Skimmed milk powder is added in amounts from 0.5 up to 3% and enables the body or viscosity of the coagulum to be adjusted and controlled.

Heating the milk between 85 and 95°C for 15–30 min is sufficient to kill all bacterial cells and spores, except the heat-resistant ones. In yoghurt such spores do not germinate and grow because of the combined effects of low pH, anaerobiosis and antagonism of the lactic bacteria. The heat treatment of milk for yoghurt manufacture achieves several purposes: it kills the natural flora; air is expelled, which gives a better medium for lactic acid bacteria; and breakdown products of some constituents of the milk are produced, including some growth factors for the lactobacilli. The precipitation of some proteins increases the firmness and viscosity of liquid yoghurts.

Homogenization is usually carried out at 55°C and pressure of 3000 psi and by reducing fat globule size the typical creamy flavour of a good yoghurt is released. The starter culture is then introduced. Control of the culture and the balance of flora and flavour are very important. When the sterile milk has been inoculated at 43–45°C it must be filled into the retail containers as quickly as possible, as the yoghurt culture grows very quickly. For stirred yoghurt the milk is incubated for 3–4 h and then cooled to about 15°C. After cooling, fruit preparations and other ingredients are added and the product is filled into retail containers. Shelf life depends upon the type of packaging, hygiene of ingredients and manufacturing process. Deterioration can occur from excessive acid production, yeast and mould contamination or syneresis. Fermentation of sugars produces carbon dioxide which creates pressure under the container lid. Containers for yoghurts must therefore prevent contamination (by yeasts, moulds and coliforms) and resist acid breakdown in citrus fruit flavoured yoghurts.

For set yoghurt, containers are first filled and then incubated in a temperature controlled room at 42°C for 4 h and then cooled to 15–20°C in another temperature controlled room and stored at 4–5°C [9].

In general polystyrene tubs are almost always used for retail packaging of yoghurt with heat sealable polyethylene coated aluminium laminated covers, although some glass containers and other materials are also used. Fresh yoghurt generally has a shelf life of 2–3 weeks at 4°C but this can be extended to a few months if the yoghurt is pasteurized and aseptically packaged.

Drinking yoghurt is a relatively new innovation and is a low viscosity drink made by blending yoghurt with fruit juice and sugar. Stabilizers such as pectin are usually added. Drinking yoghurt is normally pasteurized and filled into gable top cartons for a short shelf life product or pasteurized and aseptically filled to give a product with a longer shelf life.

Fermented meat and fish products

Fermented meat products or salamis [4, 5] cover a wide range of different products; they can be cooked, uncooked, smoked or unsmoked. Shelf lives vary from only a few weeks under refrigeration to those products that keep for many months at ambient conditions.

The early stage of fermentation is critical, in order to reduce pH quickly before pathogenic bacteria become established. Reduction in pH also reduces the water-holding capacity of the meat (thus drying it) and the production of lactic acid gives a characteristic flavour. Nitrite is present as a result of reduction of nitrate added in the sausage mix but usually a bacterial culture is added to ensure fermentation takes place properly. The presence of lactobacilli and sugars in the recipe ensure the production of lactic acid. As acid is produced the pH falls, reaching around 4.5 in 2 weeks, but these processes vary according to the product.

Spoilage may result from incorrect fermentation, growth of microorganisms, such as yeasts and moulds, and rancidity. Packaging must therefore provide an oxygen barrier as well as keep the product free from contamination. For whole sausages the packaging can be very simple; from a plastic netting to barrier plastics. Vacuum packaging in lidded or film-sealed trays is often used for sliced products.

Fish fermentation has its origin in the treatment of fish with salt [6, 7]. Fermentation is a controlled action of enzymes and microorganisms within the fish itself or from added cultures. The texture changes give a unique flavour and fermentation takes place during storage. In Japan and South East Asia many kinds of fermented fish products are available, but there are fewer varieties in Europe and the USA. Examples include fermented sushi and fermented squid, gravlax and matjes and these are often packaged in cans or hermetically sealed glass jars and bottles or plastics containers. There are also a number of fermented fish sauces, usually packed in glass bottles or plastic containers, for example Worcester sauce which is based on anchovies.

Spoilage may be from incorrect fermentation or microbial contamination although usually the toxin-producing microorganism, *Cl. botulinum*, is deactivated by the proteolytic enzymes of the fermentation. Lipid oxidation and rancidity must be prevented by the exclusion of oxygen. In addition many of these products are quite soft and can break up into smaller pieces during transport. Packaging must therefore protect against mechanical damage.

Vinegars, pickles, sauces and dressings

The preparation of vinegars, pickles and sauces has been a traditional method of preserving foods out of season. The common ingredient is acid, often acetic acid may be derived from the vinegar. This is the main ingredient responsible for the preservation of the food, although other ingredients also play a role. Most products have a final pH below 4.5, but the bacteriostatic effect of acetic acid is mainly due to the undissociated acetic acid in solution which, because it is not ionized, passes easily through bacterial cell membranes.

Vinegars are the product of two successive fermentations: a yeast fermentation converting sugars to alcohol and an acetic fermentation by microorganisms of the *Acetobacter* group converting alcohol to acetic acid. Intermediate and minor side reactions also occur which contribute to the final flavour. There are many different kinds of vinegar available including malt, distilled, spirit, concentrated, spiced and non-brewed condiment, which is really an acetic acid solution [16].

For *pickles*, *sauces* and *spreads*, other ingredients are also present. Apart from the characterizing ingredient, the product may contain other acids, condiments, such as salt, sugar, pepper or mustard, colourants and stabilizers and other additives such as permitted preservatives.

There are many different pickled products all over the world. *Fish pickles* include herring marinades, cockles and rollmops which are marinaded in acetified brine before packaging in vinegar and spices. They are usually packaged in cans or glass jars.

Pickled fruit and *vegetables* can be divided into two types of product: 'clear' and 'in sauce'. Clear pickles are usually packaged in jars topped with liquor, pasteurized and hermetically sealed. The short time lower temperature pasteurization used destroys the acetic tolerant spoilage bacteria. Higher temperatures and longer times will inactivate microbial enzymes and therefore the temperature and times used depend on the product.

Pickles in sauce are usually made by preparing the sauce separately from the particulate components and the method used depends upon the ingredients. After the sauce and the pretreated fruit or vegetables have been mixed, the product is filled into jars and hermetically sealed.

Thick sauces and spreads such as brown sauces which are used in pickle-in-sauce products, essentially consists of a finely divided suspension of fruit and vegetable particles in thickened spiced acid syrup. The ingredients are brought to the boil and simmered, and the sauce is then filled hot or cold depending on its ingredients. Normally the sauce is partially cooled before packaging.

Thin sauces are highly spiced and flavoured vinegars or vinegar/acid liquors, with no added stabilizing agent present. Manufacture is generally similar to the manufacture of thick sauce [17].

Mayonnaise, salad cream and *emulsified dressings* are oil-in-water emulsions of vegetable oil and vinegar with salt and spices, usually emulsified with egg yolk and sometimes thickened. Mayonnaises have high oil contents, for example 65% in the United States, 50–80% in Germany and 50% in France.

Salad cream is a viscous oil-in-water emulsion in which egg yolk is the primary if not only emulsifying agent. It contains vinegar and other ingredients that may include sugar, spices, mustard and colouring [17].

Emulsified sauces are manufactured by the following basic process. The ingredients except for the vegetables and oil are added and the temperature raised to 85°C to gel the thickener and deactivate enzymes. The mix is then cooked to 60°C. The egg and oil are forced into an emulsion which is then blended with the pre-mix before filling anaerobically into the final container. Non-emulsified dressings do not contain egg yolk and must be kept as uniformly mixed as possible.

Spoilage [16] can arise from:

(1) Growth of yeasts, moulds or bacteria
(2) Deterioration in organoleptic quality from the action of enzymes of either vegetable or microbiological origin or through oxidation
(3) Deterioration from trace metals causing oxidative rancidity, e.g. iron
(4) Deterioration from the physical or chemical interaction of the ingredients, e.g. maillard reactions, browning of onions
(5) Presence of foreign matter as a result of packaging or storage failures.

Packaging considerations.   Packaging used for vinegars, pickles and sauces, etc., must protect from microbiological or foreign matter contamination and protect from oxidation by keeping oxygen out. Moisture uptake should also be avoided. Glass jars and cans, often with outer cartons or shrink-wraps, are common. Seal integrity is very important to avoid leakage, oxidation and microbiological contamination. The interior of the closures and the container itself must be resistant to acetic acid. Chilled storage of 5°C is recommended.

Salad products such as coleslaws, etc. are a rapidly growing sales area. Ingredients include fruit, vegetables, nuts, cereals, pasta and coatings of mayonnaise, salad cream or clear dressings. The products are either canned or distributed chilled in plastic pots. In the canning process their acidity is important and cans must be lacquered. For chilled products good hygiene is important in preparation and will obviously affect shelf life. Common spoilage organisms are yeasts and moulds. Packaging must also protect from surface drying and texture changes. Variations in pH and moisture changes must be kept to a minimum.

Other fermented products

There are many types of fermented product; often they are specialities from regions throughout the world [8], probably the most well known is *soya sauce.* Whole soya beans are soaked in water and cooked and then inoculated with *Aspergillus oryzae.* The product of the fermentation is mixed with salt and water in fermentation vats resulting in a mash which is pressed to remove the oil. The liquid left is soya sauce [18]. There are many varieties of soya sauces and most are packaged in plastic or glass bottles although bag-in-box systems have been used in Japan. Sometimes the closures for glass bottles have a dispenser for easy pouring.

Miso [18, 19] is a fermented soya been paste made in a similar way to soya sauce. It is used in cooking and as a base for soups. There are three major types of miso; rice based, soya based and barley miso with further subcategories for each type. Basically, whichever raw material is used, the mixture is inoculated with yeasts and lactic acid bacteria and fermented in brine. After incubation the product is pasteurized either in the package or by passage through a tubular heater. Packaging is usually in barrier film bags, bag-in-box containers, glass or plastic bottles.

Tofu [20] is not a fermented product but is produced much like cheese. Traditionally, it is made by soaking soya beans overnight, draining the water and then grinding the soya beans to a pulp and cooling with water. The pulp is then removed and the resultant soya milk is coagulated with a calcium or magnesium salt. After the curds and whey are formed, the whey is drained off and the curds pressed into cloth-lined boxes. The texture obtained, firm or soft, depends on the curd size and time of pressing. Once the tofu is cut into sections it is either packed as a bulk item in pails filled with water, or packaged in retail units of water-filled plastic tubs. The resulting tofu has a moisture content of 80% and is very perishable. It is particularly susceptible to surface drying, moisture loss and microbial contamination. Its short shelf life has limited its distribution and availability in the past, but to increase shelf life some tofu is now pasteurized and increasingly it is aseptically packaged in laminated cartons. Fried and flavoured tofus are also available.

Tempeh, natto and *sufu* [18] on the other hand are true fermented soya products. *Tempeh* resembles a mouldy sponge cake-like product but has its origins in Indonesia. It is manufactured by splitting, soaking and cooking soya beans and then adding the bacterial culture, *Rhizopus oligosporus*. The inoculated beans are then wrapped in banana leaves or perforated plastic bags and incubated at room temperature for 48 h. The resulting tempeh is covered in a whitish grey mould which gives it its characteristic flavour. Freshly prepared tempeh is then sold as such or used as an ingredient. *Natto* is a very simple product in which whole soya beans are soaked and cooked and then inoculated with *Bacillus natto*. It is also used as a soup and is packed in glass or plastic containers. *Sufu* is prepared by fermentation of tofu. Different moulds can be used to produce the sufu curd which is usually bottled with brine and pasteurized.

In Europe and the United States, with the increasing interest in healthy eating, second generation products based on tofu and tempeh, such as ready meals, desserts, ice-cream, tofu and tempeh burgers, have developed. Their packaging requirements will depend on their processing as well as product characteristics.

Fats and oils

Spoilage mechanisms

Fats and oils are more susceptible to chemical spoilage than to microbial attack. The chief types of spoilage are oxidative rancidity produced by chemical or microbial oxidation, and hydrolytic rancidity, due to lipases naturally present or to microorganisms [21, 22].

Fats subjected to either or both of these changes may contain fatty oxyacids and hydroxy acids, glycerol, other alcohols, aldehydes, ketones and lactones, and in the presence of lecithin (e.g. in butter and margarine) may include trimethylamine with its fishy odour. Also, some of the pigments produced by microorganisms are fat-soluble, and therefore can diffuse into the fat producing discoloration ranging through yellow, red, purple or brown. Because of the low moisture content of fatty materials, the growth of moulds is favoured more than bacterial growth which can also cause oxidative and hydrolytic decomposition. The oxidation of fats and oils may be catalysed by various metals (e.g. copper) and by light and moisture as well as by microorganisms.

Oxidative rancidity causes objectionable odours and flavours to develop and may adversely affect the nutritional quality, e.g. by destruction of essential fatty acids and vitamins. Such oxidation is prevented or delayed by natural or added permitted antioxidants. Tocopherols, which occur naturally in some fats, have good antioxidant properties and may give effective protection, but in fats which lack this natural protection an antioxidant must be added. This is particularly important for fats which require a long shelf life, e.g. certain baking fats. Packaging materials used for wrapping fats may include antioxidants providing this does not lead to the product containing more than the permitted amount.

Hydrolytic oxidation is caused by lipases; enzymes which catalyse the breakdown of oils and fats to give soapy and rancid flavours or odours. These enzymes may originate from plant or animal sources and may be destroyed by pasteurization. Lipases of microbial origin, in contrast, require a stronger heat treatment as they are very heat-stable. Controlled action of these enzymes is useful, however, in the development of desirable flavours in fatty foods such as cheeses.

Margarine [22] is a fatty food similar to butter, but the fat is not derived solely, if at all, from milk fat. Basically margarine consists of an oil and an aqueous phase which may be water or pasteurized fresh milk or reconstituted dry milk which has been subjected to a ripening process using bacteria to produce diacetyl and other aroma-giving substances. In production, the fat blend is mixed with an emulsifier and then other ingredients such as vitamins, colours and flavours are incorporated before emulsification and packing.

Packaging requirements

The packaging materials used in margarine manufacture must fulfil two functions: (a) to protect the product from spoilage during transit, storage and use, e.g. from microbial spoilage, oxidative changes, water and oil transmission, and odour and flavour changes; and (b) to provide sales appeal and convenience to the consumer.

Margarines (and other fats such as lard) are sold in blocks or tubs. Vegetable parchment, greaseproof paper or wrappers of aluminium foil laminated to parchment or greaseproof paper are also used and it is important for all to accept printing sufficiently well to give an attractive pack. Tub margarines were originally packaged in paperboard tubs, made by laminating board to aluminium foil. More recently the development of margarines which can be spread at refrigerator temperatures and whipped margarines which contain relatively large volumes of finely dispersed air (or nitrogen) and are claimed to mix more readily in cake making and to be better as cooking fats, has led to the rapid growth of soft margarines now sold in plastic tubs [12].

The materials used for tubs include PVC, polystyrene and acrylonitrile butadienestyrene (ABS). These materials are normally thermoformed, although injection-moulded high density polyethylene and polypropylene tubs are also used. Tubs in coextruded plastics and combinations of board and plastics have also been employed. All permit printing of attractive decorations on the tub. Thermoformed PVC is the most common starting material in use for lids; in most instances the printing on the lid includes the statutory information such as the ingredients list and the name of the product. Tub margarines are sometimes over-wrapped in paper sleeves which hold two tubs together and provide a larger surface area for decoration.

A significant proportion of margarine production is used in the bakery trade. These products are specifically designed to provide the properties required by the baker, and 10–15 kg pack sizes are produced. Similarly, professional catering establishments are also important users and for these the wrappers and tubs used are of the simplest design.

Low fat spreads contain less fat than margarines and contain a higher proportion of water. A blend of stabilizer and emulsifier is necessary to retain the texture of the product which because of the higher water content is more susceptible to mould growth.

Cooking and salad oils. Various oils are extracted from their appropriate source and have different characteristics and flavours. In general the flavour of oils should be bland and mild but off-flavours can develop on storage. Oils should also have a clear and bright appearance. Frying oils and salad oils for domestic consumption were originally packaged in glass bottles but now the trend is to use plastics bottles. While glass has many advantages (impermeability, cleanliness, durability, rigidity and its lack of susceptibility to mould growth), it is brittle, and breakages in transit can ruin a complete caseload. This led to the development of moulded thermoplastic containers in high and low density polyethylene, PVC and polystyrene. The choice of material is governed by cost, chemical resistance, permeability, ease of processing and degree of clarity or opacity needed.

Oils are also susceptible to deterioration by light in the presence of oxygen, including residual oxygen in the container. Some limited protection may be gained by the use of coloured glass or plastic, but this may detract from the appearance if the customer wishes to see the true oil colour, or by designing the pack so that the top part of the bottle which shows the oil surface, is protected by a sleeve or label, leaving the clarity and colour of the oil to be seen lower down [23].

Plastics containers for oils must be reasonably rigid, resistant to microbial attack, and to transmission of water vapour and oxygen. Oils for the catering and food processing industry are usually packed in steel cans or drums or supplied in steel or mild steel tankers. All fittings must be free of copper or copper alloys to avoid oxidation. Bag-in-box systems have also been used in Japan. Storage temperatures should be maintained as low as possible, usually 10°C above the melting point of the particular oil.

References

1. *The Preservative in Food Regulations*, HMSO (1989), SI No 553.
2. *The Antioxidant in Food Regulations*, HMSO (1980), SI No 1831.
3. C.F. Niven, Influence of microbes on the colour of meats, *Am. Meat Inst. Foundation Bull.* **13** (1951).
4. M.D. Ranken, Meat and meat products, in *Food Industries Manual*, M.D. Ranken (ed.), Blackie, Glasgow (1988).
5. J.M. Davies, Meat based snacks, in *Snackfood*, G.R. Booth (ed.), Van Nostrand Reinhold, New York (1990), p. 205.
6. M.J. Urch, Fish and fish products, in *Food Industries Manual*, M.D. Ranken (ed.), Blackie, Glasgow (1988).
7. J. Nielsen and A. Brunn, Fish based snacks, in *Snackfood*, G.R. Booth (ed.), Van Nostrand Reinhold, New York (1990).
8. G. Campbell-Platt, *Fermented Foods of the World: A Dictionary and Guide*, Butterworth, London (1987).
9. K.I. Burgess, Dairy products, in *Food Industries Manual*, M.D. Ranken (ed.), Blackie, Glasgow (1988).
10. C.R. Oswin and L. Preston, *Protective Wrappings*, Cann Publications (1980), p. 41.
11. National Dairy Council, *Facts about Soft Cheese*, NDC, London (1982).
12. Euromonitor, *The UK Food Report*, Vol 1, Euromonitor, London (1989).
13. M. Hilliam, *The European Dairy Industry*, Leatherhead Food Research Association, Leatherhead (1990).
14. National Dairy Council, *Facts about Yogurt*, NDC, London (1982).
15. Dairy Trade Federation, *Code of Practice on Yogurt*, DTF, London (1983).
16. G. Campbell-Platt and K.G. Anderson, Pickles and salad products, in *Food Industries Manual*, M.D. Ranken (ed.), Blackie, Glasgow (1988).
17. K.G. Anderson, Pickles, sauces, salads and dips, in *Snackfood*, G.R. Booth (ed.), Van Nostrand Reinhold, New York (1990).
18. S.-I. Sugiyama, Fermented Soyabean products, *Food Ingredients Int.* **2** (1990), p. 19–21.
19. D. Fukushima, New processes in the development of traditional soyafoods in Japan, *Am. Soybean Assoc. Tech. Bull.* **4** (1988), HN2.
20. S.K. Karta, Critical and variable factors in tofu processing, in *American Soybean Association Singapore 1990 Human Nutrition Highlights*, **3** (1990).
21. W.C. Frazier, *Food Microbiology* 2nd edition, Tata McGraw-Hill, New Delhi (1967), Chap. 21.
22. J.B. Rossell, Fats and fatty foods, in *Food Industries Manual*, M.D. Ranken (ed.), Blackie, Glasgow (1988).
23. K.G. Berger, Practical measures to minimise rancidity, in *Rancidity in Foods*, 2nd edition, J.C. Allen and R.J. Hamilton (eds.), Elsevier Applied Science, London (1989), p. 67.

13 Juices, soft drinks and alcoholic beverages

There are many different types of beverage: alcoholic and· non-alcoholic, carbonated and uncarbonated, acid and non-acid. In most countries soft drinks, fruit juices and nectars are controlled by regulations [1, 2, 3]. Other general labelling and additive regulations also apply.

Packaging materials used for food liquids should maintain good hygiene and have sufficient mechanical strength to prevent leakage and contamination from the outside. They should also be inert and provide a barrier to light. Seals are important and low gas permeability is required. The container must also be capable of meeting the demands of the processing and filling lines [4].

Fruit juices and beverages

Fruit juice is defined as the 'fermentable but unfermented juice pressed or squeezed from the fruit excluding the peel'. Fresh juice has a very short life after extraction from the whole fruit, due to enzyme or microbial action, unless it is rapidly processed and/or preserved. Orange juice, for example, is normally concentrated to four or more times its original strength, by multistage film evaporators, reverse osmosis or freeze concentration and the concentrate is kept at subzero temperatures. Frozen concentrated orange juice (FCOJ) is transported by bulk tanker and 250 kg steel drums. Increasingly aseptic packaging of the concentrate is being used to reduce the costs of cold storage and transport. Much orange juice sold by retail is produced from concentrate. Non-citrus juice, apple, berry fruits, etc., are obtained by pressing and filtration.

Since all these juices provide an excellent medium for microbiological activity, processing and packaging require the production of a sterilized product for chilled short life or preserved long life. Aseptic or hot-fill techniques into cartons or plastic bottles for the former and glass bottles and cans for the latter are the usual techniques employed.

Aseptic ambient temperature stable juices are processed by pasteurizing the juice at 90°C, rapidly cooling and filling cold using the TetraBrik or Combibloc system. A shelf life of 6–12 months without chilling is obtainable in this way.

Where a chilled chain distribution of no more than 2 or 3 weeks is required, full sterilization is not needed provided the microbiological load is initially small.

Components and characteristics

Many different kinds of product can be obtained from fruit juices [5], e.g. squashes, nectars, cordials, carbonated drinks, but in general there are five main components to be considered in relation to packaging; acidity, enzymes, vitamin C, colourings and flavourings.

Acidity. All fruits and their juices contain organic acids that can be detected by taste, and consequently fruit juice products usually maintain an acidic character. In most fruit there is one dominant acid, and other components of the mixture occur in secondary or trace amounts. Acidity is often used as an indication of maturity, since it decreases on ripening. Blending can be used to extend the production of acceptable fruit juices over the limitations of maturity and varietal characteristics; for instance, high-acidity early-season oranges can be blended with low-acidity late-season oranges to provide a product closer to the mid-season optimum. Varietal differences in fruit also affect the final product quality.

Enzymes. Enzymes exist in all fruit juices and are also used in their processing. The most important commercially are the pectolytic enzymes. These sometimes have to be destroyed and sometimes added. For instance, the cloudiness of some citrus products is related to the presence of pectin. The natural pectolytic enzymes of the fruit, unless destroyed, degrade the pectin resulting in clarification of the product. Materials for cloudy citrus products thus require pasteurization at temperatures high enough to inactivate the enzyme (usually 95°C for 30 s, depending upon the pH). Commercial apple and blackcurrent processing, however, requires the addition of pectolytic enzymes to destroy the pectin.

Vitamin C (ascorbic acid). The best sources of vitamin C are fresh fruits and vegetables. The vitamin C content of fruits increases until just before ripening, and then decreases, due to the action of an enzyme, ascorbic acid oxidase. When fruits are cooked, much of the ascorbic acid transfers from the tissue into the liquid and may be oxidized, oxidation occurring more easily in iron, copper or badly tinned vessels. If the material is brought rapidly to the boil, the ascorbic acid oxidase is inactivated and subsequent loss due to oxidation is reduced. Losses in vitamin C also occur during storage. They are reduced by low temperature, and by preventing contact with air and exposure to light. Canning and aseptic packaging in cartons fulfils the last two of these conditions. Addition of sulphite has a preserving effect on vitamin C.

Colour and flavour. In fruit jüices and fruit juice beverages, colour and flavour are very important. Many fruit drinks contain certain legally permitted colourings which are added to overcome the bleaching effect of the sulphite used as a preservative, and to provide an attractive appearance. Flavourings are also used in the manufacture of soft drinks. These are in the form of a prepared syrup which may contain fruit with essential oils and other flavouring materials, colouring and artificial sweeteners.

Spoilage and its prevention

Yeasts are mainly responsible for spoilage in fruit juices. Different kinds may predominate in the juice, and their growth also depends upon the temperature. Spoilage of raw fruit juices at room temperature results in an alcoholic fermentation, followed by the oxidation of alcohol and fruit acids by yeasts or moulds growing on the surface. To prevent spoilage, every living yeast cell must be removed or suppressed by pasteurization, filtration and/or preservatives.

Pasteurization. In pasteurization, the temperature of the juice is raised to about 70°C by means of a tubular or plate heat exchanger, and the heated juice is maintained at this temperature for a short holding period, after which it is cooled before filling. In flash pasteurization, temperatures of up to about 80°C for 15 min are used for very short periods. At the low pH of fruit juices, commercial sterility is practically achieved.

Filtration. Filtration is normally employed to achieve brilliance and clarity and to remove yeasts. All methods use filter aids, usually fine treated diatomaceous earth which is retained by the filter. The main forms of filter are woven cloths, rotary vacuum filters and vertical metal washers commonly called 'candles'. Asbestos pads on a plate and frame are gradually being replaced by alternatives. A brief outline of the fruit juice processing from hard and soft fruit is given in Figure 13.1 [6].

Preservatives. Preservatives can be added to fruit juices and fruit juice beverages, although in general they are not added to juices, particularly the small bottles and canned fruit juices, because the contents are usually consumed as soon as the bottle or can is opened.

Preservatives are necessary, however, in concentrated drinks or squash products where a life of several weeks is expected; carbonation enhances the effect of the preservative. The main preservatives in use are sulphur dioxide and benzoic acid. Sulphur dioxide, which inhibits browning, is gradually lost even from hermetically closed containers, so that the preservative effect diminishes with time. It also has the disadvantage of bonding to aldehyde or ketone groups, and thus the total sulphur dioxide content (which is limited

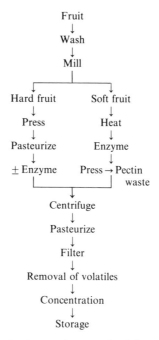

Figure 13.1   Hard and soft fruit processing (adapted from Veal [6]).

by law) may be more than that actually available as a preservative. Benzoic acid has no inhibiting effect on browning, but usually has less bleaching effect than sulphur dioxide on natural and artificial colours. Sulphur dioxide may be added as sodium sulphite, or sodium or potassium metabisulphite, and is usually used in fruit-based products. Benzoic acid is usually added as sodium benzoate to fruit squashes, etc. Flavour and packaging requirements usually determine which preservative is used. Microorganisms which are resistant to the legally permitted quantities of preservative have been found in bottling plants, but fortunately these are rare.

Carbonation enhances the effect of preservatives and also increases the acidity of the product which inhibits some but not all microorganisms. In general non-acid drinks, such as root beer, are better growth media for microorganisms than acid drinks, such as colas and fruit drinks. Moulds require oxygen and therefore do not grow in carbonated beverages, but may develop on the surface of uncarbonated drinks. Carbonation is achieved by passing the liquid through a vessel which contains carbon dioxide under pressure and for most the amount of carbon dioxide dissolved in the liquid is kept below the maximum that the liquid could absorb. Otherwise, any change in pressure would upset the equilibrium conditions and carbon dioxide could come out of solution and cause difficulties during filling.

Spoilage therefore is mainly from yeasts, especially in drinks containing sugar resulting in cloudiness and ropiness and sometimes a musty odour and taste [7].

The water used for soft drinks is purified and filtered to remove microorganisms. However, along with packaging and closures, this can be a source of contamination.

Packaging requirements

Glass bottles. The traditional glass bottles used for fruit juices and fruit juice beverages provide many advantages, as we have seen in earlier chapters, in particular inertness, easy cleaning, durability and rigidity. Glass is not susceptible to mould growth and is impermeable to odours, vapours and liquids.

Hot-filling and in-bottle pasteurization are generally employed for pure fruit juices or products which do not contain preservative. *Hot-filling* is achieved by passing the liquid product through a heat exchanger and then filling above about 70°C. The closure is then applied. Any microbiological contamination on the inner surfaces of the bottle and the closure is destroyed by the hot liquid, and adequate sterility is obtained without heating the container.

In-bottle pasteurization is carried out by heating the filled, closed bottles to 70°C and holding for approximately 20 min. The product expands and above atmosphere pressures are produced in the bottle which together with its closure must be adequate for such treatment. Both hot-filling and pasteurization are usually followed by cooling.

Glass bottles can also be covered with a polystyrene shield which enables bottles to be reduced in weight without risking breakage of bottles. Sleeves give protection and graphics can be added easily. Some bottles are shrink-wrapped with plastic sleeves.

The PET bottle and other plastics. There has been considerable growth in the PET bottle market in recent years and Table 13.1 shows the packaging trends for sparkling soft drinks from 1970 to 1985.

In Europe, PET bottles are displacing those made from PVC for products such as edible oils and mineral waters, as well as glass bottles for carbonated products. Improvements in processing technology have resulted in the appearance of stretch-blown PVC bottles, which became more competitive than the earlier non-oriented bottles.

In energy terms PVC has the edge: 1 tonne of PVC requires 2 TOEs (tonne or equivalent) compared with 3.5 TOEs for PET, but PET has better resistance to impact than PVC. For larger containers, PET is more economic in weight than PVC and is tougher. Stretch-blown PVC bottles are more liable to irregular wall thickness than those blown from PET, where the wall

Table 13.1 Packaging trends for sparkling soft drinks [5]

Year	Returnable (%)		Non-returnable (%)		
	Glass	Bulk	Glass	Cans	PET
1970	77		9	14	
1975	59		17	24	
1980	50	6	19	20	5
1981	46	8	16	20	10
1982	31	10	14	24	21
1983	23	9	14	23	30
1984	18	8	13	23	38
1985	16	8	9	24	43

thickness achieves uniformity during the stretch operation. Stress cracks are also occasionally found in stretch-blown PVC bottles and this type of bottle does not find favour with the environmental pressure groups.

PET bottles have also become even lighter than before (a 2-litre bottle usually weighs less than 40 g) and can be coated externally with PVdC to provide improved resistance to gas permeability. However, because they rely upon internal pressure to provide rigidity, PET bottles are not normally used for still (non-carbonated) soft drinks.

Polyethylene and PVC bottles are still used for squashes and cordials, but shelf life is restricted compared to glass [5].

Standard PET bottles begin to distort at temperatures above 65–70°C, which means that they cannot be pasteurized by the traditional method, but can be used in flash pasteurization.

Unlike glass the PET bottle will lose carbon dioxide with time, about 15% over 4 months [4, 8]. Hence PET is preferable for drinks with a high carbon dioxide content in large bottles, whereas for lower carbonation levels, and small or medium sized bottles, other materials may be better.

Other forms of plastic container have also been used [5, 9]. For example the Plasto-can, a coextruded plastic container with conventional aluminium easy-open can ends; the Rigello container, a multilayered polypropylene foil extrusion with a spherical bottom and tear-off cap assembled in a paperboard cylinder. New combinations of materials in can form are also being developed. Coca Cola have patented a PET/aluminium can with easy open top [10]. High barrier plastics cans which will be recyclable are under investigation. Orange juice has also been packed in clear oriented polypropylene bottles which provide good oxygen and moisture barrier properties [11]. Tamper evident pull-tab closures are used on this container.

Paperboard basket carriers, plastic clips (on bottle necks) and shrink films are used to provide multipacks holding 3, 4 or 6 units.

Cans. Fruit juices and fruit juice concentrates are frequently distributed in cans [12]. The most common are standard tinplate containers, but specially lacquered and coated cans are also used, especially for high acid products. Cans are usually hot-filled, but sometimes aseptically filled. Cold-filling after pasteurization is occasionally employed but refrigerated or frozen storage is then advisable. Products preserved with benzoic acid can also be filled cold after pasteurizing, but sulphited products are incompatible with cans. The juice tends to deteriorate in the can, due to corrosion and an increasing amount of tin and iron in the product.

In the normal hot canning process, the juice is first de-aerated to improve its flavour stability, and then pasteurized to destroy microorganisms and inactivate enzymes. After hot-filling into the cans, the lids are applied and seamed immediately before cooling, which forms a slight vacuum in the headspace as the liquid contracts. This is desirable as the presence of oxygen encourages corrosion (cold-filling operations usually involve undercover gassing, in which the head space is replaced by carbon dioxide immediately before seaming on the lid).

Carbonated beverages are susceptible to metal pick-up and are therefore packaged in lacquered two-piece aluminium cans or three-piece tinplate with side seams having a special tab design to withstand the internal pressure. Warming the filled cans immediately before packaging is important, otherwise the cans when filled with cold carbonated liquid attract a layer of condensation from the atmosphere and may corrode on the outside.

Frozen orange juice concentrate has been distributed in composite paperboard or plastics canisters approximately 170 ml in capacity [5]. There are many pack variations, including canisters with tear-off ends or paperboard spiral-wound containers with plastic linings. The concentrate is frequently left unpasteurized to provide maximum freshness of flavour, and spoilage may result if it is left unfrozen.

Bulk frozen orange juice is packed in 200-litre polyethylene drums or polyethylene lined steel drums or transported in tankers. However, aseptically produced juice (e.g. in bag-in-box systems) is replacing bulk frozen juice.

Beverage cans are also sold in [13] multipacks of four, six or more. The most common form of overwrap which assists handling and distribution is a plastic ring carrier which slips underneath the rim of the can and grips tightly throughout distribution. Paperboard multipacks are also popular as well as shrink films.

Cartons. Pasteurized fruit juice and soft drinks can be packaged very successfully in cartons with a polyethylene coating and in plastic containers [13]. These products have a limited shelf life when stored in a refrigerator. Materials selected must not absorb flavour components from the juice. In addition, acid diffusion into the plastics material can delaminate the package. Polyethylene is the most common surface contact material and is regarded as chemically stable to most food products. Packaging materials must also provide the best possible barrier to light as light affects the colour and nutritive value of fruit juices.

Aseptic filling of fruit juices and other drinks into TetraPaks and other systems (e.g. Combibloc, PurePak, Elopak) has also become popular, giving the product an extended shelf life. The product is treated by flash pasteurization, e.g. 85°C for 15 s, which destroys yeast and pectolytic enzymes. Such products have advantages over hot-filled products or nonaseptically packaged products which need a chilled distribution chain. Aseptically filled juices may also be supplied in giant bag-in-box systems and in drum containers with capacities up to 1000 litres. Such systems are used for vending and dispensing supplies on draught [14].

Storage. In general for long term storage, fruit juice after pasteurization is best kept under frozen conditions. However, for commercial reasons, cool storage at 2°C is usual. Preserved fruit juices and concentrates also benefit from cool storage, but the antioxidant effect of sulphur dioxide and peel constituents in comminuted materials has established that ambient temperatures are commercially viable. Fruit juices packaged under aseptic conditions in well sealed packs can be stored satisfactorily at ambient conditions for periods up to 9–12 months depending on the fruit concerned.

The distribution of mineral waters that have been artifically carbonated has been made possible in plastic bottles only since strong enough bottles with tight closures have become available [9]. A newer development is the canning of mineral water which is more acceptable in some environments than bottles, e.g. on the beach, at stadia and at swimming pools [15].

Beers and ales

Apart from mead, *beer* is the oldest carbonated beverage. It is not known when the first beer was bottled but records date back as far as the mid-1800s [9]. Beer is a fermented drink produced from malted barley which is milled, mashed in hot water and boiled with hops before fermentation with a special yeast strain. During fermentation the yeast converts the sugar to alcohol and carbon dioxide. After fermentation the young beer is aged in large vats, during which time the proteins, yeasts and other undesirable substances precipitate and the beer becomes clear. Flavour, aroma and body also develop. The beer is then carbonated and cooled, clarified or filtered and packaged in bottles, cans or casks. Beer for cans or smaller bottles is flash pasteurized in containers at around 60°C.

Spoilage

The boiling of the wort with hops for 2.5 h provides sufficient heat to destroy all but the most resistant bacterial spores, including those of some species of *Bacillus* and *Clostridium*. The combined action of heat and hop antiseptics destroys most of these organisms and inhibits any survivors. Beer should hinder, therefore, the growth of microorganisms because of its low pH, its content of antiseptics (carbon dioxide, alcohol and hop extracts) and its low temperature of storage. Also anaerobic conditions prevail throughout processing and storage and most beer is pasteurized or filtered. However, beer is very susceptible to spoilage from microorganisms and to physical and chemical problems. Non-microbial defects include:

(1) Turbidity due to unstable constituents
(2) Off-flavours caused by poor ingredients or contact with metals
(3) Poor physical characteristics

Microbial defects [16] can also arise but spoilage is generally due to wild yeasts and from bacterial contamination. Good hygiene practices and effective packaging will therefore reduce the problem of microbial spoilage.

As with soft drinks, beer contains dissolved carbon dioxide which creates internal pressure within the package and therefore the container must be capable of withstanding that pressure (90 psi). All carbonated beverages are susceptible to oxidative changes in flavour and so should be protected against oxygen. Packaging must be inert, i.e. there should be no migration. Beer is more sensitive than other carbonated beverages to oxygen, loss of carbon dioxide, to off-flavours and to light [17].

Most commercial pasteurization is carried out after sealing the consumer package. Beer is also sensitive to flavour uptake from the packaging materials in which it is stored, in particular metal pick-up from cans. Therefore it requires complete protection by using impermeable lacquers. High speed filling and closing processes are used and therefore the package must be free of defects and dimensionally stable.

Packaging materials

Glass bottles and lacquered metal cans are the usual containers for retail beer, although coated PET bottles are also used. Since in-container pasteurization cannot be used in the last instance, a clean-filling technique is used to handle pasteurized or filtered beer. Pasteurizable plastics bottles are under development. Retail bulk packs using bag-in-box systems are also available [14]. Wooden and steel casks are used for bulk deliveries.

PET was not used until recently for beer because the oxygen transmission rate for capacities below 1 litre is too high for the product to survive from bottling to consumption without deterioration (if the oxygen content exceeds 5 mg/kg spoilage may occur). The problem has been tackled with a new coating process coupled with the use of a large pack size of 2 litres. This has the advantage of a low surface area/volume ratio. The 2-litre PET bottles are brown-tinted and initially replaced the large tinplate party size cans. They have a great advantage in that they can easily be reclosed while the cans require a dispensing device unless they are emptied completely on one occasion.

Cider

Cider [5] is the fermented juice of the apple. During manufacture the apples are washed, macerated and pressed; the juice expressed then ferments in vats by the yeasts naturally present on the surface of the apples. Fermentation takes about 3 weeks for best results. If the process is too rapid the flavour will suffer and the cider will become acetified and dark in colour from the presence of excess air. Storage of cider should always be in a cool place in containers that are completely full or which contain carbon dioxide in the head space. Wooden vats, concrete tanks lined with bitumen based materials or plastic-lined steel tanks are used for bulk storage.

Most cider is artificially carbonated so that bottling can continue throughout the year. It may also be pasteurized or filtered using sterile filters to improve shelf life. The alternative to carbonation is natural conditioning where the cider is matured in wooden vats and bottled at a time calculated to produce the right amount of 'sparkle'. Bottling time is critical. Too early and excess gas and/or large yeast deposits will form. Too late and no gas will be produced. Retail packaging of cider is somewhat similar to beer, in cans, glass and PET bottles. Bag-in-box systems are also used.

Wine

Wine is another fermentation product made by the alcoholic fermentation of grapes, grape juice or other fruits and a subsequent ageing process. Commercially the wine may be flash pasteurized before ageing, but this is not usual, it is then cooled, held for a few days, filtered and transferred into wooden casks or containers or plastic-coated concrete tanks to mature. Final maturing takes place after bottling. The maturing vessels are completely sealed to keep out air and the maturing process results in desirable changes in body and flavour giving the wine its characteristic aroma or bouquet. After maturing the wine is filtered, casked or bottled and stored. Some wines are pasteurized in-bottle.

Spoilage may be due to faulty fermentation, defects arising from the process and ingredients used or microbial spoilage, mainly from wild yeasts, moulds and certain bacteria. The factors which affect the susceptibility of wines to microbial spoilage are as follows [16]:

(1) Acidity or pH: the lower the pH the less likely there is to be spoilage.
(2) Sugar content: low sugar wines are rarely spoiled by bacteria.
(3) Alcohol concentration: tolerance to alcohol varies with microorganism.
(4) Some bacteria require vitamin sources to grow, e.g. lactic acid bacteria. The more vitamins present in the wine (often from the yeast) the more likely the spoilage from lactic acid bacteria.
(5) The amount of sulphur dioxide present: crushed grapes are treated with sulphur dioxide to inhibit the growth of undesirable organisms which would otherwise compete with the wine yeast. The more sulphur dioxide added the greater the retardation of the spoilage microorganisms.
(6) Storage temperature: spoilage is more rapid between 20 and 35°C and slows down as temperature decreases.
(7) The absence of air prevents the growth of aerobic microorganisms, such as moulds, yeasts and certain bacteria, e.g. *Acetobacter*, but lactic acid bacteria will grow anaerobically.

Packaging requirements

Traditionally wines have been bottled in glass, but new developments include packaging in plastic bottles, laminate lined bag-in-box systems, laminated paperboard cartons and metal cans. Wine is not difficult to can and has similar requirements to juices and soft drinks in its acidity. In addition the sulphur dioxide content must be monitored and kept at an optimum level to achieve the required shelf life. Very low levels of residual oxygen must be maintained in the can and it is often necessary to pretreat the wine with nitrogen prior to filling. In addition when canning still wines, it is sometimes necessary to build up the internal pressure in the can in order to prevent collapse [12].

As with beer, bulk wines are also packaged [14] in bag-in-box systems using a laminate or metallized film bag such as LDPE metallized PE/EVA which collapses as the wine is released through a plastic tap. This minimizes air contact with the wine and provides high clarity allowing the air bubble in the bag to be seen and hence controlled. The multilayers also provide high inter-layer bond strength and improved flex-crack resistance which can be improved further using aluminium foil. The bag is contained in an outer decorated corrugated board display box.

Some countries, where wine production and consumption is high, have developed TetraBrik and Combibloc cartons for packaging wine. Transport costs can be reduced by 50% [17, 19]. The wines have been found to be equivalent in every way to those packed in glass except for a slight reduction in sulphur dioxide content. However, studies have shown that red wine in bottles is preferred to those in alternative containers, but there was no statistically significant preference in the case of white wine [20].

Distilled spirits

Spirits [17] such as rum, whisky and brandy also are first produced by
fermentation of a suitable mash, but the product obtained is then distilled
to produce a high alcohol content. Ageing takes place in wooden casks and
is a chemical rather than a biological process. Microbiological spoilage
problems are rare but packaging must be resistant to both water and alcohol
and must not alter the product flavour or the alcohol content. Traditionally
distilled alcoholic spirits have been packaged in glass, although lightweight
non-breakable PET bottles are becoming popular particularly on passenger
aircraft. Spirits and drinks with 'mixers' are also packaged in ready to drink
form in cans and small glass or plastic bottles.

References

1. *Soft Drinks Regulations*, HMSO, London (1964), SI 760 as amended.
2. *Fruit Juices and Fruit Nectars Regulations*, HMSO, London (1974), SI 972.
3. P.R. Sheard, *A Guide to the British Food Manufacturing Industry*, Nova Press, London (1991).
4. A.J. Iversen, Cartons for liquids, in *Modern Processing, Packaging and Distribution Systems for Food*, F.A. Paine (ed.), Blackie, Glasgow (1987).
5. A.J. Francis and P.W. Harmer, Fruit juices and soft drinks, in *Food Industries Manual*, M.D. Ranken (ed.), Blackie, Glasgow (1989).
6. K. Veal, Key stages in fruit juice technology, in *Food Technology International Europe* (1987), p. 136.
7. W.C. Frazier, *Food Microbiology*, Tata McGraw-Hill, New Delhi (1967), Chap. 15.
8. B.I. Turtle and B.I. Turtle, PET containers for food and drink, in *Food Technology International* (1990), p. 315.
9. L. Karfalannen, Packaging of carbonated beverages, in *Modern Processing, Packaging and Distribution Systems for Food*, F.A. Paine (ed.), Blackie, Glasgow (1987).
10. K. Kimura and G. Mitsu, Discussion on development and innovation of food packaging, *Food Policy (Jpn.)* 3 (1989), 65–71.
11. Anon, Clear OPP resin extends shelflife for orange juice, *Food Drug Packag.* 53(8)(1989), 8.
12. J.A.G. Rees and J. Bettison, Heat preservation, in *Food Industries Manual*, M.D. Ranken (ed.), Blackie, Glasgow (1988), p. 477.
13. J.V. Bousom, Carriers, beverage, in *The Wiley Encyclopaedia of Packaging Technology*, M. Bakker (ed.), Wiley, New York (1986), p. 129.
14. J.H. Briston, Recent developments in bag-in-box packaging, in *Food Technology International Europe* (1990), p. 319.
15. Anon, Perrier puts water in new 11oz cans, *Bev. Ind.* August (1989), 35.
16. W.C. Frazier, *Food Microbiology*, Tata McGraw-Hill, New Delhi (1967), Chap. 24.
17. A. Brody, Food packaging, in *The Wiley Encyclopaedia of Packaging Technology*, M. Bakker (ed.), Wiley, New York (1986), p. 163.
18. A. Markowski, Carton containers for wine, *Ind. Bevande* 18(19) (1989), 195–200, 203 (in Italian).
19. N. Buchner, H. Schlotter and H. Waterndorf, Studies on the shelflife of high quality wine in heat sealed Combicans, *Neue Verpack* 41 (9) (1988), 26–28, 33–34, 36–38 (in German).
20. G. Anelli, Containers for wine; research with voices, *Imballaggio* 38(392) (1988), 134–135 (in Italian).

14 Developing packs for food

In this chapter we consider packaging development, starting at the beginning and considering the objectives from the point of view of either a new package for an old product or a new package for a new product.

First, we must consider the reasons for initiating the development, and it is important to be single-minded in this. Package developments may take place for several of the reasons listed in Table 14.1 and some of these may be conflicting. It will be necessary to decide the most important objectives, one at a time, and to consider secondary objectives later. For example, if we want to reduce the damage rate to a particular product in a package and also to increase the shelf life of the product at the same time as we reduce unit costs, we must rank the objectives in order of importance—it will be better to consider first improving the package technically, and reducing the unit costs afterwards.

The second consideration relates to the reasons for changing a package. Some answers to the questions of when, why and how change should occur are given in Table 14.2. Most are self-explanatory, but some discussion is needed about the timing of changes.

The product life cycle

The 'product life cycle', shown in Figure 14.1, indicates that every product goes through at least four phases in its life. These are its *introduction, growth* into *maturity* and decline into *obsolescence*. After its first introduction into the market, the product passes through a very low-volume phase. During the following growth phase, volume and profit both rise to a maximum. Volume stabilizes during the period of maturity, although unit profits typically start to fall off. Eventually, towards obsolescence, the sales volume declines considerably. The length of this life cycle, the duration of each phase and the shape of the curve will vary widely for different products but in every instance obsolescence will eventually occur, for one of three reasons.

First, the need for the product can disappear, as happened to the orange juice squeezer when frozen packaged orange juice became available. Later still this also disappeared when aseptic cartons for juice were developed. Secondly, a better, cheaper, or a more convenient product may be developed to fill the same need, for instance frozen peas which had the effect of

Table 14.1   Objectives of packaging development

What can better packaging do?

Reduce unit costs
Promote wholesale, retail, and/or consumer acceptance of the product
Increase turnover, sales and profits
Improve shelf life
Reduce waste
Extend existing markets and/or make new ones possible
Provide customers with better method of using the product
Reduce the damage rate and hence complaints
Improve handling in transport or in retail outlet

Table 14.2   Changing a package

(a) *When?*
(b) *Why?*
    Product changed
    Distribution changed
    New retail outlets
    Competition altered
    Customer needs different pack
    New packaging material/method available
(c) *How?*
    Size
    Shape
    Graphics
    Better production
    More convenience

considerably reducing the need for the canned processed product. Thirdly, an existing competitive product may quite abruptly, through better marketing, gain a decisive advantage.

The part played by packaging in the life cycle of any product depends upon the factors which are critical during each phase. During development and the period of introduction, the importance of getting the packaging right is paramount. In fact with food products the main feature of new product development is often only concerned with the design or a convenience feature of the packaging. In the second (growth) phase, advertising and distribution play the major part in determining the difference between the growth rate and the rate of increase in profits. Distribution factors are, of course, vitally dependent upon packaging, and any advertising which is produced must be very much package-related, or the product will not sell. It is probably during the third phase of maturity that packaging has the least part to play. Here the effectiveness of marketing techniques is much more important. Nevertheless, one can prolong maturity by making alterations to the packaging

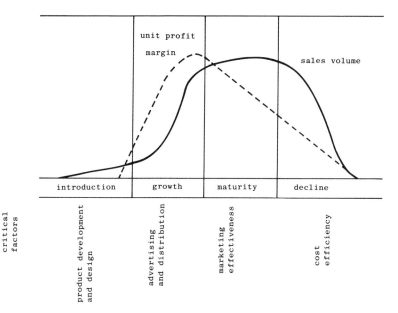

Figure 14.1   The product life cycle.

which give the product a new image when sales are beginning to fall off. Finally, in the decline phase, packaging can again play a big part in the cost efficiency of the exercise.

An important point to remember is that food products mature more rapidly now than they did 20 years ago, and life cycles are consequently getting shorter. This trend to more rapid maturity is true not only of consumer products but also of industrial products. For example, the competitive advantage afforded by nylon when it was a new product ensured a growing sale for more than a decade. In recent years, however, the tempo of innovation and substitution has reached a point where many companies will not invest in expensive research for new products which mature within 2 or 3 years, and now concentrate on making alterations to existing products which develop their use.

It is important that any company marketing a number of products should ensure that at appropriate times their use and packaging are reviewed, and those that have reached maturity or are about to decline are checked against new products in the growth phase. Action must be taken to ensure continuity. This action may well include packaging changes.

Consequently the answer to the question of when should we change a package is not related to a specific time after its introduction, but rather to the position within the product life cycle.

Planning for change

Having decided on when to change a package, we now have to consider the planning of that change, and Table 14.3 indicates the areas which have to be considered. It is not just the packaging department who are involved with this operation, but many of their colleagues within the company. Also, people outside the company, for example advertising agencies, package makers and packaging consultants, all have a part to play and the company policy on this point should be understood clearly. In particular, many companies overlook the considerable help that can be achieved by using appropriate research organizations in forwarding packaging development. Much of the considerable work required in literature research, for example, can be short-circuited through an approach to specialized organizations.

Finally, it is vitally important that package planning is related to lead times. Lead times for many packages are quite long and it is useless to start a package development project which must be completed within, say, 3 months if the lead time for one of the components concerned is longer than the 3-month period. Lead times must be known and are part of the stock-in-trade of the package development team within the company.

Basic considerations for package development [1-3]

The ultimate design in any particular package must be a compromise between various viewpoints relating to the product or the market requirement as defined by management, marketing, sales, manufacturing, R&D, quality control and so on. Since packaging has an immediate and long-term influence on total profit and sales, it should be considered in a logical manner at the beginning to achieve the best results.

All products, grown or manufactured, with few exceptions, are packaged for distribution. The basic function of packaging is to identify the product

Table 14.3 Planning

(a) Who is involved?
Product R and D
Market research
Advertising
Sales
Designers and artists
Legal department
Buying
Production management
Packaging line team
Transport
(b) Who is involved outside own company? Policy?
(c) Lead times

and safely carry it through the distribution system to the consumer. Packaging designed and constructed for the *sole* purpose of safe delivery adds nothing to the value of the product—at best, it preserves farm- or factory-fresh quality through distribution. Cost-effectiveness is the only criterion for success. If the packaging facilitates the use of the product, is reusable or has another use, some value is added that may justify greater cost. How much value for how much cost gives rise to several more criteria for judging success. Such criteria must be based on a policy and guidelines.

The way in which a package will be developed in any particular company will depend on the internal organization of that company but whoever is ultimately responsible must be aware of a number of points, the most important of which are:

(1) The current situation with present and new packaging materials in relation to properties, costs and availability.
(2) The marketing and technological developments in machinery.
(3) The current position with competitors' packaging for similar products.
(4) Modifications to both package and product which could improve the situation.
(5) Social implications, such as the current situation on waste disposal, recycling and returnable containers.
(6) The current packaging systems and their likely development.

Once these have been considered, we have reached the starting point of an interesting packaging decision chain beginning with product research and development and ending with the consumer. Three separate but interrelated development areas are involved: (a) package structural development; (b) graphics development, and (c) packaging-line engineering development. Let us look at these in turn.

Package structural development [4]

As we discussed in chapter 1, to make decisions in this area we need facts about the product and facts about the journey hazards, or the answers to what kind of protection the product needs (how much? for how long? against what?).

Since product damage is usually caused by either climatic conditions or physical environments, determining the protection needed and constructing packaging to provide it is mostly done by chemists and engineers. It follows that the technologists producing packaging structural designs must be competent to test packages against shock and vibration, evaluate the compatibility of package and product, and make test packs and check their technical adequacy against a target shelf life and distribution pattern. They must also know about materials, the sources of those materials and the relative economics of the packaging process.

We start therefore with given qualities of the product and some criteria for shelf life, plus marketing targets for content, size, number, volume or weight, and end with the production of a *complete* and tested specification for packaging that (i) protects the product for the time required, (ii) can be afforded in relation to the probable selling price, and (iii) can be handled on existing or projected production lines. 'Complete' means that it must cover all the levels of packaging needed, primary, secondary and tertiary.

We must remember that the four major packaging media compete, and they offer different advantages at differing prices. The secret of success often lies in combining the properties of two or three of those media to achieve the desired result.

Thus, the development of the design of a package operates from an input of policy guidelines, product, distribution and market criteria. Uses technical tools and skills to make, evaluate and test packages against predetermined protective and preservative needs. Produces as output specifications for economic solutions to the choice of packaging options at primary, secondary and tertiary levels.

The package development process will be structured as follows:

Step 1   This is conceptual—to identify the possible types of package that meet the requirements.

Step 2   A process of preliminary sorting—examining each possibility in some detail, checking costs, availability, protective function, adaptability to existing packaging lines, speed of filling, etc. Some packages can be rejected at this stage on an absolute basis; others because they could never be produced in time for the product launch. Generally the remainder can be graded in ranking order 1, 2, 3, etc.

Step 3   Here, certainly the best package and perhaps numbers two and three will be produced as samples for testing and comment and for work on the possible graphics. This will bring in the graphic designers who must provide paste-ups of dummy packs so that the overall effectiveness can be judged. After decisions have been reached they must then produce artwork and copy from which the printing plates can be made. We need to note that work in this area cannot start until the structural development has reached the stage where the possible location and sizes of printed areas can be broadly outlined. We must also stress that the graphics are not an isolated part of the design, but must be integrated with all the other parts of the marketing mix.

Step 4   Now we must make a study of the packaging engineering line needed. Will existing machinery be utilized or is it necessary to order new? If the former, work to prepare jigs and adapt machines must be considered. If new machinery must be obtained, both capital and running costs must be estimated.

Step 5 The results of steps 3 and 4 will now be brought together, discussed, modified and decisions taken about the final packaging.

Step 6 Finally, the package quality criteria, the defects which must be avoided and the quality control procedures must be listed and specifications for purchase and quality assurance written.

Packaging coordination

Structure, graphics and line engineering have now been brought together into a coordinated whole. Since many people have been involved in these procedures one individual will have had to do the coordinating. This needs an interdisciplinary specialist whose functions are:

(1) To respond creatively to the needs of the product manufacturer.
(2) To understand the capabilities and limitations of suppliers of packaging in respect of technical skills and costs.
(3) To communicate the necessary information and to influence decisions in order to achieve the objectives, while dealing with people over whom he has little or no direct authority.
(4) To prepare the appropriate specification.

This packaging coordinator must therefore have good communication skills and be a diplomat. Figure 14.2 illustrates the interconnections between the

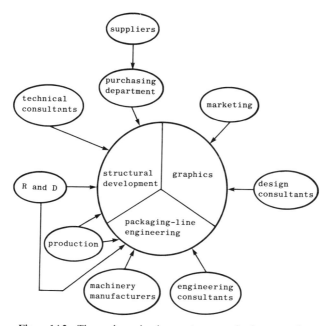

Figure 14.2 The package development communications complex.

various company functions in respect of the three areas: structure, graphics and engineering involved in design. It does illustrate a point even though the figure is an over-simplification and reality is more complex. Production, for example, will have a relationship with quality control, maintenance engineering, work planning and others. Marketing will have to take account of financial aspects, etc. The key to getting the relative roles of each correct in respect of each other is *good communications*.

Graphics

Graphics and print are two of the critical elements of today's merchandising. Since so much of our merchandising is by self-service, the product on the supermarket shelf must frequently sell itself. Advertising, if properly integrated with the packaging, will play a part in influencing choice but at the moment when the consumer takes the product off the shelf it is the packaging that may well tip the scale. The graphics of any package have three marketing elements: the brand, the product and the customer, because together they answer the question 'who sells what to whom?' Thus the graphic part of the design must convince the consumer to buy, use and re-purchase what is offered. The people who work on the graphics of a package must be creative to devise ideas for graphic treatment which will convey the desired message to the customer, outline the benefits of the product, instruct on its use, and give the necessary legal information (e.g. net weight, name of producer or distributor, etc.) required.

Packaging-line engineering

The packaging coordinator might be considered the stage director in packaging development. The important roles played by structural development and graphics development have already been outlined, but a third factor, packaging-line engineering development, can be the villain of the piece if not considered early enough.

Almost any change in a package—size, flexibility or stiffness, surface friction, etc.—can have an effect on the efficiency with which that package can be filled, closed and transported. The effect is usually in an unwanted direction unless the change has been previously checked by small pre-production trials and compensated for.

The technology of packaging machinery (see chapter 4) is the province of mechanical engineers, who have a significant knowledge of electronics, hydraulics and sometimes computer science. The importance of packaging-line engineering is not just based on the need for production efficiency—delays in solving engineering problems can prevent projects reaching fruition in time.

Cost of development [5]

However package development is achieved, there is always a cost. Even if the user company gets a packaging supplier to develop a new package, there will be the cost of defining what is required and this will involve staff time in the user company. Secondly, the user must ultimately meet the cost of development in the price for the package supplied. It is untrue to say 'package development costs nothing since our suppliers do it for us'.

Table 14.4 shows the costs that have to be considered. It is vitally important to define the requirements, to carry out an initial search on what has been done before, and to consider each of the other items right through to teething troubles that will occur on the packaging line. Each of these should have its cost estimated at the beginning so that the proper amount of time and money can be invested in getting a good result.

Table 14.4   Costs

Cost of the following
Defining requirements
Initial search
Design, function and graphics
Models and samples
Market testing
Preparing buying specifications
Preparing quality measurement service
Tooling
Teething troubles

Developing a domestic food packaging for an export market

Frequently, food producers are involved in upgrading their packaging to meet the needs of an export market. Because the hazards of transport and climate are different and the time between packaging and consumption has increased, the packaging, as well as the marketing needs, must be changed.

The following aspects have to be considered:

(1) Product requirements: Will the food need extra protection to preserve its taste, colour, moisture content etc.? Can the same primary packaging be used in the export market? If not, why not and how must it change (more protection, different language, altered recipes for use, colour changes, etc.)? If the same primary is possible can the extra protection be provided by an intermediate wrapper for 5 or 6, say, primaries or a better quality shipping container, or unit load?

(2) Distribution requirements: What, in general, do the importers/wholesalers/retailers in the new market ask for in respect of:

    (a) shipping package weight and dimensions;
    (b) retailer display requirements;
    (c) how the packages are price marked and set out in retail stores;
    (d) the need for UPC or EAN bar codes;
    (e) the types of package used in the country concerned?
(3) Consumer requirements: Is the same size primary acceptable in the new market? If not, is this because of different family sizes, need for individual portions, unit dispensing or unit price, etc.? Are there any new needs for opening, reclosing after use, disposal, etc. of the packaging? Does the market concerned have any prejudices to certain colours, words or designs? If the product has a particular selling point domestically is this valid in the new market?

References

1. F.A. Paine, *Package Design and Performance*, Pira International (1990).
2. E.A. Leonard, *Managing the Packaging Side of the Business*, AMACOM, New York (1987).
3. R.C. Griffin and S. Sacharow, *Principles of Package Development*, AVI Publishing (1980).
4. S.G. Guins, Notes on packaging design, private communication on teaching notes for MSU School of Packaging.
5. E.A. Leonard, *Economics of Packaging*, Morgan Grampian, New York (1975).

15 The economics of primary packaging

In the process of selecting the primary packaging for a particular product four basic questions must be answered:

(a) What must the packaging achieve?
(b) What types of packaging are available?
(c) What are the pros and cons of the available packaging in terms of our desired achievement?
(d) What is the resulting cost of using a particular packaging in relation to the packaging operation and the physical distribution system which will be employed to get the product to the ultimate user?

Once we have answered the first three questions, which are largely technical, the final decision will be made on economic grounds; preferably on the basis of the *total cost equation* and not on the cost of the package that the customer carries home.

Cost comparisons[†]

It is a major exercise to set out, quantify and cost all the factors relating to a packaging system [1]. In practice, therefore, decisions often have to be made on specific areas. Typically, a cost comparison between possible systems will focus on the following facts.

Packaging material (or container) prices: To be strictly comparable, quotations should be obtained at the same date (or as close to each other as possible), for the same quantity, with identical terms of sale. Primary, secondary and tertiary packaging must all be included.

Machinery cost: The capital cost will be depreciated over a number of years, in accordance with the company's policy.

Machine efficiency: This should be output achieved as a percentage of theoretical maximum output (e.g. if a machine can run at 1000/h, but gives an actual output of 800 in 1 h, then its efficiency is 80%). Losses are due to many causes, including jams, minor delays, change-over time, operator fatigue and the need to make adjustments.

[†]Throughout this chapter several examples quote costs from the 1970s or early 1980s. More up to date costs would alter the actual figures but not change the relative costs and the principles involved are the same.

Machine speed: As purchased this should always be in excess of initial requirements, to allow for growth later.

Line efficiency: If two machines, each running at 85% efficiency, are linked in sequence the line efficiency cannot be greater than 72.25% (85% of 85%), unless we arrange to hold a buffer stock between operations, so that if the first machine stops, production can continue (if the second machine stops briefly, the buffer stock will increase). In a sequence of machine operations, it is desirable for each machine to run slightly faster than the one before it. The required output may also be achieved by a single fast machine, or by two or more slower machines in parallel. These alternatives will usually result in different costs, because of differences in capital, labour requirement and efficiency. Remember that if the entire production comes from one machine, a major breakdown stops production completely, whereas a breakdown on one machine in a group of three still permits two-thirds of the operation to continue.

Labour costs: Manning requirement for packaging lines, rates of pay, including supervision and maintenance, must all be taken into account.

Inflation: It is sensible to try to make allowance for the effects of inflation. It is fairly certain that a machine bought today will cost less than a similar one bought next year. A slight labour saving this year is likely to become a larger labour saving (in money terms) next year because of increased wages. However, if a material saving is made by a pack change this year, the saving may increase, or may turn into a loss next year, depending upon the relative price movements of materials, and these cannot be reliably predicted.

Thus we can see that the assessment of costs is not a simple operation and that many factors must be taken into account. To illustrate this, we shall consider the packaging of a liquid, relatively problem-free product, in large total quantity but in small unit quantities (0.25–0.5l) [2] from the point of view of compatibility.

The relevant areas of costs are: (a) the cost of the packaging operation, filling, closing etc., *internal* to the company doing the packaging; (b) transport and distribution costs *external* to the company, and (c) costs of retailing, etc.

The *internal* costs in turn include the expenses involved in obtaining the packaging materials, the costs of those materials, the cost of storage space, the cost of storage administration, losses from capital tied-up in stock and storage space, the costs of the packaging operation, of collation and warehousing, and the cost of internal transport.

The costs of the materials will obviously be dependent on their nature and on the quantities purchased. Table 15.1 gives a summary of the prices quoted by leading packaging producers (about 1973–1974) for specific packaging. We can see from these that the costs of some packagings are far more dependent on quantities purchased than others, and one-trip packages are not necessarily always cheaper than returnables. Tinplate and aluminium cans are invariably more expensive than returnable glass.

Table 15.1   Cost prices for different quantities of various packaging [2]

Packaging	Volume of package (litres)	Cost for x thousand packs in DM/100							
		$x = 2.5$	6	10	12	15	25	50	100
Glass bottle[a]	0.33	20.5	18.5	–	17.7	16.8	14.5	14.1	13.9
Euro-glass bottle[a]	0.55	20	18	–	16.3	15.2	13.5	13.3	12.9
Lightweight glass	0.33	10	9.35	–	8.75	8.40	8	7.85	7.7
Tinplate can	0.35	–	–	22	–	–	–	21	20.5
Aluminium can	0.35	–	–	22	–	–	–	21	20.5
Plastic tub	0.25	–	–	7.3	–	–	–	7	6.7
Plastic tub	0.4	–	12	11	–	–	–	9.9	9.3
Polypak[b]	1.0	–	–	2.7	–	–	–	2.6	2.5
TetraBrik	1.0	–	–	6.7	–	–	–	6.4	6.15
Blocpak	1.0	–	–	7.8	–	–	–	7.5	7.35

[a] Multitrip returnables, remainder all one-trip.
[b] Polyethylene pouch.

The costs of the packaging operation and quantity packaged will obviously have a considerable effect on the overall costs. The total cost calculations for the packages listed in Table 15.1 are set out in Table 15.2. Here too we can see the different cost relationships applying as functions of plant utilization as well as the effect of the specific packs on the cost levels.

In addition to filling and closing costs, unit packs are then usually collated into transport units. This will produce further cost possibilities, according to whether returnable or non-returnable units are transported in returnable crates, or whether the non-returnables travel in one-trip transport outers. Koppelman [2] quotes the following as examples for the study reported here.

Assuming a quantity movement of 25 000 hectolitres per year:
    Wrap-around fibreboard cases   13.68 DM/litre
    Fibreboard trays with shrink wraps 11.90 DM/litre
    Shrink-wrapped palletized loads   9.17 DM/litre

Thus we see that the best, most economic, package for any particular purpose can vary considerably depending on the quantity packed and the utilization of the packing capacity.

In every instance a careful costing is necessary to make decisions as to how cost reduction could be achieved, or alternatively how much extra cost will be involved in upgrading the package image.

The economics of the glass primary package

Glass in the mind of the average consumer tends to have an image of high quality. It is firm and strong to touch, and in this sense glass carries the association with drinking vessels, tableware, decorative lighting and higher

Table 15.2 Filling costs for different degrees of machinery utilization

Packaging	Volume of package	Plant capacity (1000/h)	Cost at utilization of $x\%$ in DM/hectolitre							
			$x = 100$	90	80	70	60	50	40	30
Glass bottle	0.33	24	11.1	11.3	11.6	12.0	12.4	12.5	13.5	14.2
Euro glass bottle	0.5	36	3.53	3.63	3.76	3.93	4.25	4.77	5.15	6.8
Euro glass bottle	0.5	72	3.03	3.12	3.23	3.50	3.62	4.06	4.67	5.68
Lightweight glass	0.33	24	10.9	11.0	11.2	11.6	12.2	13.0	13.9	15.0
Tinplate or aluminium cans	0.35	18	6.25	6.61	7.06	7.50	8.34	9.42	10.65	12.50
Plastic tub	0.25	8	5.23	5.29	5.4	5.50	5.73	5.98	6.40	6.91
Plastic tub	0.4	8	4.45	4.51	4.6	4.7	4.82	4.99	5.15	5.4
Polypak	1.0	3.6	1.19	1.32	1.38	1.6	1.87	2.24	2.8	3.75
TetraBrik	1.0	3.6	1.96	2.10	2.24	2.6	3.04	3.64	4.55	6.08
Blocpak	1.0	3.6	1.45	1.52	1.60	1.85	2.16	2.59	3.24	4.34

quality than most primary packages. Additionally, the product packed in a glass container can be seen. This may be an advantage or a disadvantage with some products; the consumer is able to see something about quality, look into defects, and other attributes of the product. On the other hand, of course, the fact that the glass is transparent will allow the transmission of light and if the product is affected by this then its colour may change or other reactions may be induced within it. So, while transparency can be a selling factor it can be a deterioration factor for some products. Glass may itself also be coloured and this in many instances serves to identify a product. Beer is often associated with brown or green bottles. Half-white, opal and other colours can also be produced if desired. Glass is also very versatile in relation to shape: because of its thermoelastic properties, permanent shape can be induced into glass during the moulding process. Shape is an important factor not only for appearance but also in ease of handling on filling and packaging lines. Also, in many instances bottle shape identifies the bottle to the product. For example, the milk bottle is absolutely recognized and never has any label indicating its contents. With glass, of course, we also have the possibility of returnable as against single-trip packages.

Glass is perceived by the package handler as brittle, and the fact that it will break when dropped is well known. It is also, of course, subject to abrasion and the abrasion produced both in handling in the filling plant and during transport can induce weakening of the strength of the container. It is, however, a very good package from the point of view of integrity. It is a hollow blown vessel and the only entry point is with the seal or closure. No materials are used to make joints in the actual container itself. The closure is therefore the all-important point as regards package integrity. Finally, glass is almost inert and therefore compatible with virtually any food. This is an important consideration in many instances.

Factors affecting cost [3]

Raw materials.   The three principal raw materials used in making glass are sand, limestone and soda-ash. In addition to these components, glass or cullet which is scrapped during the manufacturing process (and now to some extent collected through bottle bank and other recovery schemes) is added at a level of between 15% and 30% to the mix. This increases the ease of melting and of course also reduces the initial cost. Energy is also required. The glass industry originally used mostly coal for its energy, but oil, which is more convenient, has superseded coal over the last two or three decades.

Process efficiency.   The process efficiency for manufacturing glass containers lies somewhere around 87%. Much can still be done to improve the heat transfer between glass and the moulds which form the containers, and some increase in process efficiency can be effected. It is, however, obviously quite high at the present time. Glass containers were usually formerly visually inspected after manufacture and manually packed. Visual inspection these days has been partly replaced by machinery utilizing photo-electric methods to detect faults. Some human inspection, however, is necessary for final quality assessment.

Lightweighting. A single production unit consisting of a raw materials handling and preparation plant, a furnace, say four glass-forming machines and all their associated conveying, annealing, inspecting and packaging equipment requires a considerable capital expenditure (in 1980 in excess of £10 000 000). This capital investment must obviously be utilized efficiently and hence the glass container manufacturer strives to get the maximum number of containers per ton of glass. The natural way to achieve this, of course, is by lightweighting the container. The chemical strengthening of glass surfaces to prevent damage by abrasion is another method of improving the strength properties. Ideally, the best container in terms of lightweighting is one where the height to diameter ratio is approximately 1.0, particularly if the products are not carbonated. However, all glass container design must compromise between customer requirements, cost-effectiveness and process restrictions in producing the final accepted container. Computer-aided design is now able to do this very quickly. Twenty years ago, the design of a glass container by traditional techniques would have taken 3 or 4 days. Computer-aided design reduces this to somewhere less than half an hour, depending upon the complexity.

In the first part of this chapter, we stressed the complexity of economics of primary packages and the fact that they also influenced many other costs. The total pack concept was introduced, and is well illustrated by the development of packaging for beer carried out by the Glass Manufacturers' Federation [4] a few years ago. The exercise was split into three sections: (a) container price, (b) total packaging costs and (c) brewery costs.

The model, which was developed by the GMF in conjunction with the P-E Consultant Group, was based on information provided by a significant number of United Kingdom breweries and a comprehensive evaluation of European bottling and canning technology. All the data were averaged, and whilst any individual cost component may differ from that found in a particular brewery, the overall comparison between tin cans and glass containers was believed to accurately reflect the cost situation at that time. Of course, sometimes, depending upon the distance between suppliers and the types of machinery involved, this could be reversed and tin cans would prove to be more economic primary packages for beer than glass containers. In addition one must take into account the possible advantages of the can as against the bottle. The costs are broken down in Tables 15.3–15.8. Tables 15.3 and 15.4 give the unit package price, and the total packaging cost, including the outer packaging, respectively. Tables 15.5 and 15.6 provide the brewery costs for the whole operation, and Table 15.7 gives the total overall packaging cost using the averages.

In this exercise, the filling lines were defined as modern purpose-built installations. The effective output per annum was calculated on two-shift operation per day over 240 days per year. Storage costs were based on holding stocks of empty containers for only 3 days and the distribution costs

Table 15.3 Typical container prices (10 oz volume)

	£ per 1000	Price per unit (pence)
Widemouth 10 oz glass container	27.25	
Rid-cap closure (four colours)	7.40	
Four-colour label	0.80	
Total	35.45	3.55
Standard 10 oz two-piece can in tinplate, printed four colours		
Aluminium ring-pull end	50.50	5.05

Table 15.4 Total packaging material costs including transport packaging

Glass bottle component	Price per unit (pence)	Tinplate can	Price per unit (pence)
Bottle	3.55	Can	5.05
Four bottle shrink multipack	0.16	Four can Hi-Cone multipack	0.17
Corrugated tray	0.19	Corrugated tray	0.19
Shrink film to produce a twenty-four bottle (i.e. 6 multipacks) transport unit	0.10	Shrink film to produce a twenty-four can transport unit	0.08
Total	4.00	Total	5.49

Table 15.5 Data used to establish packaging costs at the brewery

	Widemouth bottle I	Widemouth bottle II	Cans
Line speed (per minute)	600	1400	1400
Capital cost of line (£000s)	680	1400	1200
Operating efficiency (%)	73	73	73
Manning	10	12	9
Maintenance cost p.a. (£000s)	116	239	172
Container losses (%)	1.8	1.0	0.8
Product losses (%)	0.35	0.30	0.25
Effective output p.a. (millions)	100	235	235
Depreciation and interest (%)	20	20	20
Distribution distance (miles)	100	100	100

Table 15.6 Brewery costs (pence per unit)

	Widemouth bottle I (600)	Widemouth bottle II (1400)	Can (1400)
Goods inwards	0.016	0.016	0.009
Packing hire	0.789	0.542	0.435
Finished goods handling	0.045	0.045	0.033
Distribution	0.187	0.187	0.153
Overheads	0.511	0.222	0.222
Totals	1.548	1.012	0.852

Table 15.7 Total packaging cost (pence per unit)

	Material costs	Brewery costs	Total
Widemouth bottle 600 pm	4.00	1.55	5.55
Widemouth bottle 1400 pm	4.00	1.01	5.01
Can (2-piece) 1400 pm	5.49	0.85	6.34

Table 15.8 16 oz comparison

Bottle costs (pence per unit)		Can costs (pence per unit)	
Bottle	3.83	Can plus rip end plus print	6.31
Rip cap	0.74		
Label	0.10		
Total	4.67		
3-bottle shrink packaging	0.28	4-can Hi-Cone packaging	0.17
Corrugated tray	0.29	Corrugated tray	0.25
Shrink film for 1 doz. pack	0.17	Shrink film for 2 doz. pack	0.10
Total	5.41	Total	6.83

were calculated from the point at which the containers left the brewery to the point when they entered storage. It was assumed that all containers were moved from the brewery to the depot on a maximum size of vehicle and a typical distance of 100 miles was taken for the distribution. The exercise was carried out for both the 10 oz container and the 16 oz container (Table 15.8) and in both instances the widemouth rip cap bottle was used.

We will close this part of the chapter by repeating the factors which affect cost of glass: raw materials, energy, process efficiency, packaging inspection, possibility of lightweighting, design procedures and filling line technology. When all these factors are taken into consideration, we have the 'total pack concept' described earlier.

The economics of cans and canning [5]

A metal container may range from simply a means of distributing a product to a vehicle with a very precise marketing and technical purpose. The processed food can is an example of the latter, without which there would be no final product, and for success its technical function must be operative throughout the pack's life.

Thus the optimum cost for a metal container must depend on the purpose for which it is required. It must first of all fulfil the technical requirements. In the case of processed foods, it must ensure a safe pack, which will be no risk to public health, and be sufficiently strong to maintain its integrity. It must be compatible with the product being packed, whether it be a food product or a corrosive chemical, under its marketing conditions. It must be able to withstand the abuse it will receive in the market place, and convenient to retail, particularly in supermarkets. The final user must find it convenient either to empty the pack or to use its contents on occasions as required. It must have market appeal. All of these factors primarily dictate the final pack specification and thus its cost.

Over the years, considerable technical advances have been made to give a relatively cheaper metal container which still has adequate technical performance. However, technical innovation (which often increases the package cost), such as the easy-open end, has also contributed to increased sales and profitability.

A vital factor in the success of the can is the minimum cost of its use as a package. Rationalization of container size and design contribute to the development of high-speed can handling systems. Canning lines operating in excess of 1500 cans a minute are possible, and the robustness of the can, allowing it to be handled efficiently under these conditions, plainly contributes significantly to minimizing the overall cost of the final pack. Its strength and compactness further allow the outer packaging to be minimal and for it to be stored and distributed economically. The actual specifications and therefore cost of the metal container, as well as the handling equipment and processing techniques, must be carefully chosen to ensure maximum economies in use.

Finally, the promotional aspects of the container in achieving sales must be considered and investment in design both of container and decoration assessed alongside basic packaging cost and use considerations.

Thus many factors contribute towards optimizing the costs in the use of the metal container as a packaging medium. The basic pack costs must be related to the container's technical function, its efficient use and maximum convenience for marketing and final consumer acceptance.

Almost every food product one can name has, at some time or another, been packed in a metal can. Throughout the world, the metal can is used to convey and distribute the necessities and luxuries of life. Not many years ago there were some who forecast the demise of the tinplate container, claiming it would be displaced by newer plastics materials or lightweight glass, and that new ways of processing food would no longer require the can in the form we know. On the contrary, the metal food can continues to be made in ever-increasing quantities, and the new materials and processes have found their places alongside it for those products for which they are best suited.

We have used the words 'tin' and 'can' almost indiscriminately to refer to various metal containers. The words have come to mean much the same, and so the words are used almost synonymously in the rest of this chapter.

Economics of making tins [6]

The cost of making a can is made up of direct material costs, direct labour costs and overheads including indirect materials, indirect labour, depreciation, factory rent, rates and cost of energy, etc. In general-line tin box making, direct labour costs may be between 5% and 15%, with material 30–70% of total selling price. The remainder is overhead and profit. In high-speed can making, direct labour is an even lower percentage, but overheads tend to be higher because of the cost of the complex machinery involved.

Assuming that, as a rough guide, material will approximate to 50% of the cost of any can, it is thus the largest identifiable single cost. Much ingenuity has been displayed by engineers in trying to achieve the greatest possible economy in the consumption of tinplate or other box-making material, and considerable sums of money have been invested to minimize waste in making cans by high-speed methods.

The ideal package shape for economy

The problem is a matter of containing the greatest volume by the least area of sheet material. If that were all there were to it, every pack would be a sphere because a spherical shape is the most economical way of using sheet material to contain a given volume. In practice, of course, we must use packs which are either cylinders or cubes or rectangular prisms.

Assuming that the pack has no seams, no joints and no overlapping edges, in the way that a slip lid overlaps the body of a can, then a cylindrical tin is like a short section cut from a seamless tube and fitted with two ends, each of which is a simple disc of diameter just equal to the diameter of the tube. For such a container, it can be shown that the maximum volume is contained by the minimum area when (a) for a cylindrical container, height equals diameter, and (b) for a rectangular container, all sides are equal (i.e. a cube).

Such mathematical perfection is upset by practical considerations. Circular

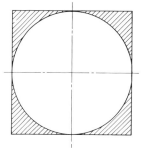

Figure 15.1 Round disc cut from a square.

components, for example, are cut for the cylinder ends from square pieces of material. This is wasteful because more than 22% of the square piece will be thrown away and less than 78% used to make a lid or bottom end (Figure 15.1). It is also necessary to allow for a seam in the cylindrical body and, with a slip-lid tin, there is an overlap of the lid on the body as well as the material in the bottom end seam. Taking account of all of this, the most economical proportions for a slip-lid tin are such that the height exceeds the diameter slightly. As a rough guide, a height to diameter ratio of 1.2:1 might be about right. The exact proportions can, of course, be calculated in each instance; they will vary slightly from small tins to large ones, and be determined by the exact amount of material employed in seams and the overlap of the slip lid on the body.

Contrast these ideal proportions for economy with those used for retail sale. Some do have proportions approaching the ideal, others are far less economic. Typical of the latter are large round biscuit tins which are popular because of their reuse value in the home. Here, considerations such as display value and utility outweigh material economy in the designer's order of importance.

Getting the most out of available material

A large part of the can maker's work is concerned with cutting round discs from sheets of tinplate or aluminium. The obvious way to do this is shown in Figure 15.2a. The strip, cut from a sheet of tinplate, is fed to a stamping press and the discs are each cut out in quick succession. In practice, the tool fitted to the press is designed to produce, say, can ends, and one almost-finished end would be formed at each stroke of the press; but it is convenient to talk in terms of discs, because we are only concerned with the plane geometry of the problem for the moment.

It is not practicable to arrange cutting so that the edges of the discs just break through the edge of the strip. Instead, a small margin must be left

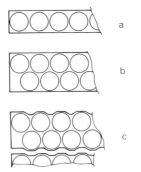

Figure 15.2 Progressively economical ways of cutting discs from a strip: (a) single row, plain
strip; (b) staggered layer on double plain strip; (c) scrolled sheared layout.

along the edges of the strip and between each disc. The material which
remains after the discs have been cut out is called 'shred'. In Figure 15.2a
the shred which remains is about 25% of the original sheet. In other words,
only 75% of the original strip finishes as discs. Such a wasteful arrangement
is to be avoided, if at all possible. Figure 15.2b shows a rather better
arrangement, where the discs have been staggered, and two rows are cut
from a single wider strip. The result here is that about 76% of the material
goes into the discs.

   The principle of staggering the pattern of discs is a sound one, but most
of the waste arises along the edges of the parallel strip. Figure 15.2c shows
how the can maker reduces this. A large sheet of tinplate is cut into strips,
with zig-zag edges as shown, and the strip is then fed longitudinally into the
stamping press. This system achieves the use of 80% of the strip. This is a
much better result, and its only weakness is that because squared sheets of
tinplate are usually involved there is still waste along the straight edges. Even
that has now been overcome by using coils of tinplate instead of sheets and
cutting the zig-zag pattern strip from continuous strip, fed from a coil. In
this way, an overall 82% usage is very nearly achieved.

The total pack concept

As we saw in the section on glass packaging, the other factors in the economic
equation can be very important. This was well illustrated by the Glass
Manufacturers' Federation exercise, which showed that sometimes a glass
container provides the more economic result although a metal package is
less expensive overall. The important thing to remember is that the cost of
the primary pack is only one item in the overall distribution cost and often
is not the decisive one. All problems must be considered individually and
generalizations can lead to error.

Costing tinplate packages for food worldwide [7]

This being a hard commercial world, the only significant economic factor in the food industry is the retail price and its formation. Very broadly, and with considerable variation, the retail price formation breaks into:

- Return to raw material supplier   10%
- Processing   30%
- Storage and movement   30%
- Wholesale and retail profits   30%

Packaging is part of the processing proportion and, again with considerable variation, it averages 6–7% of the retail price formation. Within the packaging cost, the cost of the package may reach 80%, but even so it is relatively small in the broad analysis of retail price formation. In total-project costing the unit price of a can is unimportant when compared to the losses of product arising from unsatisfactory processing, poor storage and movement, and to the influences of distribution profits.

When computers became generally available, comprehensive costing was fashionable, but it is rarely used today because the collection of data is expensive and unreliable, and because political influences are always likely to offer the unexpected event. In support of one such comprehensive costing of canned processed food, the following list of cost factors was drawn up relative to the retail price formation:

(1) *Retailing*	Profit	
	Storage	
	Display	
	Handling	
(2) *Wholesaling*	Profit	
	Storage	
	Handling	
	Collection and delivery	
(3) *Shipping*	Freight	
	Insurance	
	Handling	
(4) *Processing*	Raw material including collection	
	Storage	
	Processing energy and labour	
	Depreciations	
	Delivery to shipping	
	Profit	
(5) *Packaging*	Package design	
	Package cost	
	Package filling	

In such comprehensive costing, it is common for canning to be considered alongside other forms of packaging. It is less common for the comparison to be a simple matter of balancing the various unit package costs and filling costs. Over 90% of world food is produced in locations where there is little choice of package forms, and each location has its own pattern of relative costs. Notably, the unit costs of tinplate cans vary and in some countries the alternatives are often more attractive. If tinplate cans are necessary, it is essential for the domestic production of cans to be considered.

Food canning worldwide

The mass production of cans requires a local market for the output. The export of cans, even as flats or ovals for subsequent reforming, is not economic with present freight costs and conditions, although such exports are relatively common because medium-scale domestic production has not yet developed. Furthermore, the mass production of cans requires domestic steel sheet and tin metal at reasonable prices, and the availability of sufficient capital to ride a project over a long introductory period.

Raw material costs for any new producer of cans are unlikely to fall in the near future. How much a fall in tin prices would influence can costs is difficult to calculate. The main problem is capital involvement. To be economic in world markets, a can-making line must reach 150 cans per minute and is better at 600 per minute. Most of the world food processors, existing or potential, need about one million cans per year and would find it expensive to import empty cans.

Since the development of arc and diffusion furnaces, it has become possible for any country to remelt old ships, for instance, for good quality steel. The historic technique of deriving iron from ore and then making steel is not essential, at least until the world has used up the mountains of scrap available. Unfortunately, the engineering industry has not yet produced rolling mills related to melted-down ships but one may reasonably expect suitable plant to develop in the near future. Arc furnaces are economic down to about 20 tonnes per day of remelted steel, and it has been said that electrotinning is possible at this level when rolling mills are made available.

In theory, therefore, we might expect a rash of relatively small tinplate producers in many parts of the world. Unfortunately, world installed capacity for tinplate is presently 20 million tonnes per year and 40–50% of this is not used. Any new tinplate facilities must consequently be highly efficient to compete. Most of the potential makers of cans realize that they will have to work initially from imported tinplate sheet. If they have to work from imported tinplate, there is no need to worry too much about mass production, because can-making may be geared to the conditions of demand. A can-making line for economic cans costs something like £5 million for three-piece cans and perhaps £7 million for drawn cans in 1981 but it is possible to consider medium- or small-scale production where the circumstances are appropriate.

The future is certainly with cans using welded seams. The conventional three-piece soldered can is virtually obsolete, because cans with welded seams are more economic and more in line with market demands.

Semi-scale production

The interest in semi-scale production of cans arises from the fact that most world food production is semi-scale, seasonal and relatively remote from established can manufacture. Modern canning deals with many products, each dominating canning patterns as the seasons dictate, and is associated with can production which may not be regular and may serve a wide area of use. If such a concept is well planned it is able to provide employment for a large number of workers, and it allows *total-crop marketing* to operate.

Total-crop marketing deserves some explanation. Any crop comprises four main fractions, which are (1) best quality for fresh exports, (2) average quality for domestic fresh markets, (3) lower qualities for processing, and (4) lowest qualities for compost or animal feed. Modern thinking is that lowest qualities may be fermented or otherwise processed for upgraded discard, and may become a new line of sale for some production areas. For example, some carbohydrate crops may be fermented to increase the protein content of the discard, so that the resultant animal feed has a value in sale and thereby becomes a processed product. In many crop histories the discard may reach three-quarters of the total crop, so it becomes a serious consideration.

Semi-scale production of cans is feasible when can prices from mass production are relatively high because

(a) The size of the canning operation is small and seasonal, so cans are bought in a series of small lots, inevitably at higher prices than those for regular large lots.

(b) Political and fiscal controls often provide an artificial high price for imports of canned foods, either as protection or simply because the money for imported empty cans is not there.

(c) Distances are great between can production and the can filling operations, so that vehicles are forced to travel with empty cans over great distances, which inevitably pushes up price.

Semi-scale can production requires planning, mainly to determine where in can-making the operation should rely on imports.

A can-making line may start with flat oval bodies for reforming and ready-made ends, precut body blanks and ready-made ends, tinplate for cutting to body blanks and ready-made ends, or tinplate both for cutting to body blanks and for pressing into ends. It is not unusual for a potential can user to consider a phased programme, starting with oval bodies and ready-made ends, with a view to eventual production of bodies and ends from bought-in tinplate. This progressive procedure allows a logical flow of capital for equipment and training as the markets for canned products are developed and provide the capital. As a rule such developments favour a standard grade of tinplate and a standard diameter of can, the cans being varied in size simply by adjusting the lengths of body blanks fitting the standard ends. This sometimes results in peculiar-shaped cans, but it saves money by the standardization of ends, and since much of the potential market is domestic the cans need not conform to international voluntary standard shapes.

Can manufacture and use follows a sequence:

(1) Selection of sheet in terms of quality, dimensions and ability to work on potential equipment
(2) Slitting to blanks
(3) Forming or drawing
(4) Provision of one end
(5) Filling and provision of the other end
(6) Labelling and provision of outers.

The associated sequence for the product is selection followed by processing. Since both sequences may be varied to suit the situation, both need to be studied together, not simply examined in the light of what others have done or are doing. For example, in semi-scale production at say 30 cans per hour it is possible to use second-grade tinplate which will not run on high-speed equipment but can result in excellent cans. A can-maker could save a quarter of his input costs by using off-grade tinplate, and possibly more by accepting tinplate dimensions rejected by mass producers.

Reforming of ovals. Reforming was introduced to reduce the cost of transporting empty cans. As a rule, if empty cans have to travel 50–100 miles before filling, the packer should seriously consider making his own cans. The economic distance varies because some fillers have return-journey empty vehicles which might as well carry cans, and because some carriers can cause considerable damage to empty cans over even the shortest distances. Reforming, however, has not proved very successful. The reforming of an oval to a round takes more skill than is found in some reforming stations, and there has been a history of poor cans arising from reforming. As a rule, anyone still reforming should seriously consider setting up a sheet forming line.

Sheet forming. Sheet forming is popular for semi-scale production. Output is of the order of 25 cans per hour for labour-intensive units to 1000 cans per hour for semi-automatic production. A labour-intensive unit might use two workers per line occupying about 500 ft^2, whilst a semi-automatic line

Table 15.9 Comparison of labour-intensive and semi-automatic can production costs

	Labour-intensive	Semi-automatic
Two-shift output/year (units)	150 000	5 500 000
Installed cost (£)	8500	200 000
No. of workers (two-shift labour)	4	30
Productivity		
cans/year per £ invested	16.3	27.2
cans/year per worker	35 000	190 000
Raw material costs (£/tonne)	147–440	440
Labour cost (£/year per worker)	1000	2000

might use 15 workers per line in 4000 ft². The installation cost for a labour-intensive unit is of the order of £8000–9000 whilst the cost for a semi-automatic line might reach £200 000 (1980 figures). A labour-intensive unit is able to save money by using all grades of tinplate, including misprints, whilst a semi-automatic line is restricted to the use of high-quality tinplate. Furthermore, a labour-intensive unit is able to use casual labour whilst a semi-automatic unit must use speciality workers. Both are obviously man-paced, so the output and unit costs vary considerably according to the workers and management. Even so, experience has produced some measure of comparison (Table 15.9).

Both the raw material costs and labour costs given in Table 15.9 are unreliable statistics and are included to show that labour-intensive projects may profit from lower raw material and labour costs. A more important consideration is that labour-intensive projects fit better into regional projects for products. An output of 150 000 cans per year fits well with a project utilizing one or two tons of food product per week, which is a useful level for semi-scale food processing. The real need is for an intermediate scale of can production, about one million per year to coincide with many existing projects for products. In the developing world, food processing projects are often about five to ten tons of product per week. It is also common for such projects to be remote from can suppliers and to lack the means of using alternatives to cans.

Pressing.   Can making is one of the simplest of operations. Sheet metal is either bent or stretched to shapes which are then joined together. The bending of sheet is well within the limits of mechanized manual labour, but pressing, which is stretching, requires an input of energy. As far as is known the use of human energy has not so far resulted in effective production of can lids or drawn bodies, but there are efforts to make such production effective. A press for can ends needs to be rated at least at 20 tons, calling for a ratio of at least 400:1 for any manual device. This ratio does not seem to be available on any screw or lever presses on the market, although they would be possible using the double-lever principle. A double-lever press might provide can ends at about 100 per hour, well within the requirements for labour-intensive sheet-forming lines. However, it is doubtful that such manual presses could manufacture drawn cans.

Drawn cans.   Drawn cans were introduced not only to overcome problems of lead metal from soldered seams, but also to provide all-round surfaces for printing. It has since been shown that drawn cans are less expensive than three-piece cans in metal consumption and, since metal cost may be 60% of can cost, this is significant. The cost of a line for drawn cans is much the same as that for a line for three-piece cans and there is no logical reason why three-piece cans should continue to exist in the future. The drawing of cans as a technology has much to learn—particularly from the blown-plastics industry, which is similarly concerned with wall and bottom thickness. In theory, any metal can be drawn and ironed, but aluminium stretches better than steel and new alloys may stretch even better than aluminium. One study of steel versus aluminium revealed that the easy drawing of aluminium compensated for the lower cost of steel and the final can price for both was roughly similar.

Cans in RPF.   The important consideration in any food processing operation is the retail price formation (RPF). In this the monetary value of a can is relatively unimportant. The important value is the potential value of the can in allowing more of the total crop to bring profit, allowing processing to be carried out, providing the product in a form which is desired by the market, and allowing extension of the market. The value-evaluation is not easy and most processors of food do not carry out the necessary calculations. As a rule, they do simple calculations of cost and quantity with a high margin for errors, particularly marketing errors.

As a general rule, the ex-works value of a canned food should be 40% of the retail price. Of the ex-works price about three-quarters should be processing cost and one quarter raw material cost. Such a breakdown of retail price formation and ex-works cost gives a reasonable balance of distribution of income and thereby reduces future disputes and disruptions to trading. The 30% of retail price formation which is for processing includes packing, which many like to break down further into package cost and filling cost.

The economics of cartons

The user of cartons rarely buys the carton board direct, and his interest in its properties is limited to the overall requirements for the cartons and how they behave during and after erecting, filling and closing. The demands made on any package [8] are to contain the contents, to provide a barrier between the contents and the environment, to provide physical protection for the contents, to provide a shape convenient for distribution and sale, to add sales appeal to the product, to instruct and inform about the contents, and to be convenient for the use of the product (e.g. provide dispensing, etc.).

To this end, a food manufacturer buys cartons and service from a convertor of carton board and is only indirectly concerned with the properties of board itself. Frequently the food producer does no more than specify to his carton supplier what is required of the finished package, leaving the carton maker free to select the source of board supply. The carton maker will use his own judgment in selecting materials (Table 15.10), bearing in mind the user's requirements, as already stated.

There is also one other most important general requirement which is often overlooked. This is uniformity, for each carton must, so far as possible, be the same, in appearance, in its behaviour on packing lines and in its protective capability during transport. Translated into terms of board requirements this means a good source of board must maintain the same properties, not only throughout the making, but also from one making to the next. Consequently specifications written by carton users for carton makers usually incorporate some form of simple inspection for uniformity, printability, dimensions, strength properties, and may cover functional properties by an actual packing

Table 15.10 Examples of uses for different types of board

Type of board	Uses
Unlined chipboard	Rigid and folding boxes for soaps, detergents, hardware, electrical goods, boots, shoes, also showcards, stationery, tubes and bookbinding
Cream lined chipboard	Used for folding boxes in food, pharmaceutical and clothing industries
White lined chipboard (WLC) (2 grades, No. 1 and No. 2)	One of the most popular types of board for economical folding cartons; used in cartons for breakfast cereals, detergents, clothing, including shirts and hosiery, toilet tissues, some foodstuffs, hardware, toys and games
Triplex board	Some cheese cartons and other foods, pharmaceuticals
Duplex folding cartonboard	Folding cartons for cigarettes, frozen foods, and other foodstuffs, pharmaceuticals, cosmetics, toiletries, biscuits, cakes etc.
Duplex folding cartonboard, coated	Similar to uncoated duplex where a high finish for greater printability and superior quality is needed
Machine coated folding cartonboard	Folding cartons; markets as for WLC but where superior finish is required for greater printability and surface appeal
Solid white food board	Frozen and speciality foods, cosmetics, pharmaceuticals
Laminated or plastic-coated boards	Specialized applications, frozen foods, cakes, oils, fats; or high quality prestige packs, whisky, cosmetics etc.

line test. Remember the food packer 'tests' every carton supplied, on the packing line.

Board selection: the economic considerations

In packaging there is a constant need to reduce costs, and this is true of the carton as much as any other primary pack. This is not the same as saying that the carton at the lowest price is necessarily the most economic. If, by the unwise reduction of carton cost, one finishes up with an inferior package, the loss in production and perhaps in sales may greatly outweigh the difference in price. Moreover, in many instances it is worth paying a little more for a carton if one can be sure of getting the delivery and other services needed. All this, of course, has little to do with the properties of the board, but board properties affect the unit cost of a carton profoundly.

The packaging cost is made up of a number of parts: the cost of the carton, the costs of overwraps, the cost of the transit pack, and the cost of the cartoning (erecting, filling, sealing) operation. Of these the greatest is the cost of the carton, amounting to 75% or more for some products. The cost of the carton in turn is made up of a number of parts: board, ink, printing, cutting and creasing, glueing, etc., and of these the greatest is the cost of the board, amounting to 50% or more of the cost of the carton. Thus the boardmaker holds an important position in the economic chain. Relative costs of typical boards are shown in Table 15.11.

What is stiffness?   The most important property of board for carton making is stiffness. When any material is bent one surface is stretched and the other compressed (see Figure 15.3), and if too much force is applied one or both surfaces may break. The resistance to bending is a measure of the stiffness. The manner in which carton board is made means that the stiffness in the machine direction is always greater than that in the cross direction. The ratio between the two lies between 3:2 and 9:2, depending on the type of board-making machine.

Table 15.11   Relative costs (per tonne) of typical common boards

All 0.45 mm thick	Relative cost[a]	Grammage (g/m^2)
Unlined chipboard	1.00	330
No. 2 white lined chipboard	1.44	340
No. 1 white lined chipboard	1.61	340
Coated white lined chipboard	1.74	330
Duplex board (UK)	2.19	270
Scandinavian duplex board	2.73	280
Solid sulphate board (bleached)	2.63	372
Polyethylene-coated sulphate board	3.10	325

[a]Ratios calculated on prices ruling in November, 1980.

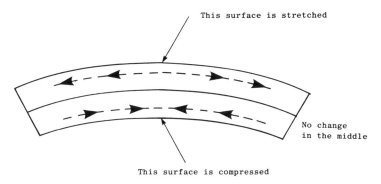

Figure 15.3 Behaviour of board on bending.

Stiffness depends not only on the type of board and the method of board making, but the number of plies, and the thickness and basis weight of the sheet. The basis weight determines the yield (the area per ton) and thus influences profoundly the unit cost of a carton. If the user is seeking value for money in carton compression strength, then stiffness is the board property by which it is most influenced. Both machine- and cross-direction stiffness are involved; usually for top-to-bottom crush resistance and flap bowing the cross machine direction is of greater importance than that in the machine direction. Besides being associated with carton compression strength, stiffness is also related to the machine performance of the pack. The boardmaker who finds a method of increasing the stiffness of board at lower cost has undoubtedly produced an attractive product.

What does stiffness cost?[9]   Since we are really buying stiffness to give the protection the product needs, it is worth inquiring into its cost. Boards differ both in price and in stiffness and stiffness is not necessarily related to price alone. Consider two actual boards, *A* and *B*: *A* is $305 \, \mathrm{g/m^2}$ at 1800 price units (PU) per tonne, and *B* is $390 \, \mathrm{g/m^2}$ at 1500 PU per tonne. First, we must remember that although we buy by weight we use by area, and we must first calculate the cost of unit area.

Board *A*: $\dfrac{1800 \, \mathrm{PU}}{\mathrm{tonne}} \times \dfrac{305 \, \mathrm{g/m^2}}{1\,000\,000} = 0.55 \, \mathrm{PU/m^2}$

Board *B*: $\dfrac{1500 \, \mathrm{PU}}{\mathrm{tonne}} \times \dfrac{390 \, \mathrm{g/m^2}}{1\,000\,000} = 0.59 \, \mathrm{PU/m^2}$

So although board *B* appears cheaper per tonne, it is actually slightly dearer per unit area. Even this is not enough, for, when we examine the boards for thickness we find that both are 0.5 mm thick, and when we measure the stiffness they are almost identical; board *A* having 13.1 Taber stiffness units and board *B* 13.0. Thus the price of stiffness is 6% higher for board *B*, even though its tonne price is almost 20% lower.

Board selection for containment, compatibility and protection

Since a large number of different products are packaged in cartons, the performance required differs from user to user and hence the board requirements may differ. We can therefore only deal with this subject in a somewhat superficial way, unless we are prepared to discuss the life history of each and every product. We must also recognize that the performance of the carton will depend not only on the properties of the board but also on the way it is made and sealed.

Board selection is assisted by the following functional advantages:

- Available in a variety of grades and calipers and surface finishes, white or coloured
- Suitable for coating and lamination
- Supplied in reels or sheets
- No standard sizes, thus allowing economic cutting
- Well-established converting processes
- Able to be printed by major processes
- Adhesion usually no problem
- Some grades suitable for direct food contact
- Opaque
- High strength-to-weight ratio compared with plastics
- Cushioning protection given by corrugated boards
- Recycled when not contaminated by plastics.

Against this should be set the following disadvantages:

- Permeable to gases and vapours unless coated
- Strength deteriorates on exposure to high humidity or moisture
- Can support fungal growth
- Inflammable
- Some grades contaminate foodstuffs.

The designer of packages needs information on the following.

(a) *Product requirements.* Is it a liquid, powder or granules? Where will it be used? How can the product be damaged? How can it be filled, dispensed?

(b) *Marketing.* What are the constraints on pack size? What are the handling considerations? What image is needed for the product? What will the sales unit be?

(c) *Distribution.* How will the product be warehoused? What marketing is required? How will it be displayed?

(d) *Legal.* What legal requirements are there on print, ink? Are weights and measures acts involved? What transport regulations apply?

(e) *Financial.* What is the pack to product price relationship to be? What are the costs in design, equipment and launch? What could storage costs for empty packs be?

Board selection for efficient running on a packaging line [8]

An increasing number of goods are now cartoned automatically, and one of the factors controlling the efficiency of this operation is the ease with which packages can be withdrawn from the magazine, erected, filled, sealed and passed through the packing line. The properties of the carton and the board are important in this respect. First, if the cartons are curled they will not feed properly and a frequent cause of curled cartons is curled board, which in turn is often related to its moisture content. The carton user does require to have flat cartons and this means he will require flat board.

Cartons on the packing line should be capable of being formed into shape rapidly and at the correct time. This is dependent on the ability of the crease to weaken the structure of the board at a specific line, which means that the crease must be easy to fold relative to the board. A critical quantity is the crease stiffness divided by the board stiffness. This ratio must not be too high: values below 7 in the cross direction and 3 in the machine direction are normally satisfactory. Higher ratios than these can lead to poor carton erection and bowing of the flaps and panels. Similarly, when the board is folded the liners or coating should not crack, nor give such a weak hinge that the board splits. This again is a factor controlled partly by the cutting and creasing technique used, but it also depends on the properties of the board.

An example of the costs of poor creasing [9]. The importance of getting the creasing right can be illustrated by an actual example from the packaging of pharmaceuticals. Two boards, apparently similar, were available but the cartons made from them were priced differently.

Cost of cartons in price units (PU)

	Per 1000 cartons	For a job of 1 000 000 cartons
Board *A*	7.87	7870
Board *B*	8.89	8890

Superficial comparison says board *A* saves 1020 PU or 11.4%. However, to find the *true* cost we must consider how well the cartons run on the filling machine. The nominal operating speed of the machine in question was 4500 cartons per hour, but the production records were:

> *Nominal speed = 75 cartons/min or 4500 cartons/h*
> No. of cartons from board *A* causing stops = 57/h
> No. of cartons from board *B* causing stops = 21/h

Now each carton causing stops wastes on average about 10 s to clear the machine, and we must take into account the cartons which would have been filled if no stops had occurred, i.e.

Production per hour (rated at 4500)

	Board A	Board B
Broken cartons	57	21
Production loss	711	265
Hence actual production per hour	3732	4214

The fixed costs of the packaging line are 60.53 PU/h, and the more cartons filled in 1 h the lower the fixed costs per carton.

Fixed costs

	Board A	Board B
Cost of packaging line per hour	60.53	60.53
Number of cartons produced per hour	3732	4214
Fixed costs per 1000 cartons	16.22	14.36

Thus taking purchase price and performance into account we have

	Board A	Board B
Carton cost per 1000 cartons	7.87	8.89
Packaging costs per 1000 cartons	16.22	14.36
Total cost per 1000 cartons	24.09	23.25

Thus we see, that although superficially the cheaper cartons are the best buy, in terms of true costs the more expensive carton gives a saving of 3.5%, which on a job of 1 000 000 cartons saves 840 PU despite the initial price disadvantage. The higher the speed needed, the more significant the stoppages become.

Making the side seam, and closing the carton. Another operation of importance to the food packer is the glueing of the carton, and the properties of the board should be such that rapid setting of the adhesive and a strong joint are obtained. If the board is too absorbent the joint may become starved of adhesive and easily break: if the board is non-absorbent to a high degree, the adhesive may not set in time: and finally, if the plies are weakly bonded although a good joint has been made, the carton may come apart, not at the adhesive interface but at the interface between the liners. Thus for high speed glueing on the packing line we need a board of the correct surface absorbency properties with the correct ply-bond strength.

Board selection for appearance and print quality

Generally, the food packer sets great store on the appearance of the carton. Although this is influenced by the printing skills of the converter, the board must be right or the carton will not look right. Further, it is essential that the inks on the carton should not rub either when the carton is going through the packing line or when it is in transit. Again, although the printer can do something to ensure non-rubbing by correct selection of inks, his task is easier if the board properties are correct. All boards should be free from defects such as specks of bitumen or particles of metal, the latter being particularly important if metal detectors are used on the packing line. The appearance of the finished carton is also impaired if the stiffness of the board is such that the panels bow, or if during transit and storage the carton becomes crushed.

A specific example (based on prices ruling in 1974) [10] The costing process may be best understood by using a specific example and showing the possible alternatives. Let us assume we want a carton to pack 450 g of a powdered solid foodstuff. Where the product is a solid, particularly where it is free-flowing, the choice of configuration requires cooperation between the carton manufacturer and the customer. First of all, it is necessary to find the volume 450 g of the product would occupy. Assuming a fairly dense free-flowing product, the volume will be about 550 ml. The next step is to calculate a base size for the carton suitable to pack it on a pallet. Presuming that the pallets are of the standard size (1000 × 1200 mm) and that B-flute corrugated containers will be used, a good fill is obtained, with a reasonable warehousing figure of 2 dozen cartons per layer in a case, with a base size of 10×3.1 cm. The third dimension is a mathematical calculation which results in 18 cm.

The standard method of measuring a carton is (1) the longest dimension round the opening, (2) the shortest dimension round the opening, and (3) the dimension between the openings. Thus this carton could be made three ways: $18 \times 10 \times 3.1$ cm, $18 \times 3.1 \times 10$ cm or $10 \times 3.1 \times 18$ cm. In addition, there are several styles of carton which could be used. It is assumed that these cartons would be machine-packed. Table 15.12 shows estimated prices for these cartons based on a 290 gsm material costing £180.35 per tonne. The first set of dimensions in column 1 does not lend itself to glueing because of the shallow depth, so this has been calculated on the basis of a tray with a three-flap lid which would be top-loaded. In column 2, the second set gives side loading and has been calculated on the basis of an economy seal-end carton. In column 3, the third set gives end loading and for comparison has been calculated in economy seal-end style. Columns 4, 5, 6, using the same end loading as column 3, show different styles: tuck ends with tucks on the same panel, reverse tuck ends, and full panel seal ends respectively. Finally, column 7 takes a different set of dimensions altogether which provide the same volume, and an economy end-load style has been used to give direct comparison with column 3.

Turning now to the factors which affect converting costs, it is obvious from this table that the number of cartons on a sheet has an effect on material costs. This is due to the fact that there is a constant width surround of waste material on all sheets; the more cartons on a sheet, therefore, the smaller the amount of waste per carton. Equally, this factor affects printing, varnishing and cutting and creasing costs, as these operations are also performed in sheet form.

Making the assumption that printing is to be done in one pass, the six-colour machine rate gives the result that cartons cost approximately

- 50p per 1000 at 20 on a sheet
- 55p per 1000 at 18 on a sheet
- 62p per 1000 at 16 on a sheet
- £1.11p per 1000 at 9 on a sheet.

Varnishing and cutting and creasing are somewhat cheaper but the same principle applies.

The next and most important factor is quantity. The figure given at the bottom of column 3 in Table 15.12 of £7.99 per 1000 is for a quantity of 1 000 000. Approximate prices per 1000 for this carton with variation in the total quantity would be:

- 100 000 = £19.00
- 500 000 = £9.00
- 1 500 000 = £7.60
- 2 000 000 = £7.30

Table 15.12 Estimated carton sizes (quantity 1 000 000; material 290 gsm; price £180.35 per tonne)

Costing data	Style 1 Tray	2 Economy side	3 Economy end	4 Tucks same panel	5 Reverse tuck end	6 Full glue end	7 Economy end
Carton size (cm)	$18 \times 10 \times 3.1$	$18 \times 3.1 \times 10$	$10 \times 3.1 \times 18$	$10 \times 3.1 \times 18$	$10 \times 3.1 \times 18$	$10 \times 3.1 \times 18$	$8 \times 8 \times 8.65$
Flat blank size	241×291	158×433	237×273	272×273	272×273	241×273	243×332
No. of blanks on sheet	16	18	20	20	20	20	16
Sheet size	981×1174	902×1311	929×1375	967×1392	967×1375	981×1375	920×1340
Weight per 1000 sheets (kg)	334	368	370	391	386	392	357
Board cost per 1000 sheets (£)	60.24	61.85	66.73	70.52	69.62	70.70	64.48
Cost of other materials per 1000 sheets (£)	13.76	14.55	15.87	16.48	16.38	16.57	14.52
Total material cost per 1000 sheets (£)	70.00	76.40	82.60	87.00	86.00	87.27	79.00
Material cost per 1000 cartons (£)	4.62	4.25	4.13	4.35	4.30	4.36	4.94
Total cost including conversion, carriage etc. per 1000 cartons (£)	8.29	8.64	7.99	8.49	8.43	8.27	9.02

It is therefore vitally important that the greatest care and attention is paid to order quantities. Most suppliers begin to charge storage for cartons after about 2–3 months, so there is no point in ordering 10 million cartons to get a low price, and then calling forward about half a million a month. However, one can take a good look at graphic designs to get as many as possible originated in the same standard colours, so that a number of orders can be combined on a group sheet. There are problems in doing this but these are comparatively minor compared with the savings. In addition to the quoted savings on increased quantity, there is a saving on plate-making. If each variety is printed as a separate job, each set of plates would cost something like £80. The printer would probably not be able to accommodate more than four varieties on a sheet. A good planner with five varieties might put three on one sheet and two on another. The plates would cost about £90 and £85 respectively, or £175 for the two, a saving of £225, since the five plates individually would cost 5 × £80 or £400.

With graphic design, more often than not, the food packer asks his carton maker for a quotation, presenting him with finished artwork. When designing the carton, the artist has no idea whether the printer will use a six-colour or a four-colour machine. Probably with a little cooperation at the initial conception of a design, the job could be run with a single pass through a five- or six-colour machine including a gloss varnish, whereas too frequently the printer is given a completed design knowing that an almost identical match will be difficult. The result is likely to be that he will use all six of his printing decks for colour, which means an extra pass for gloss varnishing, costing about £4 or £5 a thousand sheets plus a make-ready, plus an extra printing plate.

The remaining processes are diemaking for the cutting and creasing operation, which is affected by the complexity of the profile, and the number, but has little effect on a reasonable sized order; cutting and creasing itself; stripping or knocking out the cartons from the sheet; and glueing.

Stripping is relatively inexpensive, and has only a marginal effect on the price. Glueing is important, however; looking at cartons 2 and 3, 1 000 000 No. 2 costs about £650 more to convert than 1 000 000 No. 3 and more than half of this is the extra cost of glueing. The reason for this is the proportions of the blanks, No. 2 is too wide in relation to the running length but No. 3 is almost ideal. As a consequence, the running speed for No. 2 is considerably reduced.

To summarize, cooperation between carton maker and designer is essential right from the design concept stage, and the brief which the carton maker gets must cover all aspects, including in-plant handling problems.

The economics of packaging with flexible materials

As with cartons, the cost of flexible packaging materials depends on the cost of the basic film and/or coatings, etc., the costs of printing and conversion

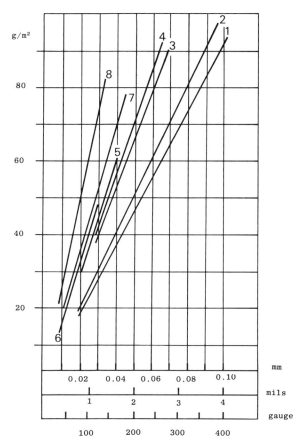

Figure 15.4 Thickness and weight of various materials. 1, low density polythene; 2, nylon; 3, cellulose acetate; 4, PVC (unplasticized); 5, cellulose film (PT); 6, polyester (mylar, melinex); 7, PVdC copolymer (saran, cryovac); 8, aluminium foil.

	grade	g/m^2	mm
MSAT cellulose film	300	350	0.026
	400	430	0.032
	450	500	0.037
	600	640	0.042

including cylinders and dies, and the costs of packaging and delivery. Similarly, the density of the base films or foils used will vary, as did the density of the various boards, and since again materials are purchased by weight and used by area this will be an important factor. Figure 15.4 shows for some typical materials the relation between thickness and weight, while Table 15.13 gives an indication of the wide range of possible materials available. The comparative costs of equal areas of some of the more common materials are listed in Table 15.14 and an indication of the potential constructions and uses of these materials is given in Table 15.15.

Table 15.13   Possible component layers in flexible laminates

Polyethylenes	Papers made of plastics
Polypropylenes	Bonded fibre fabrics
Polyvinylchlorides	Cloths and scrims
Polyvinylidene chlorides	Spun bonded fabrics
Polyvinyl acetates	Regenerated cellulose films
Polyvinyl alcohols	Cellulose esters
Polyesters	Chlorinated polyolefins
Polycarbonates	Chlorinated polyolefins
Polyurethanes	Chlorinated polyolefins
Polystyrenes	Natural and synthetic waxes
Phenoxies	Natural and synthetic waxes
Ethylene–vinyl acetate copolymers	Natural and synthetic bitumens and asphalts
Ethylene–ethyl acrylate copolymers	Natural and synthetic bitumens and asphalts
Fluoro- and chlorofluoro-hydrocarbon polymers	Natural and synthetic resins
Ionomeric copolymers	Adhesives of all types
Vinyl copolymers	Prime, key, bond- or sub-coats
Block and graft copolymers	Latex-bound mineral coatings
Papers	Latex-bound mineral coatings
Paper-like webs of mixed cellulose and plastics	Aluminium and steel foils
	Deposited metal layers

Table 15.14   Approximate comparative costs of equal areas of various flexible materials

LDPE film $18\,g/m^2$	1.0
HDPE film $18\,g/m^2$	1.1
Moisture-resistant cellulose films (depending on type)	1.6–3.3
Cellulose acetate film (0.025 mm)	2.1–2.5
PVC (unplasticized) (0.025 mm)	2.2–2.6
PVdC (0.025 mm)	3.9–4.5
Polyester film (0.025 mm)	5.5–7.5
Aluminium foil 0.0009 mm	1.8–2.0
$18\,g/m^2$ PE/300 MSAT cellulose	5.5

Economics of plastics moulded packs

Blow mouldings.   The main products from the blow-moulding process are, of course, plastic bottles, and the most important material used for this is undoubtedly polyethylene. The present applications for plastic bottles are well known and will not be considered in any detail here. A much more important consideration is the alternatives facing a buyer of plastic bottles from the inception of any project and some of the most important questions are:

(1) Has it been shown that the product can be satisfactorily packed in a plastic bottle? Which plastic?

(2) What is the potential price bracket or ceiling? Can a stock bottle be used?

Table 15.15 Construction of flexible laminates

	Nylon	Polyester	PVdC cellulose coated film	Polythene	Oriented polypropylene	Paper	Aluminium foil
Vacuum packaging deep-draw	√[a]			√[a]			
Vacuum packaging and boil-in-bag		√[a,b]		√[a]			
Boil-in-bag	√			√[b]			
Vacuum pouches and form/fill			√[a]	√[a]			
Vacuum pouches form/fill and liquids				√[a]	√[a]		
Form/fill				√[b]			
Long-life packs				√[b]		√[b]	√[a,b]
Snacks		√[a]	√[b]	√[a]			√[a]
Powders and hygroscopic products			√[a]	√[a]			√[a,b]
Confectionery, powders and hygroscopic products				√[b]		√[a]	
Shampoos				√[b]		√[a,b]	√[a]

[a] Adhesive, lamination.
[b] Polythene extrusion.

(3) What form of closure is required? In this connection, all plastic bottles are fitted with caps (either injection-moulded or compression-moulded) and possibly components. Tool costs for such components can be fairly high and economies can be made by choosing from a manufacturer's existing range.
(4) What decoration is required? Plastic bottles can be printed by two main processes, silk-screen and dry offset. Dry offset is cheaper for longer runs and multicolour work but is restricted to cylindrical or near-cylindrical bottle shapes. Is embossing a possible alternative to printing?
(5) What annual quantities are needed or estimated?
(6) Delivery rates? Economies can be achieved by laying down tools to produce at rates much greater than the annual requirement, provided delivery can be accepted and storage is available and desirable.

Manufacturers of plastic bottles usually have considerable testing and design facilities available for customer service, and on any new projects close cooperation should be maintained.

Injection mouldings

Probably the most important aspect of injection moulding as applied to packaging is that it is a more precise process than many other aspects of plastic conversion. The uses for injection mouldings fall into two main categories: (1) containers such as tubes, pots, cups and jars, and (2) components for use with either plastic or non-plastic containers, e.g. caps, plugs, pourer-spouts, sprinklers and valve assemblies. One of the important plastics materials is polystyrene, particularly the high-impact variety. For certain applications demanding greater flexibility, however, polyethylene is usually employed. Cellulose acetate is chosen for some applications where decoration and not performance is the main requirement, e.g. fancy boxes. Some of the most important points to be borne in mind when buying injection mouldings are:

(1) The relatively high tool costs for any new component; one of the manufacturer's existing range can often be used
(2) The time involved in producing new tools; usually 4–6 months
(3) The fact that the economics of production are closely related to the number of impressions of the tool and therefore rate of off-take is an important factor.

Thermoforming

This process is related to the injection moulding process but has certain important differences. The process consists essentially of forming (by vacuum or pressure) articles from a sheet of heat-softened plastic. It is a cruder process than injection moulding and is usually a cheaper method of producing certain containers and components. It should be emphasized, however, that as thinner containers are produced, they tend to be less robust. The main applications are for disposable drinking cups, cream and trifle dishes, composite ends and some advertising and display material. With the exception of the use for components, the container field use is mainly limited to where the container has a short life between point of sale and point of consumption, e.g. for ice-cream, and soft drinks for the cinema and similar trade.

All thermoplastics can be used in the process, but the main ones are cellulose acetate (where clarity is required) high-impact polystyrene and PVC.

Compression mouldings

Compression mouldings are one of the oldest forms of plastics packaging and their main impact has been in the field of screw caps and closures. The main materials used are urea and phenolic resins. The main points to note in connection with compression mouldings are that tool costs are generally lower than for injection moulding tools and that some restrictions on colour of components apply with the cheaper grades of moulding powders.

References

1. F.A. Paine, *Guidelines for Cost Effective Packaging*, Notes for Dept. of Industry and Bristol Polytechnic Low Cost Automation Centre Course (1979).
2. U. Koppelmann, On the cost structure when using packaging, *Verpackung Rundschau*, **25**(5) (1974), 416–418.
3. D.G. Osborne, Economics of the primary pack—glass containers, paper presented at *Institute of Packaging Course on Food Packaging* (1980).
4. Glass Manufacturers' Federation study: we are indebted to Mr. D.G. Osborne of Rockware Glass Ltd for the data presented here, and to the GMF.
5. Based on a presentation by Mr. G.F. Norman of the Metal Box Co. Ltd.
6. D.W. Price, in *The Packaging Media*, F.A. Paine (ed.), Blackie, Glasgow (1977), pp. 2.13–2.51.
7. Contribution from Mr. Allen Jones, consultant.
8. D I. Hine, papers on several Pira courses.
9. Data in this section from Finnboard, visual aid presentation.
10. Data in this section by courtesy of Mr. A.H. Pickford, Metal Box Co. Ltd.

16 Using barrier materials efficiently

In addition to protection against mechanical damage, many food products must be protected from atmospheric environmental influences such as rain, water vapour, oxygen and odours. Barriers are mostly used to provide resistance to the passage of gases, vapours and odours; of the gases, carbon dioxide, oxygen, sulphur dioxide and nitrogen are the principal ones concerned in packaging foods. Water vapour is the most important vapour, while flavourings and essential oils are important odours.

Barriers (Table 16.1) are also required against liquids such as water and alcohol as well as oils and fats, and in addition protect against general soiling by dust and dirt. Barriers may also be used to prevent contamination by moulds and bacteria and to keep insects and rodents at bay. Finally, certain wavelengths of light and other radiation may need to be excluded.

Table 16.2 lists the properties of many of the plastic packaging films. The rigid media (glass and metal) of course only permit entry or exit of gases and vapours through joints or closures; in flexible packs, such areas may also be more important than the material used for the barrier. Good seals are clearly essential for all barrier packaging. However, gases or vapours can enter even a sealed container if it is made of plastic, not glass or metal.

Transmission of gases and vapours through barrier materials

As long ago as 1830, Mitchell [1] studied the flow of gases through rubber. In 1866, Graham [2] presented his theory of permeability in the *Philosophical*

Table 16.1 Types of packaging barriers

Physical barriers against	Product loss, dust and dirt, bacteria, moulds and yeasts, insects, rodents, children, pilfering
Liquid, gas and vapour barriers against	Water, oils and grease, oxygen, nitrogen, carbon dioxide, sulphur dioxide, water vapour, aromas, flavours
Miscellaneous barriers against	Light (UV and visible), gamma-rays, X-rays, magnetism

Table 16.2 Properties of plastic films used in packaging

Plastics material	Density (kg/m³)	Water absorption (24 h) (%)	Water vapour transmission rate (38°C, 90% r.h.) (g/25 µm/m² d)	Oxygen transmission rate (23/25°C, 50% r.h.) (cm³/25 µm/m² d. atmos)	Printability	Transparency	Resistance to sunlight (outdoors)
Cellulose acetate	1220–1340	1.7–7.0	155–630	1800–2400	Excellent	Excellent	Excellent
Polyamides (nylons)	1010–1190	0.3–2.8	63–340	40–1400	Good	Fair–Good	Fair–Good
Polycarbonate	1200	0.15	172	4500	Excellent	Excellent	Good
LDPE/LLDPE	900–930	0.01	16–24	7100–7800	Good	Poor–Fair	Fair–Good
HDPE	945–965	0.01	4–7	2100–2900	Good	Poor	Poor–Fair
Polypropylene (homopolymer)	900–910	0.01–0.03	11	2400–3800	Poor	Poor	Poor
Polypropylene (copolymer)	890–910	0.03	–	–	Good	Fair–Good	Poor–Fair
Polyvinyl chloride (unplasticized)	1350–1600	0.04–0.4	14–80	80–300	Excellent	Good	Excellent
Polyvinyl chloride	1160–1400	0.15–0.75	80–500	80–9000	Excellent	Fair–Good	Fair–Good
Polyethylene terephthalate (PET)	1340–1390	0.1–0.2	16–20	47–94	Good	Excellent	Excellent
Polyethylene vinyl alcohol copolymer (EVOH)	1120–1210	Very hygroscopic	24–120	0.2–1.6 (0% r.h.) 13–23 (100% r.h.)	Good	Good	Good
Polyacrylonitrile copolymer (PAN)	1150	0.28	60–80	12	Good	Excellent	
Polyvinylidene chloride copolymers (PVdC)	1640–1740	0.1	0.3–3	0.5–9	Good	Good	Poor

Magazine. He was already well known for his work on the diffusion of gases through porous solids, and had found that the relative rate of escape of various gases through rubber membranes, unlike the rate of diffusion through a porous pot, bore no relation to the densities of the gases concerned. He regarded the process as one of solution, diffusion, and evaporation of the diffusing gas, a viewpoint which is substantially that held today.

Wroblewski (1879) [3] made some of the earliest measurements, and pointed out that two possible mechanisms were conceivable: either the obstacle to the passage of the gas is the surface of the membrane, the thickness being of little consequence, or the gas is able to pass through the surface of the membrane easily, and the thickness provides the obstacle. His work led him to take the second viewpoint, and he postulated that:

(a) When a film is exposed to a gas, it absorbs a definite fraction of its own volume, and this fraction (absorption coefficient) varies with temperature.

(b) The quantity absorbed by the material is proportional to the partial pressure of the gas.

(c) If one side only of a membrane is exposed to a gas and the other side to a vacuum, then a very thin layer of the sheet immediately adjacent to the gas will absorb the same concentration as if the whole sheet were immersed, while a very thin layer adjacent to the vacuum will absorb no gas.

(d) Within the material, Fick's Laws of Diffusion will hold.

These facts were confirmed and extended by other workers [4–10], and in 1941 Barrer [11] published his book *Diffusion In and Through Solid Materials*, with two chapters on gas and vapour permeability through organic membranes.

Theory

The rate at which a gas or vapour will pass through a permeable membrane is controlled by several factors, some governed by the properties of the membrane, some dependent on the properties of the gas or vapour, and some concerned with the degree to which interaction between the gas and the membrane may occur. On the kinetic atomic theory, matter may be regarded as consisting of discrete particles packed together in various ways. If these particles are arranged in an ordered fashion, the resultant solid is crystalline in nature, and if the particles are randomly arranged the solid is considered to be amorphous, or is said to exist in a disordered state. Whether it be crystalline or amorphous, it is obvious that no solid can form a completely continuous arrangement of matter, but will consist of a network containing 'pores', the size of which will depend on the nature of the membrane-forming material. This network of pores will be more or less rigid, depending upon the degree of vibration of the atoms or molecules forming it, and the number of pores of any particular size will obviously be dependent on the degree of vibration also. Thus the more rigid networks may give rise to selective permeation, allowing only the small molecules to pass, while the more elastic networks will diffuse larger molecules, owing to the greater displacement of the atoms surrounding a hole in the network from their mean position.

The process of diffusion through polymer networks may be interpreted in terms of the Eyring concept of an amorphous polymer as a tangled mass of polymer chains and holes. Segments of the polymer chains and the holes between them will be arranged in some sort of quasi-crystalline lattice. Above the transition point, the holes will be constantly disappearing and re-forming as a result of thermal motion; diffusion will take place by the movement of the diffusing molecules from hole to hole under the influence of a concentration gradient. The activation energy for such a diffusion process would, therefore, be related to the energy required to form a hole against the forces tending to hold the polymer chains together.

Four types of 'pores' are distinguishable in organic polymers:

(1) Macroscopic and microscopic cracks and spaces
(2) Submicroscopic capillaries and canals
(3) Intermolecular spaces
(4) Intramolecular spaces.

The first type of pore leads to all types of effusion, Knudsen flow and orifice flow, collectively known as capillary flow. Submicroscopic capillaries and canals are chiefly found in non-homogeneous materials, such as pigmented plastics, rubbers, and paints. The third and fourth types of inter-spaces are dependent only on the molecular structure of the membrane-forming material, and give rise to the passage of gas by the process of 'activated' diffusion first postulated by Graham [2].

The transmission of a gas or vapour through the plastic films commonly used in packaging is normally of the activated diffusion type. Under certain conditions, where cracks, pin-holes, or other flaws exist in the material, other forms of diffusion may occur. If we envisage the process as that of solution of the gas or vapour at one surface of the film, followed by diffusion through the film under a concentration gradient and culminating in evaporation from the other surface at low concentration, it is obvious that after a comparatively short period a steady state will be reached, and the gas will permeate through the film at a constant rate, providing the pressure difference between the two sides is maintained.

Under these steady-state conditions, Fick's Law of Diffusion has been found to hold, and the transmission follows the relation

$$Q = -\frac{AtD}{l}\frac{dc}{dx} \tag{1}$$

where Q is the quantity of gas diffusing through area A of the film in time t, D is the diffusion constant, dc/dx is the concentration gradient, and the thickness of the film is l.

This expression can be integrated and, where D is independent of concentration, we obtain

$$Q = \frac{AtD}{l}(c_1 - c_2) \qquad (2)$$

where c_1 and c_2 are the concentrations of the gas at the two surfaces of the film.

Gas concentrations are normally measured in terms of the pressure p of the gas which is in equilibrium with the film. Hence c can be expressed as Sp, where S is the solubility coefficient of the gas in the film (Henry's Law). Making this substitution in (2), we have

$$Q = \frac{At}{l} DS(p_1 - p_2) \qquad (3)$$

The product DS is referred to as the permeability constant P, i.e.

$$P = DS = \frac{lQ}{At(p_1 - p_2)} \qquad (4)$$

This treatment is based on the assumption that both D and S are independent of the concentration.

For gases which obey the gas laws (oxygen, nitrogen, hydrogen, etc.) this is true, and for some other gases, where the deviations from the gas laws are small, the theory still holds reasonably well. Where considerable interaction between the film and the diffusing gas or vapour takes place, the theory breaks down, e.g. for water and many organic vapours. As already mentioned, before the steady state is reached there is an interval during which the permeability increases (Figure 16.1). The situation existing during this period is described by the second form of Fick's Law:

$$\frac{dc}{dt} = D\frac{d^2c}{dx^2} \qquad (5)$$

The general solution of this equation has not yet been found, although where the diffusion constant is independent of the concentration, a particular solution for certain boundary conditions can be deduced. The case where the film is initially free from gas at one surface which is then exposed to gas at a pressure p_1, giving a surface concentration of c_1, while the other surface is held at zero concentration, has been worked out by Daynes [12] and Barrer [13]. The boundary conditions here are:

$$c = c_1 \quad \text{and} \quad x = 0 \text{ for all values of } t$$

$$c = 0 \quad \text{at} \quad x = 1 \text{ for all values of } t$$

$$c = 0 \quad \text{at} \quad x > 0 \text{ and } < 1 \text{ for } t = 0$$

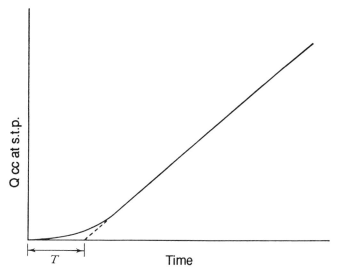

Figure 16.1   Change in rate of permeation with time.

With these boundary conditions, it may be shown that

$$Q = \frac{Dc_1}{l}t - \frac{c_1 l}{6} \tag{6}$$

It will be seen from this equation that Q will increase linearly with time, once the steady state has been reached (Figure 16.1). If we now extrapolate the linear portion backwards to the time axis we will obtain an intercept $t = T$, where $Q = 0$, and equation (6) reduces to

$$\frac{Dc_1}{l}T = \frac{c_1 l}{6} \tag{7}$$

or

$$D = \frac{l^2}{6T} \tag{8}$$

T is frequently called the lag time, and this gives an experimental basis for determining the diffusion constant D.

The variable factors associated with permeability measurements

These factors may be conveniently divided into (a) variables of the film, and (b) factors affecting the diffusion constant and the solubility.

   (a) *Variables of the film:* The main film variables are the area exposed and the thickness. It has been experimentally proved that, as predicted by Fick's Law, the permeation rate is proportional to the area exposed. In the steady state, the quantity of gas or vapour transmitted is, generally, inversely proportional to the thickness of the film but there are certain exceptions to the general rule. These are principally coated films, where not all of the measured thickness contributes to the same extent and the barrier properties increase as the thickness of the coating increases.

(b) *Factors affecting the diffusion constant and solubility.* These include pressure, temperature, the nature of the film forming material, and the nature of the diffusing gas. Firstly, with the permanent gases, the permeability constant is independent of the pressure of the diffusing gas. This is also true in many instances with other gases and vapours, providing there is no marked interaction between the film and the diffusing material. Where gases or vapours interact strongly with the film material, the permeability constant is found to be pressure-dependent. Secondly, Edwards and Pickering [6] were the first to show that the relation between the rate of permeation and the *temperature* was not linear. They found that for the system hydrogen and rubber, the permeability constant P increased rapidly as the temperature increased.

Dewar [7] reported that the logarithm of the rate of permeation gave a linear relation when plotted against the temperature. In fact, he had discovered part of the relationship which was finally demonstrated by Barrer, who showed conclusively that the correct relation between P and temperature was

$$P = P_0 \cdot e^{-E/RT} \tag{11}$$

This equation holds for both elastic and non-elastic materials. P_0 is a temperature-independent factor, while E may be regarded as a characteristic energy of activation. Both D and S also obey the Arrhenius relation:

$$D = D_0 \cdot e^{-E_d/RT} \tag{12}$$

$$S = S_0 \cdot e^{-\Delta H/RT} \tag{13}$$

In these expressions E_d is the activation energy for the diffusion process and ΔH corresponds to the heat of solution. Consequently

$$E = \Delta H + E_d \tag{14}$$

The diffusion constant always increases when the temperature increases. For permanent gases ΔH is small and positive, hence S increases slightly with temperature. For easily condensible vapours ΔH is negative, due to the contribution of the heat of condensation, and thus S will decrease as the temperature rises.

We may summarize, therefore, as follows: where no interaction occurs between a diffusing gas or vapour and a barrier film, the permeability will generally be exponentially dependent on the temperature. Moreover, the permeability constants of the most resistant barriers will be highly sensitive to temperature changes, while the poorest barrier film will often be almost independent of temperature.

A third factor is the *nature of the film-forming material*. In the Arrhenius relationship $D = D_0 \cdot e^{-E_d/RT}$, E_d can be regarded as the activation energy for diffusion. D_0, the temperature-independent factor, can be regarded as a measure of the 'looseness' of the structure of the polymer, or a measure of the number of 'holes' in the lattice at any instant of time. The activation energy for diffusion must be associated with the energy required to separate sufficient atoms to form a hole, or in other words will be a measure of the cohesive energy of the polymer. Thus, if we have two polymers such as polyethylene and rubber with similar cohesive energies, the latter having a much looser structure due to the lack of symmetry in the molecule, a larger diffusion coefficient is to be expected. Similarly, where two symmetrical polymers are concerned, polyethylene and polyvinylidene chloride, the latter has the higher cohesive energy and hence its diffusion constant is smaller.

Similar considerations will apply when polymer materials are plasticized because, as would be expected, the addition of a plasticizer decreases E_d and increases D_0. Where we are concerned with water vapour diffusing through a hydrophilic material, the water which is absorbed by the film plasticizes it, decreasing the activation energy of the diffusion and increasing the 'looseness' of the structure, thus permitting a higher rate of permeation. Similarly, if a hydrophilic film in a moist condition is being used as a barrier to a gas such as oxygen, then the plasticizing effect of the moisture on the film will again increase the size of the holes, and the gas will diffuse far more rapidly than it would if the film were dry.

We would also expect that the rate of diffusion would be dependent on the size of the gas molecule, and this, in fact, has been found to be the case, particularly where the complicating influence of the solubility can be practically eliminated. Muller [14] studied the permeability of cellulose triacetate, polystyrene and polyvinyl chloride to the inert gases (helium, neon, krypton and xenon), where the interaction between gas and film must be practically zero. He found that the permeability constant (which was numerically very little different from the diffusion constant) fell sharply as the diameter of the diffusing molecule increased. Xenon, with a molecular diameter of 3.5 Å, had a very low permeability, and molecules of a size larger than this have little or no chance of permeating.

The introduction of a pigment or filler into a plastic film can lead to an increase in the transmission rate, if the quantity of pigment employed exceeds a critical amount. The particle size of the pigment is obviously of importance here, as also will be the shape of the particles. Investigations have shown that the permeability of various pigmented systems is increased or decreased according to the nature of the pigment, the amount present, and the size and shape of the particle.

Studies have also been made of the effects of cross-linking polymers, but most of these have been carried out on rubber/gas systems. High-energy irradiation also causes cross-linking in polyethylene films. A low irradiation dosage corresponds to about one cross-link in every 600 monomer units, whereas a higher dosage is roughly equivalent to one cross-link in 30–60 monomer units. The higher dosage reduces the transmission rate to approximately one-half. It was not possible in these experiments to estimate the diffusion constant, but it would be expected that the solubility would not be affected markedly by cross-linking, hence the reduction in P is due to a reduction in D while S remains approximately constant.

The degree of crystallinity of the polymer also affects the rate of permeation. Thus the permeability of polystyrene in the oriented condition is lower than that of the unoriented material. Comparison of amorphous polyethylene films (produced by casting from a hot solution of xylene on to an armour glass plate, followed by shock cooling by immersion in chilled water to 'freeze in' the amorphous structure) with films which are cooled slowly over a period of 3 or 4 h, also showed that the more crystalline annealed film gave a lower permeability than the amorphous material. Stannett [15] and his co-workers also worked on polyethylene, and found that the permeability steadily decreased with increasing crystallinity. They found that a short period only was required (between $0°C$ and the melting-point of the polymer) to establish an equilibrium crystalline content. Nitrogen, oxygen, carbon dioxide and water vapour, together with methyl bromide and n-hexane, were studied. The decrease in permeability could be accounted for by a reduced diffusion constant, because of the more tortuous diffusion path which the diffusing gas molecules have to take. Another possible explanation for the reduction in D is that crystallites act in a similar manner to cross-linking, restricting the motion of the chains involved in the diffusion process.

The fourth important factor is the *nature of the diffusing gas* or vapour. The solubility of the gas in the film material will depend on their mutual compatibility, and generally speaking the principle of 'like dissolves like' is followed. The difference between gases and vapours is, of course, arbitrary and is primarily related to the ease of condensation at the temperature concerned. Easily condensible vapours are generally more soluble in polymers, as would be expected. It might be anticipated, therefore, that for any particular film the permeabilities of different gases are dependent on those physical constants of the gas related to the ease of condensation, such as the *critical temperature*.

Table 16.3 gives figures for the *atmospheric gases*, nitrogen, oxygen, and carbon dioxide and for water vapour collected from various sources. It also shows the ratio, taking the nitrogen permeability as 1, between the permeability constants for these various gases, and it can be seen that whatever film material is involved, oxygen permeates about four times as fast as nitrogen, and carbon dioxide four to six times as fast as oxygen, but there seems to be no consistency at all for the ratio between water and nitrogen. It would seem, therefore, that the permeability constant of a particular gas/polymer system is the product of three factors. One of these factors is determined by the nature of

Table 16.3 Permeability of films to gases and water vapour

Material	Permeability (ml/m² per MPa per day)[a]				Ratios (to N_2 permeability as 1.0)		
	N_2 at 30°C	O_2 at 30°C	CO_2 at 30°C	H_2O at 25°C 90% r.h.	P_{O_2}/P_{N_2}	P_{CO_2}/P_{N_2}	P_{H_2O}/P_{N_2}
Polyvinylidene chloride (Saran)	0.07	0.35	1.9	94	5.0	27	1400
Polychloro-trifluoroethylene	0.20	0.66	4.8	19	3.4	24	95
Polyester (Mylar A)	0.33	1.47	10	8700	4.5	31	27 000
Polyamide (Nylon 6)	0.67	2.5	10	47 000	3.7	15	70 000
Polyvinyl chloride (unplasticized)	2.7	8.0	6.7	10 000	3.0	25	3800
Cellulose acetate (P912)	19	52	450	500 000	2.8	24	2700
Polyethylene							
($d = 0.954 - 0.960$)	18	71	230	860	3.9	13	47
($d = 0.922$)	120	360	2300	5300	2.9	19	44
Polystyrene	19	73	590	80 000	3.8	32	4200
Polypropylene ($d = 0.910$)	–	150	610	4500	–	–	–

[a] All permeabilities calculated for a 25 μm thick film. Units are millilitres per square metre per megapascal per day.

the gas and may be called F_i, meaning it is the film-determining factor for polymer i. The second is determined by the nature of the gas, and is called, say, G_k, k denoting the specific gas. The third will account for any interaction between the gas and the film, and may be called γ_{ik} for polymer i and gas k. In other words,

$$P_i = F_i \cdot G_k \cdot \gamma_{ik} \tag{15}$$

γ is approximately equal to 1 where little or no interaction occurs, and becomes larger the greater the degree of interaction. On this basis it can be seen from Table 16.3 that the interaction between water vapour and several of the films involved must be quite high. Because of the unfamiliar units used in this table, two other tables (Tables 16.4 and 16.5) are given, which show the resistance relative to a common material, LDPE, taken as unity.

The first study on the effect of mixed gases on one another was carried out by Alexejev and Matalski [16], who studied the permeation of carbon dioxide,

Table 16.4 Resistance to water vapour transmission at a given yield

	WVTR (1 = 800 MN/s per mol = 18 g/m² day, tropical)	Yield (1 = 42 m²/kg)
Plain cellulose film PT	0.005	0.73
Semi moisture-proof cellulose film QMS	0.015	0.67
Cellulose acetate (25 m)	0.020	0.95
PVC meatwrap (SAF)	0.08	0.90
15–18 lb dry waxed sulphite paper	0.10	0.53
Nylon 66	0.10	0.81
PVC produce wrap (SAC)	0.13	0.90
Oriented polystyrene (25 m)	0.15	0.90
24–30 lb wet waxed sulphite paper	0.2	0.34
PVC for overwrapping (SAD)	0.2	0.72
Polyester (112 m)	0.4	1.33
Nylon 11	0.5	0.40
Unplasticized PVC	0.8	0.70
LDPE (25 m)	1.0	1.00
HDPE (12 m)	1.0	2.00
PE-coated kraft paper (25 g/m²)	1.5	0.23
Nitro-coated cellulose film (MS)	1.5	0.67
Chill cast polypropylene (25 m)	1.5	1.00
Biaxially oriented polypropylene (25 m)	1.5	1.36
Copolymer solution coated cellulose MXXT (S)	2	0.65
LDPE (50 m)	2	0.50
HDPE (25 m)	2	1.00
Coextrusion coated OPP	2	1.00
Monoaxially oriented polypropylene	3	0.60
PVDC coated polypropylene	3	1.32
PVDC copolymer	3	0.54
Dispersion coated cellulose film MXXT (A)	4	0.64
Briphane	6	0.53
PCTFE	10	0.45

Table 16.5 Resistance to oxygen transmission ($1 = 8400 \, ml/m^2$ per atm. per day at 25°C)

15–18 lb dry waxed sulphite paper	0.01
24–30 lb wet waxed sulphite paper	0.01
LDPE (25 m)	1.0
HDPE (12 m)	1.5
PE coated kraft paper ($25 \, g/m^2$)	1.5
Cellulose acetate (25 m)	2.0
PVC meatwrap (SAF)	2
Chill cast polypropylene (25 m)	2
LDPE (50 m)	2
HDPE (25 m)	3
Coextrusion coated OPP	3
Monoaxially oriented polypropylene	4
PVC produce wrap (SAC)	5
PVC for overwrapping (SAD)	30
Unplasticized PVC	50
Nylon 11	60
Polyester (12 m)	80
Nylon 66	100
PCTFE	160
PVDC coated polypropylene	500
PVDC copolymer	600
Plain cellulose film PT	700
Semi moisture proof QMS	800
Copolymer solution coated cellulose MXXT (S)	900
Dispersion coated cellulose MXXT (A)	1000
Briphane	1200

oxygen, acetylene, nitrogen and air, both alone and as mixtures through rubber membranes. They established that for wide differences in composition, the rate of permeation of a gas mixture was equal to the sum of the rates of its constituents. Other workers have confirmed this with other film/gas systems. Thus, it would appear that a gas diffuses through a membrane independent of the presence or absence of other diffusing gases.

Where one of the diffusing molecules interacts strongly with the film material, the permeation rate of both may be changed, but this is due primarily to the plasticizing effect of the interacting gas increasing the 'looseness' of the film structure. Stannett and co-workers [15] made studies using nitrogen, oxygen and carbon dioxide with various plastic films. Providing the gases were mixed adequately, no difference in the permeability was found whether the gas diffused alone or in the presence of another gas. They also made a study of the system water vapour/gas through three types of film, two of which were hydrophobic while the third (nylon) was hydrophilic. With the hydrophobic films no effect was found, but with nylon, the rate of gas transmission increased as the relative humidity increased. This is explained on the basis of the sorbed water acting as a plasticizer and increasing the 'looseness' in the structure.

Estimating the type of barrier required

Laminated and coated materials.   Laminated and coated materials can be
considered as an arrangement of membranes in series. In the simplest instance
of a laminate consisting of two barriers of thickness x_1 and x_2, it can be shown
that if the permeabilities of the two constituents are P_1 and P_2 respectively,
and the total thickness of the film $L = x_1 + x_2$, then the 'impedances' may be
added, or

$$\frac{1}{P_{lam}} = \frac{x_1}{L}\cdot\frac{1}{P_1} + \frac{x_2}{L}\cdot\frac{1}{P_2} \tag{16}$$

The equation will hold, providing both films show no interaction with the
diffusing gas. Where one or both do not meet this requirement, then the
resistances are no longer additive.

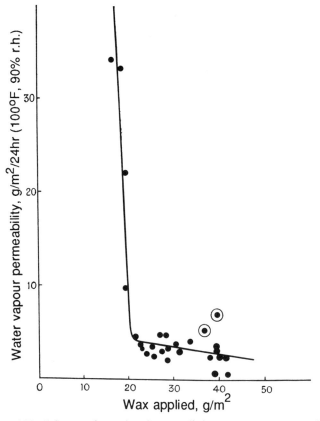

Figure 16.2   Influence of quantity of wax applied on water vapour transmission.

Coated and laminated papers. Where paper constitutes one of the 'barriers' in equation (16), a simple substitution shows that for water vapour transmission the presence of the paper should, theoretically, have little or no effect. In practice, however, the nature of the surface of the paper will have a profound influence on whether the amount of polymer applied will be effective in providing resistance to water vapour.

The first essential for any coating is the production of a complete pinhole-free film over the whole surface of the sheet. Figure 16.2 [17] shows the effect of increasing wax application in the lamination of a chip board to a white paper liner with a micro-wax composition. It can be seen that until about $20\,g/m^2$ have been applied, film formation is incomplete. It has also been shown that to obtain approximately the same water vapour permeability from laminations of glassine-to-glassine, glassine-to-greaseproof and greaseproof-to-greaseproof required 3–4, 9–10 and 12–13 g/m^2 of straight paraffin wax respectively. It will fairly easily be realized that glassine is smoother than greaseproof which, in turn, will be smoother than chipboard and plain paper. In addition, the properties of absorption will also be playing a part.

Where waxed papers are concerned, there will be a number of factors controlling the water vapour permeability, and among the most important of these will be the type of paper used, the condition of the surface, the quantity of wax applied, and the method of application. The last consideration can obviously be split into several variables, among which are the temperature of the wax, the uniformity of application across the web, the speed and tension of the web, the chilling temperature, and the time between the application of wax and chilling.

Coated papers are liable to be two-sided as far as water vapour permeation is concerned. Figure 16.3 is produced from Stannett's data [15] and shows the difference between exposing the paper to the high relative humidity and exposing the polymer. These results can be explained by assuming that the paper fibres penetrate into the polymer and when the paper side is exposed to the high relative humidity, the fibres swell and act as wicks carrying water vapour more easily, and thus reducing the effective thickness of polymer film. When the polymer is exposed to the high relative humidity, the fibres are not so swollen and hence the transmission rate is less.

The smoothness and absorbency of the surface of the substrate is also of considerable importance in determining the effectiveness of a given quantity of polymer. Figure 16.4 shows this in relation to a cellulose film, glassine paper, and a glazed imitation parchment paper coated with various amounts of polyvinylidene resin applied from a dispersion coating, using an airbrush applicator. The intercepts on the x-axis can be interpreted as a measure of the quantity of polymer necessary in each instance to fill up the irregularities in the surface, and provide a complete film without pinholes. After this point, the maximum advantage of the polymer applied is obtainable. It must be remembered that these figures are specific for the particular samples used, and

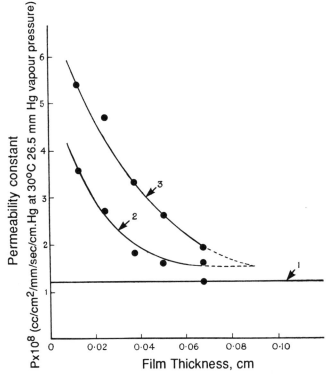

Figure 16.3   Effect of direction of moisture movement on coated paper. 1, free film; 2, polythene to high relative humidity; 3, paper to high relative humidity.

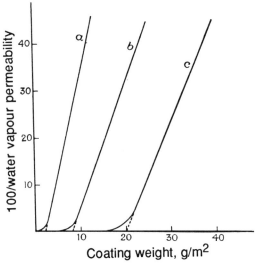

Figure 16.4   Effect of substrate on resistance of coated materials: (a) cellulose film; (b) glassine paper; (c) glazed imitation parchment (GIP) paper.

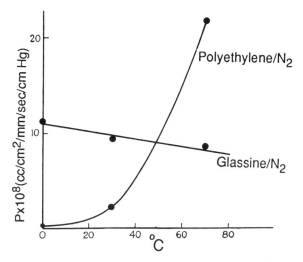

Figure 16.5   Temperature dependence of rate of permeation of nitrogen through polyethylene and glassine.

variation in the surface from one reel of paper to another can be quite considerable; e.g. glazed imitation parchment papers giving results much more similar to the glassine curve shown in Figure 16.4 than to curve (c).

Although its water vapour permeability is relatively high, it is interesting to note that a good quality glassine can be a very effective gas barrier, and, for certain gases, at higher temperature is equal to or better than some plastic films. Since transmission through glassine takes place principally by a 'capillary process' and not by 'activated diffusion', the temperature-dependence of the rate of permeation is not high, and as Figure 16.5 shows, it decreases slightly as the temperature rises. Hence, a laminate of polyethylene coated on to glassine could operate with the polyethylene controlling the rate of oxygen transmission at temperatures of 50°C and below, while at 50°C and above, the glassine would become the principal 'barrier'.

Protection of a moisture-sensitive product

Several methods have been published on the use of product, sales, packaging, climatic and other data to estimate the type of moisture barrier required for the protection of a moisture-sensitive product [18–21]. None is entirely satisfactory for all circumstances, and in order to make any assessment, it is necessary to have the following data available:

(1) The proposed weight of product per unit pack, W grams.
(2) The bulk density of the product, i.e. the volume occupied by a specific weight of the product. This permits the calculation of the dimensions of the package required (surface area of package S cm^2).

(3) The moisture content of the product at the time of packing, $M_0\%$.

(4) The critical moisture content of the product, $M_c\%$.

(5) The number of unit packs per outer case.

(6) Details of the method of distribution, i.e. whether the product will remain sealed in outer cases for a considerable period before the unit packs are removed for sale, or whether the unit package will be exposed as a unit after a shorter time in transit.

(7) The shelf life required, D days.

(8) An estimate of the average conditions to be experienced in the sales area, t (°C) and H (%r.h.).

It is also useful, but not absolutely necessary, to have data on two other points:

(9) The relative-humidity/moisture-content relationship of the product.

(10) The type of pack it is desired to use, e.g. a plain or multi-ply bag, a bag inside a carton, an over-wrapped carton, a lined composite container, etc.

Method of calculation. As a preliminary, it is useful to calculate the permeability of the material required assuming that (a) it is unaffected by folds and creases, and (b) all joins, seals and closure are 100% efficient. The problem is then to calculate the water vapour transmission rate of the package material under the conditions of use and to convert this figure to either standard temperate or tropical conditions.

The moisture content of the product may change from M_0 to M_c before the product is unusable. Therefore, the total permissible moisture uptake is

$$\frac{M_c - M_0}{100} \times W \text{ grams}$$

This quantity of moisture may be transmitted through $S\,\text{cm}^2$ in D days. Therefore, $1\,\text{m}^2$ will transmit

$$\frac{M_c - M_0}{100} \times \frac{W}{S} \times \frac{10\,000}{D} \text{ g/m}^2 \text{ per 24 h at } t°C$$

with an average r.h. differential of

$$\frac{2H - (H_0 + H_c)}{2}$$

After correcting this figure to either standard temperate conditions (25°C and 75% r.h. differential) or to standard tropical conditions (38°C and 90% r.h. differential) we obtain a measure of the problem. This approximate method of calculation provides a useful breakdown of problems involving moisture protection and can often indicate an economic solution where this is not obvious at first sight. Where information on the permeability of folds, creases, and closures is available, due allowance may be made for these in the calculations.

Packaging barriers and their relation to moisture changes in foods

The activity of water in foods can range from about 0.05 to 0.95, and the water vapour pressure in the surrounding atmosphere ranges normally from 20% to 100% of the saturated value; the two are rarely equal. Generally the food is not in equilibrium with the atmosphere when it is first exposed, and the difference in vapour pressures constitutes a potential which drives the system to equilibrium by transfer of water to or from the food. Packaging opposes this tendency by inserting an effective 'resistance' (R) between the food and the potential gradient. Perfect packaging would completely isolate the food from any external environment, but such packs are expensive, and usually unnecessary. A package which impedes the change, retards the equilibration process to such an extent that deterioration does not occur before the food is consumed. The subject is reviewed by Hine [43] and he lists the following factors which must be taken into account when predicting shelf life:

(1) The mechanism of deterioration of the product
(2) The agents responsible for control of the rate of deterioration
(3) The quality of product in the pack
(4) The desirable shape and size of the package
(5) The 'quality' of the product when packed
(6) The minimum acceptable quality of the product
(7) The climatic variations likely to be encountered during distribution and storage
(8) The mechanical hazards of distribution and storage that may affect the integrity of packs
(9) The distribution unit, whether this be an individual pack or a collation of such packs in a transit pack
(10) The barrier properties of packaging materials against the agents causing product deterioration
(11) The influence of conversion of packaging materials into packs on the barrier properties
(12) The significance and distribution of defects in barrier performances of production packs.

The task of a shelf life prediction technique is to take this data, together with marketing requirements, to give an estimate of the time before the product is unacceptable. The definition of what is acceptable is crucial to this process. Setting this limit follows from an understanding of item (1).

In the simplest cases of controlling weight-loss (e.g. in bread, or vacuum-packed cheese) the loss of small amounts of water does not make much difference to the activity level, and the potential gradient (ΔP) is virtually constant. The change in water content (q) then follows a simple Ohm's Law relation, $q = t \cdot \Delta P / R$, and the shelf life is proportional to the resistance of the pack.

Retention of texture such as a crisp texture due to a high water content as in salads or to a low water content as in biscuits, is more complicated. The loss or gain of some of the water causes a change in the activity, and the potential gradient falls as equilibrium is approached. The food has a limited 'capacitance' (C) for water and to a first approximation the change can be seen as an exponential process, having a time constant equal to $C \times R$ (during which time 63.2% of the equilibrium change occurs).

Such foods have a limited tolerance for change in water content before there is a serious loss of crispness, and the problem is to adjudge the shelf life (t') at which the change in water content (q') is still just acceptable. Since commercial practice aims at a limited period between production and consumption, the problem is one of finding the probable potential difference in practice, and selecting the appropriate 'resistance'. The resistance needed can be quite high if the product has a low tolerance for change and the intended shelf life is long: a metal can or a glass jar with good closures would be necessary. For more tolerant foods with well organized distribution systems this would be excessive protection, and a flexible wrapping can be used as the barrier. The same mathematical model applies.

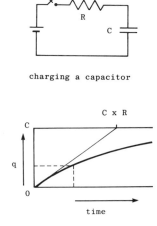

charging a capacitor

package   analogue

Figure 16.6 Analogy between electrical and packaging capacitance (see text).

The relation quoted is only approximate, the relation between water content and activity actually being sigmoid and not linear. Theoretical equations [22] have been published for homogeneous adsorbents, but they do not give a good fit with mixtures. It is sufficient to use an empirical equation [23, 24] and this in turn leads to a simple approximation [25] for the shelf life $C \cdot R/t = C/q - q/C$ which, although it gives a poor fit above $q = 0.4C$, is good enough since few foods can tolerate such a large change [26–28] without spoiling (Figure 16.6).

To illustrate the use of this equation consider the following problem: water biscuits are normally packaged at a moisture content of between 1.2 and 1.6% on dry weight. Ten packages were weighed immediately on leaving the packaging line and after 10 days storage at 25°C, 75% r.h., with the following results:

Package no.	Initial weight	Weight after 10 days storage
1	210.5	211.6
2	211.0	212.6
3	214.0	214.9
4	212.3	213.6
5	209.7	211.0
6	213.1	214.5
7	210.8	212.0
8	212.7	214.1
9	213.0	213.9
10	209.9	211.1

Three individual biscuits were extracted from the centre of another package taken at the same time; they were each weighed rapidly, allowed to come to equilibrium with the atmosphere in the storage chamber (which they did in 5/6 days) and their equilibrium moisture content on dry weight determined by drying at 105°C. The results averaged 12.4%. Given that the critical moisture content of the biscuits (i.e. the maximum possible without customer complaint) is 3.4% estimate the average shelf life at the storage conditions using the formula

$$\frac{CR}{t} = \frac{C}{\Delta} - \frac{\Delta}{C}$$

The estimation proceeds as follows:

(1) Calculate the average initial weight of the packages: 211.7 g
(2) Deduct the average weight of packaging: 8.8 g
(3) Hence the average weight of biscuits at 1.4% moisture = 202.9 g
(4) Calculate the dry weight of biscuits: 202.9 × 98.6/100 = 200.0 g

(5) Subtract each initial package weight from the corresponding weight after 10 days to give the uptake in 10 days.

1.1
1.2
0.9
1.3
1.3
1.4
1.2
1.4
0.9
1.2 } Average 1.2 g in 10 days, hence 0.12 g per day

(6) Calculate the mean weight gained by biscuits at equilibrium with 25°C/75% r.h. from experiments with single biscuits:

$$12.4 - 1.4 = 11\%$$

$$11\% \text{ of } 200 \text{ g} = 22 \text{ g}$$

(7) Use the equation to determine CR from the above data:

$$CR/1 = 22/0.12 - 0.12/22$$
$$= 183 - 0.0005$$
$$= 183$$

(8) Calculate the critical moisture gain in grams from the permitted moisture uptake in %.

$$3.4 - 1.4\% = 2\%$$

$$2\% \text{ of } 200 \text{ g} = 4 \text{ g}$$

(9) Insert this extra data in the equation to estimate the shelf life (T) in days:

$$183/T = 22/4 - 4/22$$
$$= 5.5 - 0.2$$

$$T = 183/5.3$$
$$= 35 \text{ days}$$

At high water activities, any moulds and bacteria soon flourish on foods, at low activities oxygen can be absorbed and react to develop rancid flavours. At activity values near 0.5 the Maillard reaction can cause discoloration. In between these there are two areas which offer longer shelf life. Each of the hazards can be eliminated by specific chemical preservatives or by physical deactivation, but the packaging still helps to keep the water activity within bounds. Where diffusion of water through the food is controlled by the activity gradient, it may be sufficient to let the surface of the food dry to an activity of 0.75 while the inside remains moist. The surface is then too dry for rapid mould growth. The corresponding electrical model is two resistance in series, with a specified potential required at the junction. Too high a resistance in the package could then shorten the shelf life (by mould growth) just as too low a resistance could shorten it by browning or overdrying. A wide range of packaging materials [29] has been developed to offer the necessary choice. The barrier properties of available materials are indicated broadly, by family, in Figure 16.7 [30].

The range of flexible wrappings spreads over five decades of resistance to water vapour and to oxygen. In many packaging applications atmospheric oxygen is also a factor [31] to be controlled, so the families of wrappings (Figure 16.7) are shown in two dimensions by their respective resistances to water and oxygen. The fact that families are indicated by continuous lines does not mean that continuously varying values are available: commercially the values are isolated, but they can be extended by using films of different thicknesses. Except for laminates and coated films, resistance is roughly proportional to thickness. In using the figure, those resistances to water vapour which are too low to be considered can be eliminated. Similarly, oxygen resistance can be taken into account and one side eliminated. This leaves a 'window', or rather, a funnel, since the window 'closes' with extending time. The wrapping is selected by the cost and other properties from the

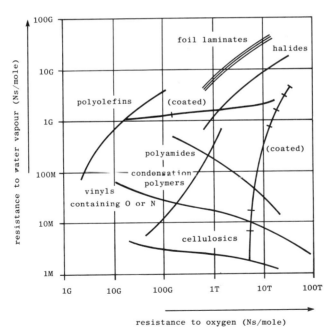

Figure 16.7   Emballistics—selecting plastic films for packaging. Barrier properties of 25 μm film (note SI units, log scales). Data from Oswin [23].

families still visible. In the examples shown in Figure 16.8, (a) shows the simple instance of biscuits; in the more complex examples, too high a resistance is seen to be as damaging as too low; (b) shows a chart for bread and (g) illustrates the difficulty of packing red meat. The generalization is, of course, imprecise but it serves to select materials to be field-tested.

Temperature has so far been ignored in the model, but it is an important factor affecting potential gradient, resistance and spoilage rate. The effects are not identical, and as Figure 16.9 shows there can be three regimes of spoilage

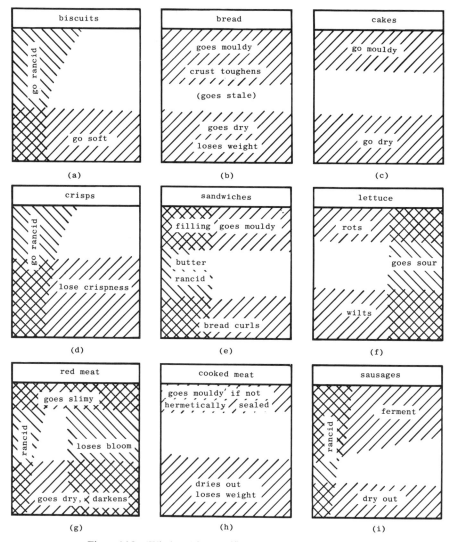

Figure 16.8 'Windows' for specific products. After Oswin [23].

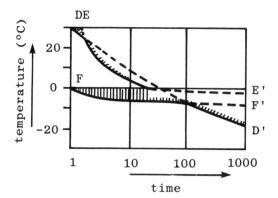

Figure 16.9 Spoilage at different temperatures. Data from Oswin [23]. DD′, spoilage of a moist food by simple dehydration; EE′, mould growth; FF′, spoilage of texture by ice crystal formation.

dominant at different temperatures. The curve *DD′* represents the time for spoilage of a moist food by simple dehydration: *EE′* represents mould growths, and *FF′* spoilage of texture by ice-crystal growth.

In such a complicated field as food preservation, it would be unwise to select a packaging material solely by this process, but the mathematical model provides enough insight into shelf life estimation to enable past experience to be extrapolated. It also provides a check to see whether the selected wrapping is used in a way which enables it to give its full potential benefit. If the right flexible wrapping is chosen, and used correctly, considerable saving can result.

The measurement of gas transmission rate

To understand the principles by which gas permeabilities are measured it is necessary to appreciate the meaning of two terms which are constantly used— *total* and *partial pressure* of gases in a mixture.

In a given fixed volume, the total pressure is that pressure which results from all the constituents of a mixture of gases, while the partial pressure of any specific component gas is the pressure which would result if that particular component was to occupy the same volume. For instance, in considering a mixture of 55% nitrogen, 30% oxygen and 15% carbon dioxide at a total pressure of 1000 mbar, then the partial pressures of the three components are:

- Nitrogen   550 mbar
- Oxygen   300 mbar
- Carbon dioxide   150 mbar

The rate at which a specific gas moves through a non-porous barrier material is a function of the partial pressure differential of that gas across the barrier, and not the total pressure difference between the two sides.

There are three basic methods [25] of determining the gas transmission rate of a material: by a pressure increase method, a volume increase method or by measuring the change in concentration.

Pressure increase method.   In this method it is usual to test with a total pressure difference across the barrier and with the same gas on each side of the specimen. Hence the (partial) pressure difference of the gas on the two sides of the specimen is equal to the total pressure difference, and gas passes from the high-pressure to the low pressure side of the barrier.

Volume increase method.   In the volume increase method the change in volume (at constant pressure) due to the permeation of the gas through the barrier is determined. Various types of apparatus have been designed and these generally show slightly less deviation of results ($\pm 4\%$) than pressure increase methods, but the equipment is usually more complex because of the necessity to adjust pressure before measuring the change in volume. Compared with the use of pressure increase or concentration increase methods, volumetric methods are little used.

Concentration increase methods.   In the concentration increase method two gases are used, a reference gas and the test gas of which the transmission rate is to be determined. A partial pressure difference across the barrier material with respect to the test gas is created without any difference in total pressure. Such methods are also referred to as *isostatic* methods. Because the pressure on each side of the specimen is the same, there is no need for a rigid support of the specimen. Furthermore, since the method of measuring the concentration of the test gas can be specific to that gas even in the presence of other gases or vapours, equipment can be developed in which the relative humidity of both the test and the reference gases can be controlled. This is not possible where equipment has to be evacuated, and is of prime importance when determinations are carried out on barrier materials having a hygroscopic base, such as regenerated cellulose. The oxygen permeability of cellulose film, for example, at various relative humidities has been measured and above 65% r.h. the rate of transmission increases rapidly. A permeability of some 3000 ml/m^2 per day atm. at 100% r.h. contrasts with about 10–20 ml/m^2 per day atm. for a dry film.

Various methods of measuring the concentration change are employed, such as chemical analysis, gas chromatography, thermal conductivity measurements or radioactive tracer techniques. It is usual to employ a sweep gas technique although when testing materials of very low permeability the reference gas side of the cell can be isolated for a convenient period and the concentration of test gas found by an integrating technique when the gas flow is restored. It is necessary to control the flow rate of the sweep gas to within close limits, and an accurate flowmeter is needed to measure the rate.

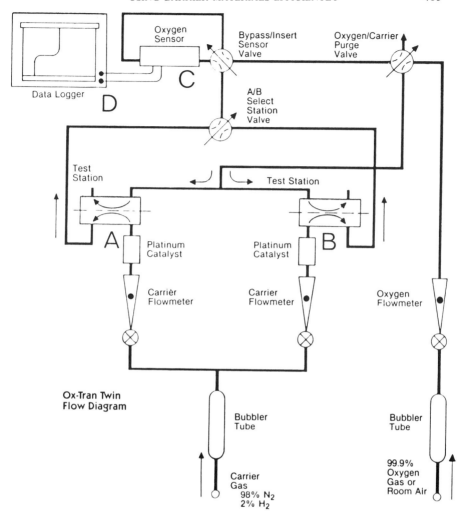

Figure 16.10 MOCON's Coulometric Oxygen Detector is a patented nickel–cadmium graphite electrode arrangement saturated with basic electrolyte to stimulate a high rate of electro-absorption of oxygen. The closed system produces a current output directly proportional to the amount of oxygen passing through the packaging material being tested which is displayed directly on the recorder. For film testing, flat 4×4 inch test specimens are clamped into special $50\,cm^2$ diffusion cells A and B. Both sides of the film sample are initially purged with an oxygen-free carrier gas to remove residual oxygen from the system and to desorb oxygen from the sample. When a stable zero has been established oxygen is then introduced into the upper half of the diffusion chamber. Carrier gas continues to flow through the lower half, carrying into the coulometric detector, C, any oxygen molecules which have diffused through the barrier. The strip chart recorder, D, indicates an increase in current through a special load resistor which is directly proportional to the oxygen transmission rate. After a short time, the recorder trace indicates visual verification of the equilibrium transmission rate of oxygen through the sample barrier.

Table 16.6 Relation between permeability to gases of various flexible materials (nitrogen transmission rate taken as unity)

Material	Oxygen/nitrogen ratio	Carbon dioxide/oxygen ratio
LDPE	2.9	6.5
HDPE	3.9	3.3
PVdC copolymer	6	5.2
PET	4.4	7
UPVC	3	8.3
PS	3.8	7.9
Coated OPP	4.6	2.7
PA	3.8	4.2
Average (range)	4 (3–6)	6 (3–8)

Oxygen transmission rate is probably the most important measurement required and the well established MOCON equipment, which is specific for oxygen or air, is the most used means of measuring it (Figure 16.10). The method is standardized in ASTM D3985-81. Since there is a relationship between the transmission rates of nitrogen, oxygen and carbon dioxide through the same barrier material, once one is known the others can be interpolated. Nitrogen is transmitted at about one-third the rate of oxygen and carbon dioxide about 6 times as fast (Table 16.6).

Measurement of water vapour transmission

The standard dish method for the determination of the water vapour transmission rate of flexible sheet materials is detailed in BS 3177 for the UK and in the USA is given in TAPPI Standard T448-M49. In the dish method, a water-absorbing material (anhydrous calcium chloride or silica gel) is separated from a controlled high humidity atmosphere by the material being examined, which is sealed over it in a dish. These methods, in which the moisture gain is determined by direct weighing, have several practical disadvantages, not the least of which are the length of time (between 1 and 14 days) needed to make a determination and the lower limit of the useful range (about 1 g/m^2 per day). When dynamic equilibrium conditions of moisture transfer have been established within and across a homogeneous non-porous barrier material, then the quantity of water vapour permeating through that barrier is given by

$$Q = \frac{P(p_1 - p_2)A \cdot t}{l} \qquad (10)$$

where Q is the quantity of water vapour, t the time, $(p_1 - p_2)$ the difference in water vapour pressure across the test specimen, A the area of test specimen, P the permeability constant of the material at a given temperature and l the thickness of the specimen.

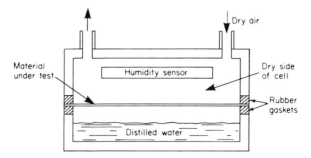

Figure 16.11   Schematic diagram of cell for rapid WVTR measurement.

It should be noted that most methods determine the water vapour transmission rate $Q/(A \cdot t)$ under specified conditions of temperature and water vapour pressure difference, and not the permeability constant, P, the value of which varies appreciably with temperature and (for barrier materials with a hygroscopic substrate) the moisture content of the base material.

The standard dish method has two main disadvantages. First, the method is slow, and second, the lower limit of $1\,\mathrm{g/m^2}$ per day renders it unsuitable for many of the better materials, particularly laminates. Over the last 25 years equipment has been developed to improve the sensitivity of measurement, while reducing the period of test to a few hours or less.

Most rapid methods of determining WVTR [25] depend on detecting small changes in the relative humidity of the atmosphere on the dry side of the barrier. Consider a test cell (Figure 16.11), the lower section of which contains water to give a saturated atmosphere (100% r.h.). The test specimen is clamped between the two halves of the cell to provide a leakproof seal. The complete cell is maintained at constant temperature. The upper section of the cell, containing the humidity sensor, is dried down by purging with dry air, and when sufficiently dry, the air inlet and outlet tubes are closed to isolate the section. The movement of water vapour through the test specimen now raises the relative humidity of the air surrounding the humidity sensor, the time for a given rise can be recorded and from this the WVTR calculated.

The humidity sensing elements vary; the infrared diffusion meter, resistance changes and the current flowing as water is decomposed by electrolysis in the so-called 'coulombmetric' cell have all been used.

Pests

Insect infestation

Many stored products, especially cereals, dried fruit, nuts, spices, cocoa and coffee, have a sufficiently low moisture content to preclude the possibility of mould growth. They are considered non-perishable and are stored for considerable periods. It is in this type of product that insect infestation is most serious. The extent of damage caused by insects is difficult to assess, but estimated losses of as much as 20% of particular crops in warm climates are not unusual. The kinds of insects that do most of this damage are moths and small beetles. The life histories of two of these insects will serve to illustrate the nature of the problem, but it must be remembered that there are many different species involved and their habits vary considerably.

Moths. The Mediterranean flour moth or mill moth (*Ephestia kuhniella*) is now cosmopolitan and found on all kinds of cereals, flour, dried milk and cocoa. It is a small slate-grey moth, 15–20 mm across the expanded wings and about 5–10 mm from head to tip of abdomen. During the day the moths are sluggish and remain resting in sheltered places. At night they become active and egg laying occurs. One female can lay up to 250 eggs. The eggs are laid singly or in small clusters lightly fastened to the surface on which they are deposited, usually on or near the future food supply, and often through the meshes of a textile sack on the grain or beans inside. The moths prefer a still atmosphere and dislike draughts. They live only 9–14 days and never take any food or water.

The eggs hatch after about 7–14 days (at 26°C (80°F) in 5–9 days) and the tiny white caterpillar-like larva immediately searches for a crack in the surface of a grain or cocoa bean and there begins to feed, spinning round itself a silk tube. The larva must quickly find easily obtainable food, for it soon dies of starvation and is unable to eat through the hard undamaged shell of a bean or grain. The larval period lasts for 10–22 weeks according to the temperature; the larva grows and casts its skin to suit its size during this time and is finally about 17 mm long and white to pinkish in colour. When fully grown the larva eats its way out of its feeding place and searches for a crevice in which to spin a cocoon and pupate. Wherever the capterpillars go they trail a silk thread behind them and the mass of webbing so produced is obvious. The larvae may wander for 2 days or more before they finally spin a cocoon. Within the cocoon a brown pupa is formed and after about 11–16 days (9–12 days at 26°C) the adult moth emerges. At least one and probably two generations will be produced in a year. Eggs laid in September or October hatch larvae which may not pupate for 5–6 months, but in heated buildings five or six generations may be produced in a year.

Beetles. Beetles have a similar life cycle to moths; they lay eggs which hatch into larvae which feed, grow and pupate, the pupae later hatching into adults. There are many hundreds of different species which vary considerably in their habits and food requirements, but the following life history will serve to illustrate the factors affecting beetle infestations. The leather beetle (*Dermestes maculatus*) is a dark-brown to black coloured beetle 5–9 mm in length; it is covered in black hairs on its back and white ones below. It is cosmopolitan in distribution. The optimum temperature for development is 18–20°C. The female begins to lay eggs 10–15 days after emergence and will only do so at temperatures above 16°C. The females must have water to drink even in an atmosphere with a relative humidity as high as 73%, and they do not lay eggs in the absence of food. The adult female may live more than 92 days and lays up to 845 eggs. The incubation period of the eggs varies from 2 to 12 days according to the temperature, and the larvae which emerge are very active and move away from light.

The larva grows through several moults, feeding on proteinaceous material such as leather goods, hides and skins, glue and bristle, dried fish, feathers, and sometimes grain and cacao beans. The larval stage lasts from 35 to 238 days according to temperature. When fully grown the larvae leave their food to find a suitable place to pupate. They will bore a pupal chamber in almost any compact substance that happens to be near at hand, and this indiscriminate boring causes damage to all kinds of materials that are not used as food. Commodities in which pupae are known to have bored before pupation are cork, hardboard, books, tobacco, tea, linen, cotton, woollens, salt, sal-ammoniac, plaster moulds, flexible asbestos and lead.

Bearing in mind the details of these life histories, it is easy to see that storage conditions and package design are important factors in controlling insect infestations. In warehouses scrupulous cleanliness maintained by frequent vacuum cleaning or sweeping of floors, walls, ledges and beams, and the immediate destruction of waste, prevents the accumulation of sources of food so essential to the newly-emerged larvae. Premises that are cool, dry and well-ventilated help to discourage insects, and low temperatures greatly reduce their rate of multiplication. Neat stacks away from walls and with adequate gangways permit frequent inspection and cleaning. Intake and outturn on a strictly rotational basis avoids the accumulation of old and neglected stock. Precautionary insecticidal spraying of walls and floors is advisable when susceptible materials are stored and when an infestation is detected, immediate action should be taken to utilize the material and to fumigate the store, but fumigation must be accompanied by thorough cleaning and is a task that should only be undertaken by experts.

When considering the type of package best suited to keep out insects, it is obvious from the minute size of the newly hatched larvae and their preference for cracks and crevices that small holes and poorly fitting closures provide easy access. Paper sacks provide a better protection than do fabric sacks, but stitching holes are a possible entry site for insects, and taping these reduces the risk of infestation. Loose folds, poorly glued seams, and the corners of cartons are all danger points.

There is no material that is absolutely insect-proof in all conditions; as previously stated, some larvae can bore into lead when searching for a pupation site. But nevertheless, in general, the stronger and thicker a packing

Table 16.7   Rate of penetration of typical packaging materials by a mixed population of insects[a]

Packaging material	Thickness (mm)	Average no. of weeks before penetration[b]
Cellulose film	0.023	3
	0.036	3
	0.041	$3\frac{1}{2}$
Polyethylene film	0.038	3
	0.050	3
	0.100	3
PVC/PVdC copolymer film	0.038	3
	0.050	4
Polyethylene terephthalate	0.025	6

[a] 11 types of boring insect including *Rhizopertha dominica*.
[b] Envelope tests: food inside the envelope.

Table 16.8   Rate of penetration of paper-based packagings by grain beetles (*R. dominica*)

Material (folded and creased) in a funnel test	Average no. of weeks to penetrate
2 plies sack kraft paper	2–3
3 plies sack kraft paper	3
5 plies sack kraft paper	6
Bitumen union kraft paper	1

material is mechanically, the more resistant it will be to insects. Smooth surfaces are preferable to rough ones, as they do not afford a purchase for the insect to begin gnawing on, and an avoidance of unnecessary folds and creases reduces places likely to harbour insects. The relative strengths of some common packaging materials to insect penetration are listed in Tables 16.7 and 16.8. It is also possible to apply insecticidal coating to some materials, which affords considerable protection.

Termites.   In considering insect damage to stored goods, no mention has so far been made of termites, as these insects present a totally different problem. There are many kinds of termite, but all are social insects living in large colonies in warm climates with a mean annual temperature above $10°C$ ($50°F$). They feed mainly on wood; but paper and board are also readily eaten. The colony lives always in darkness, tunnelling in the ground or in timber, some species making huge mounds, and in others moving over inedible barriers (such as concrete foundations) in earthen tubes, always hidden from the light.

There are two main groups of termites. In the first the colony must retain tunnels to the soil in order to obtain enough moisture. The others (drywood termites) form colonies in wood where they cannot obtain moisture from the soil. Thoroughly applied timber preservatives and continuous coats of paint help to protect wood against termites, but paper and board cannot easily be protected in this way as their strength is inadequate. Carefully constructed warehouses, with termite shields on the foundations and netting over windows to prevent the entry of flying drywood termites, together with frequent inspection, afford the best protection against termite damage.

Rats and mice

Rats and mice are a frequent cause of damage to stored food. Nearly all packaging materials can be attacked by rodents; even tins have been eaten though by rats. Although they eat a considerable amount of food, far more is wasted because of spillage from damaged packs and also general fouling. Many attempts have been made to find a repellent chemical with which to treat paper or jute sacks to deter rodent attack. But these have all been found almost useless because treated materials that are repellent to the rat when biting his way through the sack have no effect on his capacity for tearing with his claws. As even tins can be eaten through and repellents are useless, it is obvious that the best protection is afforded by keeping storage places free from infestation. Constructional details that keep out rats are the covering of gullies and drainage fittings, basement windows and cellar gratings with rat-proof covers such as galvanized wire netting of 12–15 mm mesh. Doors and thresholds should fit well and be covered with metal sheeting. All rubbish should be destroyed quickly so as to deny them shelter. Finally, freedom from infestation can be ensured by careful observation and prompt destruction wherever rodents are found. Poisoning, gassing and trapping can all be effective if properly applied by experts.

The importance of clean, frequently inspected, storage conditions cannot be overstressed. In all the examples of biological deterioration that have been discussed, no amount of careful packaging, expensive fungicides or insecticides can replace good housekeeping in warehouses, transport vehicles and retail stores.

The compatibility of foods with their packaging

Few materials suitable for food packaging are completely inert towards food and those materials that are inert often have to be used in conjunction with others that are not.

The care that is taken to produce wholesome and attractive foods needs to be matched by the care taken in the production of the packaging used to contain and protect them. Recognizing this, responsible suppliers of packaging materials and food manufacturers have for many years worked together for the benefit and protection of the consumer. Most countries have food regulations in this respect and those of the Food and Drugs Administration

(FDA) in the UK and of the European Commission (EC) in Europe are probably the best known. Generally in the USA the need is expressed by the requirement that 'food shall be of the substance and quality' demanded. Only packaging materials that do not adversely affect either the flavour or the wholesomeness of the food are allowable.

Practical, economic, and other factors such as convenience and effectiveness of presentation, have, of course, influenced the choice of material. For some food packaging situations it has been convenient to express the properties of a satisfactory packaging material in terms of a voluntarily agreed standard, e.g. vegetable parchment for wrapping dairy and other food products. The BPF– BIBRA Code of Practice for plastics for food contact applications is another example of self-imposed recommendations that are voluntarily observed.

Since we receive almost all our food in packaged condition, we need to know the extent of any interactions between foods and the materials or containers in which they are packaged. Clearly, any interactions must be small. If they were large the container would cease to be an efficient container, e.g. wet foods are not packed in unprotected paper, nor acidic foods in unprotected metal cans, for obvious reasons. Small interactions can be detected as appropriate by the senses of taste and smell, or the sensitive instruments available to the analytical chemist, or both. These interactions may result from the movement of volatile or non-volatile substances from the packaging material into the food. It is also possible for the reverse to occur, i.e. movement of a food component into the packaging can result in a nutritional loss. The volatile materials can move (migrate) without physical contact between the product and its package but for non-volatiles there must be contact. If any of the migrants are toxic, harm to the person ingesting the food or medicine could result.

Volatiles

Volatiles from the food.   Although loss of volatile substances from a food may lead to loss of desirable aroma and flavour characteristics, it does not provide any health risk. Selecting a packaging material that possesses not only the physical properties necessary for it to contain the food for the intended storage period, but also the barrier properties necessary to prevent loss of flavour and aroma during that period, is a matter for the food manufacturer to decide on the basis of quality. He must decide the degree of change that can be tolerated, and how to control it. Here compatibility is concerned only with the quality, not the health safety, of the product in question.

Volatiles from the packaging.   Volatiles from packaging materials may be odorous or non-odorous. They may or may not give rise to changes in the flavour of food that is kept in their vicinity, according to whether they are absorbed by the food and whether they are detectable by smell or taste.

There are two situations to be considered. In the first, if volatile odours are present at an unacceptable level in the packaging material, then under proper conditions of quality control, such material will never be taken into use for food packaging. The test methods are organoleptic, since odour and taint can be perceived only by the nose and tongue. If the odorous volatile material or materials coming from the packaging are chemically identifiable and not too numerous (e.g. are common solvents, up to four or five in number, and not the inexactly definable mixture that constitutes the odour from unbleached kraft paper, or strawboard) then under certain conditions gas chromatographic analysis of the odour from the packaging material can be used as the test procedure.

The necessary conditions are (1) that taste panel testing and gas chromato-graphic testing have been carried out in parallel on a sufficient number and range of packaging material samples to enable the numerical chromato-graphic results (usually mg solvent/m^2 packaging material), to be correlated with panel judgements of acceptability, and (2) that a proper chromatographic procedure is used. This second condition may sound a little strange, but anyone with experience in this field will be aware of the difficulties of obtaining satisfactory agreement between the laboratories of suppliers and buyers of packaging material.

An important example of the second situation where the volatile material is not odorous is that of vinyl chloride monomer derived from polyvinyl chloride. The quantity of this volatile considered sufficient to make a PVC container incompatible with food is far too small to be detected by odour. Conventional head space gas chromatographic analysis, using a flame ionization detector, is just adequate for some PVC samples, but detection by mass spectrometer is necessary for others, and is essential for the reliable analysis of the majority of samples of foods that have been in contact with or in close proximity to PVC-based packaging.

An EC directive on this subject specifies the limits as follows:

1 mg/kg in articles and materials intended for food use. In packaged foods there must be no trace of vinyl chloride monomer when examined by a method of analysis having a detection limit of 0.01 mg/kg.

Other monomers from plastics are also under scrutiny, and similar safe-guards will certainly be imposed if toxicological and migration studies show them to be necessary.

Non-volatiles

Wooden articles, paper products, woven vegetable fibres, and from earlier times, animal skins, horns and other natural containers have such a long history of use as food containers that the probable transfer of some of their non-volatile components into the foods placed in them has largely been ignored. Only recently has the compatibility of any of them with food been questioned. Even now the most important question about paper and board products concerns their extractable heavy metals; a question that is particularly relevant for recycled paper, which, it is urged on economic and ecological grounds, should be given fresh consideration by those involved in food packaging.

Ceramic articles also have a long history of use as food containers, and relatively recent concern about the possibility of the transfer of lead and other toxic metals (notably cadmium) from glazes and decorative finishes has brought the compatibility of this type of container into question.

Plastics have chemical structures (and additives) that are different from those of the animal and vegetable world and a comparatively short history of use. Their compatibility with food has therefore been closely examined. Interest here is not confined solely to plastics containers and plastics packaging films; plastics are also involved in metal can lacquers, in glass bottle closures, and in many laminates.

Plastics, as synthetic polymeric materials, inevitably contain some or all of the following classes of substances:

(1) Residues of polymerization initiating and controlling chemicals
(2) Aids to converting/processing/fabricating
(3) Additives to modify the physical properties of the basic polymer, e.g. to give it light and heat stability, flexibility, etc..
(4) Pigments and fillers (extenders)
(5) Low molecular weight polymeric material.

A list of the chemicals covered by these five classes would be a very long one, and the analysis of foods for the presence of traces of these chemicals present as a result of migration is clearly impracticable. In most instances it is impossible. Assurance concerning the compatibility of plastics and foods must therefore be sought by some means other than direct analysis of food that has been exposed to plastics.

The means to date has been the analysis of so-called food simulant liquids that have been suitably exposed to plastics. Thus, for *neutral, aqueous, non-alcoholic* foods, water is the simulant. For *acidic aqueous, non-alcoholic* foods, 3% acetic acid is the simulant. For *alcoholic foods*, 5 or 15% alcohol in water is used (or the actual percentage in the food if higher than 15%), and for *fatty foods*, an edible oil such as olive oil or a synthetic triglyceride, is the simulant (see Table A.5 in Appendix 1).

Legislation based on this or similar systems of test liquids has been enacted in many countries, the EC and the USA bring the most promiment (see Appendices).

References

1. J.K. Mitchell, *Am. J. Med. Sci.* **7** (1830), 36.
2. Graham, *Philos. Mag.*, **32**, 401 (1866).
3. Wroblewski, *Ann. Phys. Leipzig*, **8** (1879), 29.
4. Hufner, *Ann. Phys.* **34** (1888), 1.
5. Reychler, *J. Chim. Phys.* **8** (1910), 617.
6. Edwards and Pickering, *Sci. Paper Bureau of Standards* **16** (1920), 327.
7. Dewar, *Proc. Roy. Instn.* **21** (1914–1916), 813.
8. H.A. Daynes, *Trans. Inst. Rubber Ind.* **3** (1928), 428.
9. Kanata, *Bull. Chem. Soc. Jpn.* **3** (1920), 183.
10. Sager, *Bureau Std. J. Res. Wash* **19** (1937), 181.
11. R.M. Barrer, *Diffusion In and Through Solids,* Cambridge (1941).
12. H.A. Daynes, *Proc. Roy. Soc.* **A97** (1920), 273.
13. R.M. Barrer, *Trans. Faraday. Soc.* **35** (1939), 628.
14. Muller, *Kolloid Zt.* **100** (1941), 355.
15. V. Stannett *et al., TAPPI* **39** (No. 11) (1956) 741; V. Stannett *et al., TAPPI* **41** (No. 11) (1958) 716; V. Stannett *et al., TAPPI* **40** (No. 3) (1957) 142; V. Stannet *et al., TAPPI* **40** (No. 7) (1957), 564.
16. Alexejev and Matalskii *J. Chim. Phys.* **24** (1927), 237.
17. F.A. Paine and Y.E. Turner, PATRA Interim Report, No. 89/1/54.
18. C.R. Oswin, *Protective Wrappings,* Cam. Publications, London (1954).
19. F.A. Paine, *Chem. Ind.* **52** (1957), 1656.
20. R. Heiss, *Chem. Eng. Tech.* **28** (1956) 763.
21. C.L. Brickman, *Pack. Eng.* **2** (7) (1957), 19.
22. S. Brunauer *et al., J. Am. Chem. Soc.* **62** (1945), 1723.
23. C.R. Oswin, *JSCI,* **65** (1946), 419.
24. T.P. Labuza, *Food Technology* March (1968), p. 15.
25. J.A. Cairns *et al., Packaging for Climatic Protection,* Newnes-Butterworth (1975), p. 75.
26. R. Gane, *UK Progress Reports on Dehydration No. XI,* HMSO, London (1942).
27. T.P. Labuza *et al., Trans.* A.S.A.E. (1972), 150.
28. L.V. Burton, in *Food Processing Operations,* M.A. Joslyn ed., vol. 3, AVI (1964), pp. 473, 477.
29. C.R. Oswin, *Plastic Films and Packaging,* Elsevier Applied Science, London (1975).
30. M. Salame, AMI. Chem. E. (Maths and Eng. Sci. Div.) Conference preprints (1970), p. 290.
31. T.P. Labuza, in *Water Relations of Foods,* R.B. Duckworth (ed.), Academic Press, New York (1975), p. 462.
32. Guidance on avoiding odour from packaging materials used for foodstuffs, British Standards Institution P.D. 6459 (All British Standards are available from British Standards Institution, 2 Park Street, London W1A 2BS).
33. Methods of test for the assessment of odour from packaging materials used for foodstuffs, BS 3755.
34. Vegetable Parchment (for wrapping dairy products and other foods), BS 1820.
35. Permissible limits of metal release from glazed ceramic ware, BS 4860 (1972).
36. EEC Directive 76/893/EEC 23 (1976).
37. EEC Directive COM (74), 2173.
38. K. Figge *et al., Mitt. Gebiete Lebensm. Hyg.* **69** (1978), 20–41.
39. G. Haesen and A. Schwarze, Commission of the European Communities, Report No. Eur., 5979.
40. L.H. Adcock, W.G. Hope and F.A. Paine, *Plastics and Rubber: Materials and Applications,* **4** (2) (1980), 37–43; **4** (5) (1980), 71–77.
41. F.A. Paine, ed., *Fundamentals of Packaging,* Blackie, (1962); revised edition, Institute of Packaging (1981).
42. J.H. Briston and L.L. Katan, *Plastics in Contact with Food,* Food Trade Press (1974).
43. D.J. Hine, in *Modern Processing, Packaging and Distribution Systems for Food,* Blackie, Glasgow (1987), p. 62.

17 Specification and quality control

It is essential to distinguish very clearly between the measurement of quality and quality control. The best results are only possible with the full realization that the control of quality for any type of product is a team operation involving everyone concerned from the raw material to the ultimate consumer. So-called quality *controllers* can only operate by providing a quality *measurement* service on which managers and operators can make decisions. Secondly, quality control costs money (Figures 17.1 and 17.2) and many smaller firms think they cannot afford this expenditure. There is a misconception here. Formal quality measurement carried out on a statistical basis with a special inspection department is often not possible at an economic cost in a small company. Nevertheless, there must be a conscious measurement of quality even if it is done by the production operatives. The sooner it is recognized that control must be exercised, and indeed can only be exercised, by the men operating the machines, the sooner we can get good results. No firm can afford not to control its quality (whether it does it on a statistical sampling basis by a specific department is neither here nor there—it should evolve its own method within the limits of the organization). Any progressive company must always have production control decisions based on a predetermined quality standard which is known by everyone, and this control will be initiated by information

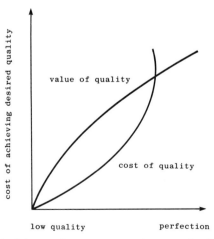

Figure 17.1 Better quality costs more and commands a higher price.

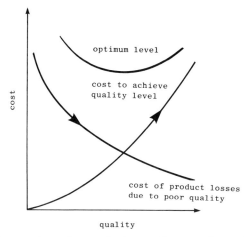

Figure 17.2 Quality costs (cost of preventing low quality, cost of appraisal, internal failures, external failures).

from some form of quality measurement. Many smaller companies are not even aware, at present, that they control their quality quite informally. The provision of a quality measurement service should be as economical as we can make it depending upon the operation and the scope involved.

Successful food manufacturing involves two necessary, inseparable and almost equally important requirements: (a) producing a good product which is both nutritious and tasty; (b) delivering that product to the consumer in as near perfect condition as is possible. The first requirement is in the realm of food science, but the second depends almost entirely on packaging. Although food manufacturers will only be successful if their packaging is right, it is equally clear that the best packaging in the world will only sell a product to each consumer once; although there are a considerable number of consumers, they have a habit of informing each other about bad products very rapidly.

What is quality? [1]

First of all we must ask ourselves what we mean by quality and which particular aspects of quality concern us in packaging? Quality can be defined in many ways, but probably one of the best is its definition in relation to 'perfection'. Perfection is something which we can very rarely, if ever, attain and the quality is the distance away from perfection that any particular specimen happens to be. This is the *level* of quality and must be determined by the particular product involved. It is obviously not necessary to produce a level of quality higher than that required to do the job. The second important aspect of quality (and this is particularly important in terms of packaging standards) is the *consistency* of quality throughout the job. For example, if we assume that perfection is rated as a figure of 100 and decide that for our particular purposes we need a package at a level of 85, then we want all the specimens that are submitted to be between, say, 80 and 90 and not varying to such an extent that we may have some at 95 or 96 and others as low as 75. In other words, quality, in the sense of uniformity, must be present irrespective of the level that is required. In setting packaging standards, it is necessary therefore to answer these two questions. Firstly, what is the level of quality we need to do the job, and secondly, how much variation about that level can we tolerate?

It is of course, very important to realize that quality is something that is built in at the time of manufacture. Consequently, it is impossible for what is often called 'quality control' in a food manufacturing unit actually to control the quality of the packaging material which they receive from their suppliers. They are only able to determine whether the supplier has reached the required standard. In other words they are a quality *inspection* department. True control of quality can only be done on the machines which are actually producing the packages or packaging material.

There is one possible exception to this, where the quality inspection also involves removal of defectives, particularly where reusable containers are concerned. For example, let us consider the use of milk bottles. These make a considerable number of journeys before being taken out of service and they must, therefore, be thoroughly cleansed every time they return to the beginning of the cycle, and chipped, broken and other defective bottles removed. In this instance the inspection not only checks the cleansing but also sorts out any defective or dirty containers. This type of sorting is best done by machine at the exit to the cleaning cycle. In any event, in such operations where containers are reused, the presence of special quality inspection problems must be countered by using special techniques. A similar set of circumstances arises in the reuse of outer containers in fibreboard, where the containers on return must be inspected to see that they are fit to go out again, and if not, removed from the system.

The next question which arises if we are going to take a sample is, how big should that sample be? The answer to this is related to the cost of inspection. How many inspectors can we afford to use in the operation and what is the minimum number required to give sufficient information to enable us to make a realistic judgement?

Once again, we must remember the two aspects of quality for which we must cater: first of all the level of quality required, and secondly, the variation about that level throughout the batch. Setting a level which is higher than necessary normally means that costs will be increased for two reasons. Firstly, the maintenance of the higher quality level on the production machines will mean that more rejects will be produced, and secondly, the requirements for a higher quality level will almost inevitably mean that the variation will become more obvious over a smaller range than if the level were lower.

Quality control

Quality control is an expression which has now been used for many years. Whenever we use it, it is pretty certain that we do not mean what we say. The expression is shorthand [2] for a longer sentence, something like. 'The measurement of quality by statistical methods to assist the control of production.' You cannot write this on a door or even say it regularly; it is too long. So we speak first about statistical quality control, then quality control, or even QC. By using these abbreviations we can easily fool ourselves into believing the shorthand and not the original meaning.

Another term frequently used, or misused, is *specification*, and this also means different things to different people. There are at least six possible motives for writing a specification, and each of these can give a different result. The six are:

(1) To invite tenders on a like-for-like basis
(2) To improve the product
(3) To assist the supplier to judge what he has made
(4) To allow the supplier to know more about what he should be trying to produce
(5) To make one's own staff better judges of what they accept or use
(6) For use in case there is an inquest later on.

However, the only valid use for a specification is as a document jointly drawn up by the user and supplier so that each knows as clearly as possible what is required. Because to reach such a situation requires considerable liaison between technical personnel on both sides, by the time it is possible to write the statement of intent such an understanding will have been reached that the document is only necessary as a record of the position and can therefore be used best not for inquests but as a basis for improvement.

Process sequence control

Every batch of packaging material made on a machine has a beginning, a middle and an end. As a first principle, process sequence control suggests that, having made the material in a certain order, we should endeavour to keep it in that order for all subsequent operations. Variations that occur in the properties of the material as it is made take place relatively slowly over fairly long periods of time. Other machines subsequently used for converting such material are capable of adjustment if necessary to deal with different properties in the material. Changes in such properties taking place relatively slowly can lead to gradual adjustment of machines. However, if the reels or sheets of material are presented to a machine in a random mix with sections which were produced miles or possibly days apart during the making operation, then the converting machine may well be unable to cope with the wider variation, and adjustments cannot be made. Machines by their very nature like to have consistent material. (They may like it consistently different from that with which we would like to provide them but, nevertheless, they are consistent.) However, *identical* machines do not exist. Any two machines for the same task will have (at least slightly) different sets of requirements. We are faced, therefore, with the apparently simple need to number in the order of making the pallets or reels coming off the machine making the material.

The converter also must deal with certain problems in warehousing, for obviously now the reels or pallets numbered in sequence must be placed into store in such a way that they can be brought out in order. There may, in fact, be a bigger problem in warehousing than any sequential numbering off the machine making the material.

A converter will often have more than one process, e.g. a carton maker may well be printing board and then, after having printed it, will want to move it on to a cutting and creasing operation and then on again to a further finshing operation. At each one of these stages it would be important to keep at least the piles of board being used in the same order, for each one of the machines concerned may well be able to deal only with slight changes in properties, and would not like changes in properties which are larger.

Measurement, the assessment of quality [3]

Those who have to measure, in one way or another, need to ask 'how exact, how accurate?' Many of the basic concepts inherent in a consideration of precision are revealed by a brief study of the history of length measurements.

The first recorded measurements of length were based on portions of the human body—the finger, knuckle, palm, foot and cubit are ancient examples. Because these varied from person to person, one unique cubit was found necessary. Thus the Royal Cubit based on the (then) king's arm was established. Since the king could not be available at all times to loan his forearm, a master copy of the Royal Cubit was made on an appropriate wall. Further, since this was sited at one static point, additional copies were made and located at other accessible places throughout the kingdom.

The accuracy to which any measurement could be made depended on that achieved in copying the master cubit, in transferring the size to subsequent copies, and then transferring the size of the copy used to the part to be measured, together with the errors inherent in subdividing the unit. In those days measurements were made for the purposes of trade and building, and, generally, fine measurements were not required. The search for better standards continued; a woodcut dated 1575 shows a method of determining the value of one foot from the mean measurements of sixteen good men entering church. Part of an act of Edward I in 1305, establishing a number of standards, read 'It is ordained that three grains of barley, dry and round, make an inch', which definition remained on the statute book until 1592.

Progress to date has led to the establishing of the yard as 0.9144 metres exactly (Weights and Measures Act, 1963) whilst the metre was defined by an International Conference in October 1960 as 'equal to 1 650 763.73 wavelengths, in vacuum, of the radiation corresponding to the transition between the levels $2p^{10}$ and $5d^5$ of the krypton-86 atom'. This can be determined to an accuracy of one part in 10^8. Thus, to the best of present knowledge, our measurement of length can be related to a standard that can be reproduced very accurately (given facilities) throughout the world and which is invariable with respect to both geographical location and time.

Some standards, such as length, are related to supposedly invariable natural phenomena; others are physical entities, e.g. the standard international prototype kilogram for mass. This can change slightly, and the British copy (No. 18) was in fact found to exceed the international prototype by 51, 71 and 59 μg in 1924, 1948 and 1960 respectively. The range of 20 μg in these determinations is equal to two parts in 10^8, accurate enough for any exercise with which we might be concerned.

It scarcely needs remarking that without measurement there is no science and no technology. One of the important aspects of scientific work is to quantify those variables found to be of consequence. The working accuracy of the equipment used to do this is clearly a matter of moment. It follows that traceability of standards is of prime importance. Every measuring standard should have been confirmed by comparison with a standard of higher quality, and so on, until the international standard is reached. Measuring devices should be checked against appropriate standards or by other means, for errors such as those in magnification, linearity of magnification, dynamic response, discrimination, constancy, datum and zero errors.

In this context, manufacturers' trade literature on measuring instruments can mislead. Is a length-measuring device that will 'discriminate to 0.000 01 inch' better than one which is accurate to 0.000 1 inch'?

It is impossible to over-emphasize the importance of instrument calibration. An instrument that is not telling the truth may cause wrong and costly managerial decisions. Indeed, such an instrument is sometimes worse than having none at all.

Often, measuring instruments can be checked by using them on standard test pieces and comparing the result with what is obtained using a similar fully calibrated instrument. Since this method would normally simulate the usual operating conditions, it has some advantages over other calibration methods, though it also has drawbacks.

Quite apart from possible sources of error in the instrument itself, the techniques used in measuring can themselves introduce errors. Some instrument errors such as backlash, lost motion, and hysteresis can be overcome by correct procedures. Parallax, arising from placing the eye either off an optical axis, or not immediately above a scale, can introduce small but sometimes significant errors. Many mechanical measurements rely on devices which should operate normal to the surface and, if they do not, 'cosine errors' may be introduced. In machines with moving parts, the effects of roll, pitch and yaw must be considered. Very accurate dimensional measurements will be affected by surface texture variations. Similarly, geometric shape, the practical meanings of straightness or parallelism, etc., will have to be taken into account. Care must be taken not to infer more from results than what they really say, e.g. straightness of generators in two directions at right angles does not define a plane, and a plane figure having a constant diameter is not necessarily round.

Consideration of errors arising from measuring techniques points to the need to establish (a) the appropriate number of significant figures for recording the test results, and (b) whether, in determining the last figure, the result should be 'rounded' up or down.

Despite taking every precaution with measuring equipment and techniques, there is no such thing as a completely accurate measurement. It is possible, however, to find by experiment how accurate any measuring process is. This determination covers all aspects, the instrumentation deficiencies and observer errors. What has to be done is to repeat the measurement a number of times with several different observers. From the results, it is possible to calculate a mean value and a figure called the *standard deviation*. This is a measure of the variability of the results and is designated σ (sigma). When this has been determined for a given set of conditions, it is possible to state the mathematical risk of the inaccuracy of a particular measurement exceeding a certain value. These chances are those indicated by what the statisticians call the *normal distribution* and are represented diagrammatically in Figure 17.3.

To summarize:

(1) Measurement is needed to quantify quality
(2) No measurement is completely accurate

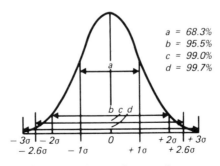

Figure 17.3   Area under normal curve.

(3) Inaccuracies can be minimized by properly calibrated and sufficiently accurate equipment used with a proper understanding of the inherent problems in measuring

(4) The probable extent of inescapable inaccuracies can be calculated where necessary.

Measurement used for quality aspects

Measurements are used to turn into numbers some aspect of the quality, size or quantity of a product. Here we are only concerned with those measurements relating to quality. In every manufacturing process, a particular quality is either inferred or specified. Often this will be in the form of a material deemed by the buyer to be 'satisfactory'. Measurements are made by the manufacturer in order to establish or quantify the quality level of his product, to determine whether the production process is running properly, and to make certain that bought-in materials or components are up to the necessary standard to ensure that the product can be made satisfactorily. Measurements cannot of course, make anything and are an additional overhead cost. However, without them the chance of producing poor quality work is increased. Poor quality itself has a cost to the firm. The individual items that contribute to this cost can be summarized under two headings as follows.

(a) *Tangible costs*
    Factory accounts:
    (1) Material scrapped.
    (2) Labour costs to produce scrap material.
    (3) Cost of maintaining additional production capacity created by defectives.
    (4) Excess inspection costs.
    (5) Cost of investigation into cause of defects.
    Sales accounts:
    (6) Discount due to 'seconds'.
    (7) Cost of dealing with customer complaints.

(b) *Intangible costs*
    (8) Delays and stoppages caused by defectives.
    (9) Loss of customer goodwill.
    (10) Loss of internal company morale due to friction created between departments.

The problem is to fix inspection levels so that the overall total cost of inspection and poor quality is a minimum.

In most production processes measurements are made at three main stages, 'goods inwards inspection', 'process control', and 'final inspection'. Some of the procedures involved have been developed by statisticians using probability theory, so a brief explanation of such ideas is needed. Perhaps the most important concept which has to be accepted is that nothing is perfect and that there is variability in all things. We should find it easy to accept this simple truth. For instance all the people reading this book are of different heights and different weights. Again, 12 consecutive cars of the same model from the production line of one manufacturer will have slightly different faults and so will vary one from another.

Where it is possible to measure some aspect of a material and express the result as a number, there is a remarkable mathematical law which predicts how the variations in the values will be distributed. If the values are grouped for size and plotted in the form of a bar chart or histogram, the result will approximate to a bell shape as shown in Figure 17.4. This is called the 'normal' or 'Gaussian' distribution and applies in many situations where measurements can be expressed in numbers which can fluctuate both sides of some mean value (so-called measurements of variables).

The greater the number of measurements, the nearer to the theoretical ideal shape the plotted distribution is likely to be. Thus if we took the heights of a small sample of 20 men taken in a bus, some vague indication of the bell shape

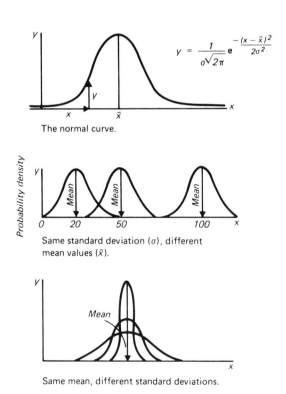

$$y = \frac{1}{\sigma\sqrt{2\pi}} e^{-\frac{(x-\bar{x})^2}{2\sigma^2}}$$

The normal curve.

Same standard deviation (σ), different mean values ($\bar{x}$).

Same mean, different standard deviations.

Figure 17.4 Aspects of the normal curve.

would be apparent. However, if we were able to take a bigger sample of, say, 1000 men drawn at random from the population, a much closer approximation to the curve would be obtained. Statisticians describe the shape of the curve for any set of figures by two calculated quantities, the algebraic mean ($\bar{x}$) and the standard deviation (σ).

Another form of measurement is often used where objects have to be assessed visually by an inspector and accepted or rejected (a so-called subjective assessment). Here the inspector looks at a fairly large number and rejects a certain percentage. Often for reasons of cost, time, or even lack of suitable quantity measurement, this visual method is the only acceptable one. Where decisions are based on choosing one of two attributes (accept or reject, go or not go, etc.) the 'binomial' statistical distribution is used, not the 'normal' one. The mathematics associated with this form of distribution mainly concern fixing the correct number of items to inspect and where to set the rejection level.

For reasons beyond the scope of this chapter, yet another distribution, the so-called Poisson distribution, is used in nearly all practical cases. This works very much like the binomial distribution when the rejection level is less than about 10%, and is easier to handle mathematically.

Sampling

Many quality control activities involve some aspect of sampling [4]. The art of sampling is to select a sample from a population so that the quality of the sample is representative of the quality of the population from which it was taken. The sample should be taken in as random a manner as possible to give each member of the population an equal chance of being selected. Confidence in the sample result increases with sample size, but large samples cost money and an economic compromise is often necessary.

Fortunately there are several sets of standard sampling tables [5] available to give guidance on sampling plans for particular populations. These tables not only give a series of sample sizes and associated accept/reject numbers for a range of acceptable quality levels, but they also contain graphs, called operating characteristics, which indicate the probabilities or risks of any particular percent of defectives in a population passing through any particular plan.

There are three methods of sampling given in such tables. The first is called single sampling, because the result is decided on the examination of a single sample. The second is called double sampling, and here a decision can be reached following the examination of a smaller first sample, provided that the quality is either very good or very bad. If the number of defectives found lie within certain intermediate numbers in the table, this indicates borderline quality, in which case a second sample is taken and the decision is based on the results of the combined sample. The third type of sampling is called multiple sampling. This is really an extension of double sampling, where successive samples are taken until a decision is finally reached to accept or reject.

Whenever we use statistical ideas we are trying to extract the maximum information from the minimum of inspection. We may want to do this because a large volume of inspection is too expensive, or because there is not enough time to do the inspection, or because the inspection is destructive (such as the manual or machine erection of cartons when used as a final inspection). Once it is found either adequate or necessary to measure samples only (of a product) then statistics can tell us the most economic way to do this consistent with a certain specified risk. For instance, if we have measured a number of samples of some material in respect of water vapour transmission rate (WVTR) we can derive a mean value ($\bar{x}$) and the 'best estimate of the standard deviation' ($\bar{\sigma}$). From these figures we infer the population's mean WVTR ($\bar{x}$) and its standard deviation (σ). Knowing the size of the sample taken, and using the special statistical tables, we can work out for any specified risk (say one in twenty on average) how far the mean of the sample might vary from the mean of the population.

We have now considered the equipment, the techniques, and the problems associated with accepting results based on the measurement of samples. It still remains to think about what we actually measure. The phrase used earlier was 'some aspects' of quality. What aspects? There is no point in measurement for its own sake; whatever property is measured must be related in some way to eventual product performance. There must be 'correlation' between measured property and final usage, e.g. the WVTR of a material can be shown to be related to the shelf life of many foods that might be wrapped in it. There are pitfalls in establishing and interpreting correlations; the choice of the most suitable tests needs particular care and attention.

Having considered the aspects so far covered and started to make measurements, money and effort will have been wasted if the results are not made available to the appropriate staff at the right time. Some way of permanently recording the results in an accessible form should be established. In brief, good housekeeping can provide an invaluable fund of knowledge for future use, whilst the choice of the best way of displaying results, such as charts, can ensure that decision-making during the production process is systematic, with minimum risk of error.

Finally no system of measurement will work as it should unless the staff (both those measuring and those whose work is measured) understand the reasons why and the methods used. The involvement of all staff in considering and, if appropriate, trying to improve quality is important. In many industries it is found that the major financial losses attributable to poor quality are caused by defects in the product which were operator-controllable. Pareto diagrams of the kind shown in Figure 17.5 can be used to isolate those defects which are most costly to the firm. In nearly all instances it has been found that between two and four types of defect account for two-thirds of the quality

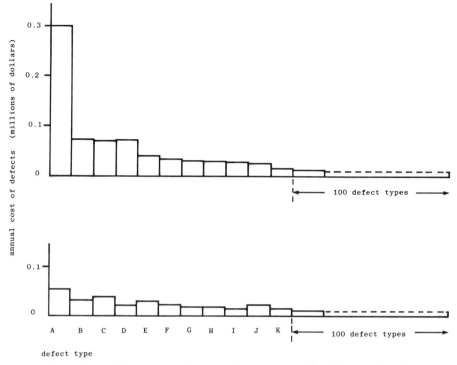

Figure 17.5 Pareto histogram showing annual costs due to defects in a metal-casting and machinery plant. Top, first year of analysis (gross cost of defects $871 000); bottom, third year analysis following corrective action (gross cost of defects $389 000).

costs. Thus this is the area in which the largest and most immediate gain can be achieved.

A 'quality' programme, properly mounted, will include explaining to operators what the aims and objects are. When this has been done with enthusiasm and drive, astounding results have been achieved. The Zero Defect programme in the United States, and Quality Circles in Japan, offer many examples of striking improvements resulting from stimulating cooperation from the work force.

Factors affecting quality in packaging

Package performance

Modern packaging lines, whether they are fully or semi-automatic or manual, often contain highly complex pieces of engineering equipment. It is important that for packaging efficiency the packaging material or containers to be processed are of the right, and consistent, quality. Substandard material or containers will cause hold-ups and reduce the speed of the packaging operation. Packaging material suppliers and container manufacturers have always exercised some control over the quality of their products, but formerly this was mainly concentrated on the visual characteristics of the material.

Nowadays much more emphasis is often placed on functional aspects, both during manufacture and in the preceding design stage. There are four key decisions on controlling the quality of any packaging or packaging container:

(a) What are the packages intended for?
(b) What properties of the materials used control the requirements?
(b) How can we measure these properties?
(c) When and how will the operators of packaging machines use the measurements we have made?

We will consider each of these later in this chapter.

In all types of packaging, quality depends on the quality of raw materials, the skill and care taken in production, and the efficiency of removing substandard materials. The *quality of raw materials* must be controlled by a specification, a 'sampling' procedure, and a 'what to do if not correct' arrangement. The *quality in process* is controlled by the operators; a quality measurement service can only provide data. The *removal of substandard material* is assisted if we record when it was made, separate it out as soon as possible, and keep the processed goods in order (process sequence control). To illustrate these principles, let us consider in turn how the quality of cartons, of glass containers and of fibreboard cases may be measured and assured to the user organization.

Quality measurement and control of cartons

The quality of a batch of finished cartons depends on the three main factors listed above, namely the quality and uniformity of the raw materials, board, inks, etc. used in manufacture, maintaining quality in process, and efficient removal of substandard material. The first stage in quality assurance is the specification and control of the raw materials to ensure that they meet the manufacturing requirements both of the converter's machinery and of the carton user. In the production processes, the skill and care of process and machine operators are the main factors, standard measurements made by the quality measurement service providing data to assist the operation. Finally, at all stages in production, substandard material must be separated from the mainstream and sorted to recover good work.

Three points must be made clear to all personnel right at the beginning:

(1) The only people who can control quality are those who operate the processes that produce the cartons. The quality measurement service (QMS) only makes measurements and provides data for production personnel.

(2) A quality measurement service is not installed to provide snares and traps to production personnel so that complaints can be made. It is an acknowledgement by management that variables exist; the measurement service will help production to control them better.

(3) Before and during all production steps, all personnel must have a clear idea (and ready access to a true example) of the end product required by the customer.

Setting packaging standards.   Like most products, packages and containers have both measurable and non-measurable features, and these must all be taken into account in setting standards. We use a system of defect levels (critical, major and minor) and divide the variables and attributes into those concerned with function and those concerned with package appearance. Any defect which causes a package to fail in any part of its function is regarded as critical, and the aim should be not to deliver such material to the user. As far as appearance is concerned, the standard here may vary from product to product and market to market.

It is essential in any package-making operation for all departments in the manufacturing unit, and for those sections of the manufacturing plant who use packages, to have a clear unequivocal target for both package performance and appearance. Realistic tolerances about the standard and appropriate inspection methods must be set up in advance.

So that everyone concerned in the operation understands clearly what is required, four 'species' of carton are defined:

(1) 'Perfect' cartons. The word 'perfect' in this context means that the result is *centrally placed as regards colour* between the light and dark limits, is dimensionally correct, and contains no cutting, creasing, printing, board or other defects outside the limits which normally score 1, i.e. a minor defect. Such cartons are the target at which everyone is aiming.

(2) 'Normal' cartons. These contain no defect above a specific limit set for any particular fault. Every 'normal' carton would be acceptable to any customer and represents the quality of work which the factory 'normally' produces. A 'normal' carton is therefore one which contains *no functional* defects that would interfere with its use on a packing line or after filling. No visual defect should be present at a level greater than 5 (i.e. a major defect) as measured by the QMS. Moreover, the total score of defects on any individual carton should not exceed a number to be agreed with the customer concerned.

(3) Rejects and (4) 'grey' cartons. Because we are dealing with variable materials and using machines that can vary and operators who are human, it is impossible to keep 100% of production within the 'normal' limit. So each customer will receive a number outside the 'normal' standard which are missed during inspection. This number must be agreed with each customer. There is never any difficulty in deciding

when cartons are obvious rejects. The difficulties always arise in that area between the upper limit of 'normal' and the lower limit of scrap.

Cartons that are to be rejects, i.e. spoils, can also be defined as cartons that either have one functional defect or one visual defect of a score greater than 10 (critical defect), or one where the total score of defects exceeds an agreed upper limit. This system adds precision to the difference between spoils and acceptable cartons.

Control of raw materials. The final arbiter of the suitability of any raw material is the machine on which that material is converted to another form, whether this is for making cartons or for packing them. It is essential, however, to lay down a fairly complete specification to control the raw materials by means of laboratory tests that relate to both the carton manufacturer's and the carton user's requirements. Most tests give numerical results and, once the difficult stage of correlating the results with production behaviour has been carried out, agreed tolerances can be written into the specification.

Certain tests are hard to assess numerically, for example the viewing of sheets of board under oblique lighting to examine their surface for correct formation or for extraneous material. If such tests are used in a specification, care must be taken to see that both carton manufacturer and board supplier are working to the same scale. Interchange of personnel between the two companies, to set up agreed visual standards, is one of the best ways of surmounting these difficulties. The most vital point of any specification is that it must be agreed by all parties concerned in its use.

It is equally important that the specification is used properly and revised as required, so that it is a means of communicating the requirements of the job, and not a document kept out of sight until someone wishes to prove why the job was wrong and where the fault lay. Raw material suppliers and carton makers, as well as carton makers and users, are all in partnership. Quality assurance should be a means of improving relations and products by better communications, not a means of proving who was wrong by documentation.

Sampling is one of the main problems in running an adequate quality control system on raw materials like carton board. If this is to be done in the carton manufacturer's plant, it involves great handling problems. Pallets and bales have to be opened, sample sheets taken and then reclosed for storage. The simplest way to overcome this is to have the board manufacturer carry out the sampling during production. The sample sheets are taken at regular intervals throughout the run and then despatched in a separate bale either along with or just before the bulk delivery. It is also very useful if the mill supplies a copy of the test results recorded through the run. This usually gives a broader picture of the delivery than the carton manufacturer could obtain by testing the advance samples, and can be used to give an idea of the likely quality of the finished work.

A carton maker is in a 'bespoke' business, every job having its own specific requirements. Hence inspection must cooperate with production to decide whether a particular property of the material would cause trouble with a particular job for which the material is intended. If something is found wrong with a supply of raw material, the problem of maintaining delivery dates also arises. In many instances, there is insufficient time to reject a delivery of raw material completely and arrange a replacement before delivery of the cartons is due. Here, also, good quality measurements can help to make the best of a substandard batch of material.

The problems during a production run can be made easier if board mills number the pallets and bales in manufacturing sequence, and the converter puts these through the machines in the same order. If this is done as a matter of course, whenever trouble is encountered the faulty part of the delivery can be isolated.

Quality in the processing departments. The objective of these departments is to prepare the surface from which the carton will be printed and the forme which will cause it to be cut. Several ancillary factors are involved that concern both artwork and design, and these also are covered under this heading. Obviously, a considerable amount of in-built quality is produced in these departments, for any defect on either printing plates or cutting and creasing formes will appear on every sheet of board converted. The departments involved are the design studio, artists and retouchers, the camera department, the letterpress and litho platemaking operations and gravure cylinder-making departments, as well as the box setting department.

It is important for all these to have as clear an idea as any other department of what the customer requires, so a target should be readily accessible to each of them. It is also important that any changes which occur after the initial order has been placed are notified to all sections and not just to the one apparently concerned. Very often a change in artwork, for example, may affect the box setting and even the construction of flaps etc., since the change in requirements for ink can affect the glueing arrangements in finishing.

Finally, at the proofing stage for each of these operations, the responsibility for checking the result against the target must be clearly defined. Each section should cooperate in producing the controls needed to ensure that every point in its brief has been covered before the work passes on to the next section.

Quality measurement during printing. The object of this operation is to control the quality of print in terms of general appearance and colour. This is achieved by providing the machine operator, at regular intervals, with certain information related directly to quality. He can then take whatever action is needed to maintain control.

As stressed earlier, one of the main requirements of the quality measurement service is that it should be possible to measure each fault and give a score that will differentiate between minor, major and critical levels. Note that this measurement is applied only to specific measurable defects, for there is as yet no absolute way to measure print quality: this relies on the eye and judgement of the customer. For the carton manufacturer, the target must be established with the customer in the form of an approved proof, sketch or specimen cartons, and any 'improvements' on such a standard could be faults.

Obviously the main types of defect occurring will vary according to the printing process used (letterpress, lithography or gravure) but it is still essential to find a means of measuring them. Certain defects can be measured and scored directly, e.g. colour can be measured by a densitometer, and print register and the size and number of spots on the print can be measured with a scale or rule. Other defects, however, e.g. streaks, smudges, set-off, etc. cannot be scored directly, but only by comparison with a set of standards previously given graded defect ratings. The quality measurement department and the machine men concerned must have access to a copy of the 'target' for the job so that they can check that the job appears correct immediately after setting up the press.

To ensure that control remains in the hands of the production personnel, the quality measurement service normally comes into operation only when a 'pass sheet' has been signed, i.e. when the production department indicates that the press is ready to run. A duplicate of this pass sheet is supplied to the quality measurement service, and after checking against the 'target' this is used as the standard the printing department should maintain throughout the production run. On this pass sheet the densitometer settings for each colour are made at the standard to be maintained.

During the run, sample sheets are taken from the machines at regular intervals. These are checked and all defects marked and scored, the scores being entered on a record chart. This chart and the checked sheet are returned to the machine and from these the operator can see the general trend of the job, each specific fault, and the score it has been given. From these he can decide on the action to be taken and whether anything can be done to eliminate the faults. The record chart is kept at the machine so that supervisors can easily see how a job is running.

After the first few checks, the control limits for the job can be set and these also are entered on the chart. If the checks made by the quality measurement service are combined with more frequent sample sheets taken and retained by the machine operator, then whenever a defect appears a check can be made to find its approximate time of starting. Once this has been done the suspect work can be marked and channelled aside for inspection and a decision on what is acceptable.

Quality measurement during cutting and creasing operations. The object of this operation is to control the box cutting and creasing quality by the criteria of functional efficiency, dimensions and appearance. A quality measurement service operating similarly to that in the printing department separates the work reaching the cutting and creasing presses into two categories: (a) material containing virtually no printing defect, and (b) material that has been marked for attention because it contains printing defects requiring inspection and possible separation. After cutting, any defective material must be segregated from the bulk of the work for complete inspection and removal of any reject cartons.

Sampling off the machine is carried out at regular intervals, and a complete cut and creased sheet removed and inspected for dimensions, register of cutting to print, quality of cutting, quality of creasing, and burst scores. The individual blanks are removed from the sheet one at a time, checking for ease of separation and quality of cutting. Dimensions can be checked with a rule or template. The quality of creasing can be checked by folding all creases through 180° and checking for liner cracking and roll formation of the back.

Wear on the creasing platen can be checked by measuring the folding torque of selected creases on a crease stiffness tester (suitable equipment has been produced by Pira). The standard for this may be taken from the pass sheet, which is signed by the supervisor or foreman when it is decided that the job is fit to run. The inspected carton blanks are returned to the machine operator with any faults marked and scored, and with a record chart showing the scores given to each individual blank and the total score for the sheet.

Again, any action to be taken must be decided on by the production staff, as the sole function of inspection is to provide an information service to the machine operator. Any defective work found during inspection will be channelled off with defective printing work for inspection and separation of the faulty material.

Quality measurement in the finishing section. The main objects of quality measurement in this section fall into four categories:

(a) To control any finishing process in relation to defects in production.
(b) To prevent rejectable material (either from the finishing process or that has been missed at any of the in-process inspection points) being despatched to the customer.
(c) To produce a record that gives an accurate assessment of the quality of the material delivered.
(d) To produce a truly representative sample of the material delivered.

To ensure that these aims are achieved satisfactorily, an inspector is assigned to every glueing machine or packing point, taking sample cartons as often as is consistent with efficient inspection of each carton. This frequency generally falls somewhere between a 0.2% and a 0.5% sample, depending on carton size. We should aim to examine every two-hundredth carton if possible. The sample cartons are inspected for all defects and a record kept of the findings in each individual check. If the fault is due to the machine from which the samples are being taken, the operator is informed immediately and action taken to correct it. However, if the fault is from printing or box cutting and has escaped earlier inspection stages, then the cartons adjacent to the inspected one, or the already packed cases that may hold suspect material, can be put aside for checking. If the record of defects kept in sequence throughout the run is combined with consecutive numbering of the cases as they are packed, then at any time, either before despatch or after a customer complaint, any part of the order can be related to the inspection record. Additionally, the inspected cartons may be packed separately in marked cases as a sequential sample for checking at the user factory before the main delivery is used.

The relation between the user and the supplier. Packaging can never be effective unless there are good contacts and understanding at all levels between the suppliers of the package or packaging material and the users. A number of relationships may exist between members of the chain from raw material suppliers to final consumer. For example, a board maker supplies board to a converter to produce folding box-board cartons, which are then supplied to a food manufacturer to pack cakes or biscuits, and he in turn supplies a wholesaler who distributes them to a retailer, who sells them to a consumer who opens the package and consumes the product. There are a number of contacts here and each one will have a bearing on the packaging requirements. The best result is always achieved when every link in the chain is considered and the packaging requirements of the product have been specified adequately and taken care of at all stages.

While we must not forget that there is a chain of users and suppliers, it is not necessary for each of them to repeat every test. Furthermore, it must not be thought that the relation between packaging supplier and food manufacturer is just that of buyer to seller. They may argue or disagree during initial bargaining, but once it has been decided that a particular food manufacturer is going to buy packaging material from a particular supplier and agreement to work together has been reached, the supplier–user relationship must become a partnership working to maintain production and distribution of the packaged food at the correct level of quality. To achieve this, not only has understanding to be reached on all the things that happen when the flow of packaging is proceeding correctly, but also on what to do if things go wrong. Obviously the procedures for delivering cartons, bags, bottles or wrappers in the correct quantities at the right time and the right quality level, must be decided. The tolerances that are allowable on various dimensions, for example, must be agreed, and the methods by which the package quality will be measured must be fully understood. The procedure for taking repeat samples must be considered in instances where a first sample does not indicate satisfactory performance, and also the action to be taken when a batch delivered is substandard must be decided. This same 'partnership' arrangement is also necessary between the converter of the packaging material into containers and the supplier of raw material and, at the other end of the scale, the retailers' and consumers' viewpoints must also be taken into account. In such a complex operation, it is easy to see how problems can arise, particularly if specifications are used not as statements of intent, but as weapons to prove the other side was wrong, or where the relationships between the various parts of the chain are less than true partnerships.

Specifications.   Two principal types of specification are used in packaging— quality (or material) specifications and performance specifications. A *material* specification states what materials shall be used for producing the package and how these materials are tested or measured directly. A *performance* specification, on the other hand, allows some freedom in choice of material but considers the package in its complete state and specifies a test (or tests) related to its use and which it must pass adequately. Most good specifications for packaging contain both material and performance aspects. Whichever way specifications are written, they may be produced for several reasons.

We are looking for optimum packaging. The methods of deciding this vary in different organizations depending on their size, nature, staff, and the services they carry out. However, whatever the evaluations (or guesses) which are carried out before deciding the packaging required, there is one final common requirement for all: to make sure that the appropriate people in both the supplier and user factories are in no doubt at all as to the decisions that have been made. We are largely concerned, then, with the need for communication.

Communication regarding specifications is required in three areas: within the packaging supplying company, within the packaging using company, and between the two companies. Let us consider the last one first. It will be of little benefit to either company if the understanding of what packaging is required is held only in the supplier's sales department and the user's buying department. In both companies a number of other departments and people will have a contribution to make and will need to know the decisions made. These needs will differ and this should be taken into account when writing the specification. The right reason for producing a specification is therefore to communicate as clearly as possible to one person in a chain of operations what is required of the next link and in the final use of the article. In order to do this a package material or container specification may be divided into three parts.

The first part is a statement by the user on what he proposes to do with it. In the case of a carton maker, for example, the specification for a board supplier should be a statement as to the use of the board. The method by which the board is to be printed should be stated and whether it is a two-colour, two-pass job or a one-pass, four-colour job, whether the printing process is letterpress, litho, gravure or flexography, whether the board is to be cut and creased after printing, whether it is to be embossed or varnished, and, in short, everything that will happen to the board along the route between arrival at the carton plant and its use as a carton. This will enable the board supplier to understand what the material has to accomplish.

The second part of the specification is a statement drawn up between the technical departments of both supplier and user, giving what is believed to be appropriate test methods and test levels for making sure that the consignment of material or containers will come up to the standard needed.

The third part of the specification is concerned with the procedures to be adopted (under all circumstances) for examining each batch when delivered, deciding whether it is satisfactory or not and presenting the various options that may occur when a batch is not within the standards stated in the second part. We all know of material which did not meet specification but performed satisfactorily; we have all come across the batch of containers apparently not up to standard but giving a completely satisfactory pack. On the other hand there are also occasions when material which fully met specification behaved unsatisfactorily; or when containers completed as specified did not run well on a filling machine. The third part of the specification must take care of such situations and lay down the procedure to be followed by the two parties when difficulty arises. Remember, once the contract for producing something has been agreed, the supplier, the converter and the user are partners. They no longer operate on their own. The total objective of all is to keep a packaging line running satisfactorily.

How to make complaints useful. From time to time in any operation, examples will be found of material which causes trouble for one reason or another. Such material is usually the subject of a complaint. Making a complaint to a supplier always has the effect of relieving the feelings of the complainant, but in order to make the best use of complaints, it is essential that the information required by the packaging material supplier should be given to him. For example, it is much more helpful to say: 'Here is an example of a box we do not like and we found some 3% of these in a batch of 200 samples that we took at random', than it is to merely send a specimen back saying: 'Please do not send us any more of these'.

With regard to functional complaints, very frequently the particular piece of packaging material or container which caused the trouble was destroyed in the machine and the samples sent back on either side of it may or may not have been ones which could cause trouble; hence the information they give is extremely scanty. One thing, however, can be taken for certain; if substandard material which does not function properly is delivered to a food manufacturer, then he will complain and send examples of the complaint back to the supplier. When he does so, the inference is 'Please do not send me any more like this', and while such information can be useful in a negative sense, it is possible to be far more positive. Instead of only complaining the food manufacturer should also send back samples when his packaging lines are running perfectly with the statement 'Please send them all like this'. It is much better to know what is required than it is to know what is not.

Quality control in a glass container factory

The demand for better quality glass containers has led to the development of what is called the *total quality control* technique. This concept recognizes that control must be exercised over every aspect of glass container production, and not just inspection of the end product and rejection of the defects. The underlying principle is that to be completely effective control starts with the design of the container, continues through the various production stages and ends only when it has been delivered and recognized as satisfactory by the customer. The emphasis is therefore placed on the prevention of defects, with the elimination of substandard containers as a secondary function. Everybody in the business is concerned with controlling quality. A very similar approach is used in other fields such as the carton-making operation previously described. Once again, the quality measurement and control function has three main responsibilities: to ensure that raw materials are right; to ensure that the containers being manufactured are right; and to ensure that the finished goods are correct and capable of being used by the consumer satisfactorily.

As with cartons, quality requirements vary considerably, depending on the type of product to be packaged, and also from one packer to another within the same product group. Glass jars for baby food have different requirements in respect of filling and labelling from those of cosmetic bottles. One must be filled and labelled automatically at very high speed, while with the other we are very much concerned with appearance characteristics as these make a greater contribution towards the marketing of the product. At the beginning, one must establish a reference standard against which samples from routine production can be prepared. There are several methods of specifying such standards, some of which are mentioned below. These can be used in isolation or, more usually, in combination.

(a) *A specification drawing.* A drawing can be prepared to show the general profile of the container, usually as a front elevation. Often there is also a plan view and sometimes a side elevation in the case of a complex shape. A typical glass specification drawing is shown in Figure 17.6. The finish detail is shown on a separate drawing. The drawing should give tolerances for all dimensions which are important to the performance of the container, and include such items as capacity or content. Other dimensions intended mainly to give additional guidance to the manufacturer when preparing the moulding equipment, etc. are shown as nominal. Tolerances should always be derived in consultation with the container manufacturer and there must be a common understanding on exactly how they are to be interpreted. The dimensional specification should also be discussed with the supplier of any packaging machinery through which the containers will have to pass. It

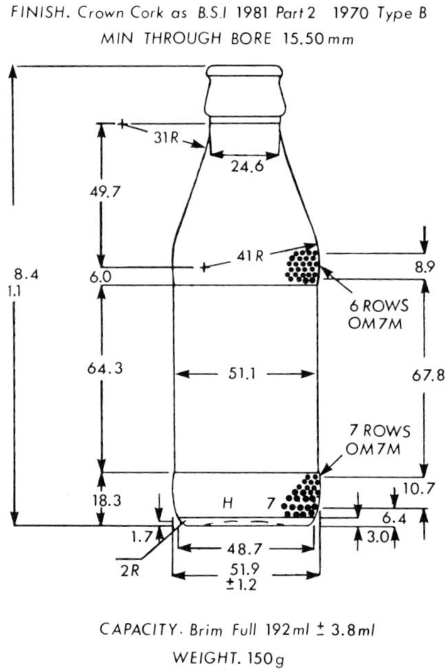
FINISH. Crown Cork as B.S.I 1981 Part 2 1970 Type B
MIN THROUGH BORE 15.50 mm
31R
24.6
49.7
41 R
8.9
8.4
1.1
6.0
6 ROWS
OM7M
64.3
51.1
67.8
7 ROWS
OM7M
10.7
18.3
H
7
6.4
3.0
1.7
48.7
2R
51.9
±1.2
CAPACITY. Brim Full 192 ml ± 3.8 ml
WEIGHT. 150 g

Figure 17.6 A glass container specification drawing.

is not sufficient merely to forward a few sample containers, because while these give a useful impression of the general shape of the container, the individual dimensions can fall anywhere within the overall specification and will rarely be representative of the complete population of containers to be supplied in bulk.

(b) *A prescribed test on an associated standard.* Here the container is tested in a prescribed manner and it must satisfy a certain minimum standard. Such tests normally concern the strength of the container.

(c) *Defect groups and associated acceptable quality levels.* Some faults or defects are more important than others. Certain defects are so important that they cannot be tolerated at all. Others can be tolerated provided there are not too many present while others of lesser importance can be accommodated in rather larger quantities. The defects can be arranged in different groups according to seriousness and each group can then be assigned an acceptable quality level (AQL). The AQL defines the worst percent defective which can be tolerated as an average for the process over a period of time. Usually there are three main defect groups called critical, major and minor. If a container is found to possess more than one defect it is normally classified as one defective unit, recording only the most serious defect present.

(d) *Limit standards.* A limit standard defines the departure from perfection which can be tolerated before a particular feature is to be regarded as defective. The limit standard can be an actual example of a defective container exhibiting the fault to the permissible standard, or it can be illustrated by means of a diagram or photograph. Sometimes a limit standard can be defined in units of length, depth or area. Where a degree of colour is involved, then reference limit standards for colour can be established.

Control zones in glass container making. Figure 17.7 illustrates the various stages at which control is exercised during glass container manufacture.

Collecting quality information costs money and any information which is collected should be fully utilized. When measurements are made they should be recorded as such and not translated into simple 'pass' or 'fail' terms, as much of the positive value of the information is then lost. Once acquired, the information can be used in several ways. The most obvious use is to indicate when the container has departed from specification and has to be rejected. A more profitable use, however, is to feed the information back to the machine operator to indicate how the process is performing within the specification (the limits of which are defined by the operator's gauges). This will enable him to take corrective action when a trend towards the specification limit is detected.

Quality control information can be useful for the mould engineer who is responsible for the maintenance of the mould equipment. It can also be of interest to the container designer. If a particular container is continually giving failure results near to the borderline for a strength test, then the design may require modification to give a more reasonable safety margin. Finally, the information may be useful should there be a query on the containers when delivered.

The way in which information is recorded is important. It should be logged in a manner which is easy to read, and report forms and log sheets should be designed so that trends show up quickly without having to study masses of figures. If measurements of deviations from the ideal or mean can be made, they are often simpler to record than absolute values, but measurements requiring calculations to be carried out before action can be taken should be avoided where possible. If the data can be presented pictorially then this is often easier to interpret.

Figure 17.8 shows a tally diagram of frequency distribution for measurements of a certain dimension on a bottle. Each time a particular measurement occurs, a cross is placed opposite that value on the horizontal axis. It is immediately obvious that the dimension is being produced towards the minimum of the specification, although the process capability for this feature is well within the specification. If some change could be made to the process to move the distribution to the dotted position, then the quality would be

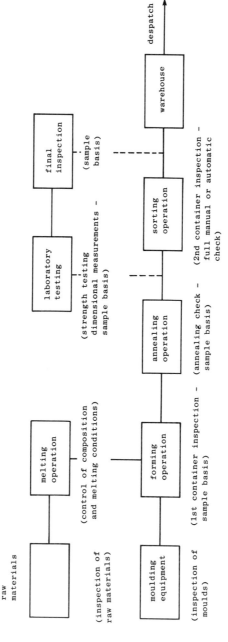

Figure 17.7 Manufacturing control zones in the glass container industry.

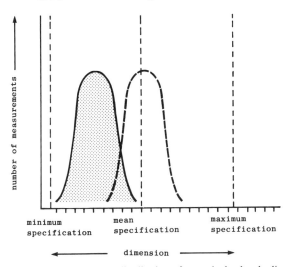

Figure 17.8 Tally or frequency distribution of a particular bottle dimension.

improved and the chance of bottles below the minimum specification would be greatly reduced.

Another method of presentation is to use a control chart. This is a graphical record of how the process is changing with time. The control chart has an advantage over the tally diagram in that the order of making the measurements is retained so that trends can be observed.

Control charts. There are two main types of control chart. The first type concerns variables and is employed where the feature concerned is a measurable one. This type of control chart might be used, for instance, for controlling the filled contents of a container during a packaging operation. Figure 17.9 shows how action limits for the chart can be developed from a tally diagram constructed from routine records. The action limits are normally drawn so that 95% of the results would be expected to fall within them, provided the process has not changed. Such limits are known as 1 in 40, or, in statistical language, 2-standard deviation (2σ) limits. Provided only one point in 40 falls outside these limits, then no action is needed, but if two points fall outside in quick succession then it indicates the process is going out of control. Sometimes the points plotted are the means of samples of five consecutive containers taken from the production line. This chart of averages can be supplemented by a corresponding chart showing the ranges of the five samples.

The second type of control chart concerns attributes and is employed where the feature concerned is either right or wrong, good or bad. Figure 17.10 shows an attribute chart which could be used to record the number of defectives in

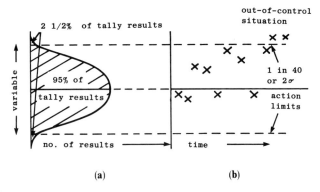

Figure 17.9   Control chart for variables: (a) tally diagram; (b) control chart.

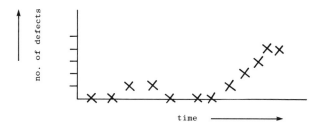

Figure 17.10   Control chart for attributes.

successive samples of five containers when a relatively high number of defects is present. Where there are only a small number of defects present then a larger sample would be necessary so that there would be an average of one or two defects per sample. The vertical axis of the chart would then be expressed as a fraction defective scale. Again, action limits could be incorporated as required.

Quality checks on corrugated cases

As with cartons, it is not possible for the case user to exercise quality *control* on his incoming cases. Control requires feedback of information during production, with machine adjustment based on that information to ensure that the specified parameters remain within the required limits. This can only be done by the case maker. The case user carries out quality *assessment* to determine whether the cases received conform to the required standard of manufacture and have not sustained significant damage during transit.

The need for quality assessment.   Many case users, especially smaller firms, believe that quality assessment is an exercise which they do not need or cannot afford. This is a misconception. In practice they will already be carrying out quality assessment in some form, even if it is only that the operators on the packing lines pick out those cases which they consider badly damaged or know from experience will not run on that line. When one considers the losses due to damage in transit, down time on the packing lines due to faulty cases, and the possible loss of customer goodwill, it should be apparent that no firm can afford to be without quality assessment.

The degree of assessment exercised varies, according to the needs of the firm, from the single visual pick and reject system on the shop floor to precise testing carried out on a rigid statistical basis by the properly trained staff of a separate 'quality control' laboratory. The needs of most firms will be somewhere within these extremes and each firm has to devise the system best suited to their particular requirements.

Whatever the system used, there are certain fundamental requirements to enable it to function efficiently. These are (1) specification for the cases, (2) standards by which to assess compliance with the specification, (3) inspection methods which ensure adequate examination of consignments, (4) recording of the results of inspection, and (5) feedback of those results.

Specifications.   The specification should include all the points which affect the performance of the particular type of case. A guide to the items which might need to be included is given in British Standards Institution PD 6112, *Guide to the Preparation of Specifications*, from which one can pick out the items which are required. Case specification can be based upon (a) supplying as per samples originally provided and agreed, in which case the user should keep some specimens himself, (b) material and constructional requirements, or (c) performance requirements. A mixture of (b) and (c) is usually used and is better than (b) alone as there is then also a check on the quality of conversion of the board used.

Setting standards.   As stated earlier in this chapter, two aspects of quality must be considered, the level and the permissible variation about that level. Like all other products, corrugated cases have some features which are measurable and others which cannot be measured; both of these must be considered when setting standards by which the cases will be accepted or rejected. It is also necessary to consider the acceptable frequency of occurrence of departures from the standard which is permissible. This will depend upon the type of defect. Such acceptable frequency levels should be the subject of agreement between the supplier and purchaser.

Inspection.   Two types of inspection may be carried out. Firstly, all items received are examined and those which are substandard rejected. While such a method might seem best from the user's point of view, this is only true when the quantities involved are small. To carry out a 100% inspection on a large number of cases would prove impossible because of the labour and time involved. Furthermore, it becomes increasingly difficult to notice the faulty case when it comes along and some will not therefore be found. The second method involves sampling by the withdrawal of a small number of cases from the consignment in as random a manner as possible. These cases are examined in detail and the whole consignment is accepted or rejected upon the results of the examination of the sample.

Defects of corrugated cases. Defects can arise during the manufacture of the board, during its conversion to the case or in transit to the user. It must not be forgotten that damage can also be caused by careless handling or adverse storage conditions in the user's premises, so that inspection should be carried out without delay after receipt of the consignment. It is then much easier to determine whether defects fall into the first three categories mentioned and are therefore the responsibility of the supplier or whether the user is at fault, and there is no liability on the part of the supplier. A number of defects which occur in corrugated cases are given in *Fabrication Manual for Corrugated Box Plants* issued by the Technical Association of the Pulp and Paper Industry of New York. Some of these are included in the following list of defects.

(A) *Faults in the material* (arising during manufacture)
    (1) Corrugations not uniform in height
    (2) Flutes are formed too low
    (3) Leaning corrugations (i.e. not symmetrical)
    (4) Liners not completely glued
    (5) Blisters in the liner
    (6) Liners scored or wrinkled
    (7) Wash boarding
    (8) Warped board
    (9) Wet board
    (10) Board has objectionable odour

(B) *Faults arising during conversion*
    (1) Wrong grade of board
    (2) Case dimensions incorrect
    (3) Trimmed edges ragged, perhaps with trim still attached
    (4) Slots ragged, perhaps with trim still attached
    (5) Slots of incorrect depth
    (6) Creases not properly formed
    (7) Case not square—skewed creases
    (8) Delamination (due to faulty glueing in manufacture)
    (9) Corrugations running in wrong direction
    (10) Faults in manufacturers' joint:
        (a) joint too wide or too narrow

(b) joint skewed
(c) joint insufficiently glued
(d) joint has too much glue—excess causes case blocking
(e) jointing tape wrong length, out of position or cut crooked
(f) tape has insufficient adhesion
(g) stitches broken or malformed
(h) stitches wrongly positioned
(i) incorrect number of stitches
(11) Printing faults:
    (a) wrong colour
    (b) poor coverage
    (c) print smeared
    (d) print out of position or colours out of register
    (e) poor definition
    (f) set-off—ink transferred to inside of next case
    (g) board crushed under print areas
    (h) ink has objectionable odour

(C) *Damage during transit and storage*
(1) Cases crushed in the bundles
(2) Edges crushed by strapping
(3) Cases dirty or contaminated
(4) Cases creased across panels
(5) Cases warped or bowed
(6) Liners scratched or scuffed
(7) Cases punctured
(8) Flaps torn
(9) Cases too dry or too damp

Method of visual inspection. The initial inspection of an incoming consignment can be purely visual, but it must be conducted in a systematic manner. The following method is suggested.

(A) *Examine the bundles of cases, as received*, for strapping or impact damage and for exposure to rain or damp conditions.
(B) *Withdraw one case* from near the middle of each bundle sampled, marking the place from which it was taken and recording from which bundle it was taken.
(C) *Examine the case in the flat and check*:
(1) *Print* is—correct design
    —right colour
    —correctly positioned
    —not smeared
    —legible and clear
    —does not smear when rubbed.

(2) Case is cleanly and squarely cut to correct design and style.

(3) Creases are correctly placed and there are no unwanted creases.

(4) Slots are correctly placed with respect to the creases.

(5) Washboarding—the impression of the flute tips should not be unduly visible on the liner.

(6) Board is not bowed or warped.

(D) *Open the case and check*:

(1) No splits in inner liner.

(2) Slot over joint is square and of correct width.

(3) Joint lap is of correct width (usually 32 mm minimum).

(4) Joint (see section H for examination of joints).

(5) No objectionable odour.

(6) No print set-off.

(7) No dust or dirt.

(8) Bend flaps through 180° and check that board does not split.

(E) *Cut a piece of board* about 80 × 30 mm from the centre of one panel (using a sharp knife), and check:

(1) The correct materials are used.

(2) The flute size is correct.

(3) The flutes are properly formed.

Then pull the components apart and check if the adhesion is good.

(F) *Close the bottom flaps and check*:

(1) Flaps meet or overlap as required.

(2) Internal dimensions are correct.

(G) *Close top flaps and check*:

(1) Flaps meet or overlap as required.

(2) Case is square.

(H) *Examination of joint*

(a) *Glued joint*: Tear the joint apart. Failure should be by tearing of the *board* and not of the *glue*. The glue should extend the full length of the flap.

(b) *Stitched joint*: check

(1) There are the correct number of stitches.

(2) Stitches are correctly and evenly spaced.

(3) Stitches are central to joint lap.

(4) Stitches are correctly formed.

(5) Stitches are not rusty.

(c) *Taped joint*: check

(1) Correct tape has been used.

(2) Tape is the correct size.

(3) The tape is centrally and squarely placed.

(4) When the tape is pulled off, failure is by tearing of the board and not by adhesive failure.

Classification of defects. Three classes of defect can be identified.

(A) *Critical defects* which will prevent the case from fulfilling its function of protecting the contents.
   (1) Incomplete adhesion of liner.
   (2) Dimensions out of tolerance.
   (3) Weight below the minimum specified.
   (4) Basis weight of a liner or medium below minimum.
   (5) Loose manufacturer's joint.

(B) *Major defects* which reduce case performance to a marginal level so that it is likely to fail under stress, although it may perform adequately under ordinary conditions of storage and transit.
   (1) Incompletely glued joint, incomplete tape joint or inadequately stitched joint.
   (2) Deep slots running into the edges between ends and sides.
   (3) Outer flaps not meeting by a gap exceeding 5 mm.
   (4) Faulty creases.
   (5) Excessively high or low moisture content of board (below 5% or above 10%).
   (6) No non-skid treatment when specified.

(C) *Minor defects* which impair the appearance of the case but not necessarily its function.
   (1) Rough-cut slots or flaps.
   (2) Washboarding, which can cause bad printing.
   (3) Visible specks or non-paper content in the outer liner, such as tar, asphalt, metal foil or dirt particles.
   (4) Bad printing for any reason.

If any of these checks indicate that the cases are not up to the standard required, then this fact should be reported. Further tests in the laboratory may then be necessary to confirm non-compliance with the specification. If, for instance, it is suspected that the board is underweight, a rough check can be

Table 17.1 Weight/thickness relation for corrugated board components

Thickness of component (liner or fluting)		Weight	
in	mm	lb/1000 ft^2	g/m^2
0.0075	0.19	26	127
0.011	0.28	42	205
0.0125	0.32	47	230
0.017	0.44	64	313
0.023	0.59	90	440

made by measuring the thickness of the components and using Table 17.1. An accurate check can only be carried out by formal test requiring air-conditioning and an accurate balance. If the specification lays down requirements for the physical or chemical properties of the board, then tests must be carried out by the required method. Such tests could be done by the user's own laboratory, an independent test house, or the case maker's or board maker's laboratory. Upon completion of the examination the defects should be graded and the consignment judged according to the permissible limits of each grade of defect.

Some of these defects, such as those arising from a machine maladjustment, are likely to appear in clusters so that a consignment or part consignment could be produced out of specification. Other defects occur randomly, and have causes such as a brief interruption of the glue feed or a tear in a roll of linerboard. The random defects are less likely to be picked up in a standard sampling plan than are systematic defects. If it is suspected that a defect is systematic, then the cases on each side of the original position from which the defective was withdrawn should be examined. This presupposes that the location of the first samples was recorded. If these adjacent cases also show the same defect, then the defect may be systematic or could be due to a temporary process fault which was not noticed. Further sampling of the consignment is then indicated.

Quality assurance

So much, then, for the mechanics of measuring quality and controlling it in the packaging manufacturing operation. The question is how the user of these packages and packaging materials can be certain that he always receives containers of the required quality without doing too much work and spending too much money. There are several alternatives. First of all, the user could rely solely on the integrity of the supplier, making no technical contacts and not checking the incoming goods in any way. Providing that there are no problems, this arrangement will be satisfactory, but immediately trouble does occur (and sure enough it will) it usually takes a long time to resolve.

At the other end of the scale, the packer could set up a complete check on everything that is delivered. This is generally wasteful both of manpower and of time and is not usually conducive to good relations with suppliers. It also gives a false impression of quality assurance since, in most instances, it is not possible for the packer to check sufficient samples to be sure of the results. Probably the best method is to work with the supplier, or suppliers, and arrange with them that the process checks they are carrying out are likely to have the desired effect, and to keep one's own acceptance checking to the minimum, merely to see, essentially, that the right type of containers have been delivered. The packer should attempt to negotiate an arrangement whereby he

BLOGGS LTD. – PRODUCT PACKAGING SCHEDULE	No. 14/D/3
PRODUCT: WIZZ–EXPORT	Date issued: Feb. 1st, 19●

1. Contents
Nominal 500 gms. All formulations.

2. Primary packaging

Carton Size	–152.5 mm × 76 mm × 216.5 mm.
Style	–One piece, cut away inner flaps.
Board	–Type No. 2 white lined chipboard.
Mill and substance	–Blacks Ltd. 375 g/m².
Caliper	–500 μm.
Print	–Five colour lithograph.
Varnish	–Lithograph on all printed areas except small print.
Adhesive sealing	–P.v.a. Smiths B7/91. Bottom application.
	P.v.a. Smiths C4/26. Top application.

3. Secondary packaging

Case	Content	–12 cartons.
	Style	–Top opening, inner flaps meeting.
	Size	–(internal dimensions) 462 mm × 310 mm × 215 mm.
	Substance	–200 K/112 SC/200 K/g/m².
	Print	–One colour. Flexographic.
	Adhesive sealing	–P.v.a. Whites 96B.
	Sealing tape	–Reinforced gumstrip—Greys "strongstrip" 50 mm wide.

4. Palletization
Eight cases per layer.
48 cases per pallet.

5.

Authorization *Responsibility*	*Department*	*Signature*
Schedule preparation	Research	
Material availability	Purchasing	
Manufacturing capability	Production	
Product packaging costing	Accounts	
Total acceptance	Marketing	

Although the packaging schedule illustrated is in a packer's plant, clearly a packaging material or container manufacturer could have a similar packaging schedule for his product.

Figure 17.11   An information specification.

BLOGGS LTD.–QUALITY SPECIFICATION	No. 14/D
TITLE: PAPERBOARD CARTON— PACKAGING MATERIAL APPLICABLE TO: NOMINAL 500 gms. WIZZ	Date issued: March 1st, 19 ●

A. Supplier
See approved list.

B. Carton terminology (general)

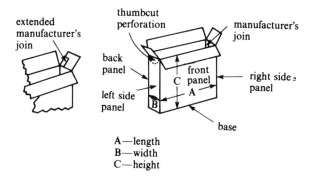

A—length
B—width
C—height

C. Carton style
Cut away inner flaps.

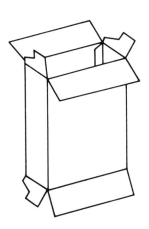

Definition
One piece, side seam glued up paperboard carton with four flaps at each end.
Two outer flaps at each end to equal length and square cut.
Two inner flaps at each end, cut away design, not meeting.   *Continued*

D. Materials of construction

1. *Paperboard*   500 μm No. 2 white lined chipboard (375 g/m² Blacks Ltd.).
2. *Adhesive*   Normal moisture resistant type used in carton manufacture.

E. Carton construction

Carton blank (printed surface uppermost, reading correct).

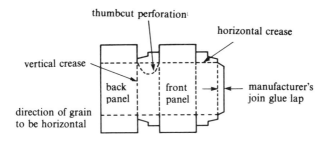

Manufacturer's join—The glue lap to be to the right of the right side panel and glued to the inside of the back panel.

F. Dimensions

Dimensions are measurements from centre of one crease to the centre of the next crease, measured on the inside surface of the carton blank.

Length   152.5 mm
Width   76 mm
Height   216.5 mm front panel *214.5 mm sides and back panel.

Manufacturer's join glue lap to be 12.5 mm wide.

*The front panel creases to be staggered equally at top and base. See Bloggs Drawing No. 14/D March, 1975 for detailed dimensions of carton blank.
(Note—not shown here).

G. Prefolding

To be prefolded along each vertical crease during manufacture to facilitate easy erection.

H. Method of flat fold

Folded glued carton when viewed from the back to have back panel to the right of the manufacturer's join with print reading correct and join central.

I. Printing

1. *Design*

   (a) Supplier's design proofs to be supplied for approval by Bloggs and each design to be given a Bloggs Design Number.
   (b) Design location to be shown on supplier's design proof.
   (c) Print bleeds on to the flaps to be within 2 mm of flap creases and cut edges.

2. *Inks*

   (a) Type—to be low odour, rub resistant and light stable.

Continued

(b) Colour—see range of approved colour samples.

(c) Appearance—to be sharp and clear with the register free from readily noticeable inaccuracies.

3. *Varnish*

To be overprinted with a rub-resistant low odour varnish on all print areas except print on inner flaps adjacent to horizontal creases.

4. *Identification*

(a) Manufacturer's Code and Die Station Number to be printed on bottom flap of left side panel. Manufacturer's name flash may only appear on this flap.

(b) Bloggs Design Number and Print Edition Number to be printed on bottom flap of right side panel.

J. Packaging

Cartons to be banded in units of 100.

Cartons to be edge packed into corrugated cases with three units of 100 per case. The units to be banded with bands of glazed kraft paper at least 100 mm wide and fixed by adhesive tape. The positions of glue laps on adjacent bundles to be reversed. Cases to be palletized five layers high as shown in diagrams.

Pallet load to withstand multi pallet stacking (maximum four pallets high). Carton load to be tight-strapped by Whites 64 to pallet in four positions as shown in illustration.

Protective material to be used between strapping and carton load.

A close-boarded wooden expendable pallet should be used with dimensions that match the pallet load. Irrespective of pallet bottom design it must have at least three stringers each at least 125 mm wide.

Stacking pattern

See illustrations.

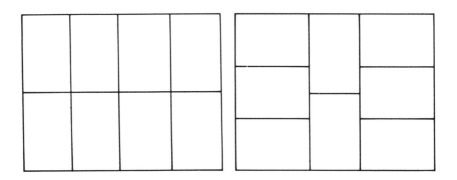

K. Identification

A printed carton to be fixed on each side of pallet load and each pallet to be clearly labelled on each entry face with the following:

SUPPLIER:

BLOGGS ORDER No.:

BLOGGS DESIGN No.:

QUANTITY:

DATE OF MANUFACTURE:

DATE OF DESPATCH: *Continued*

Finished pallet

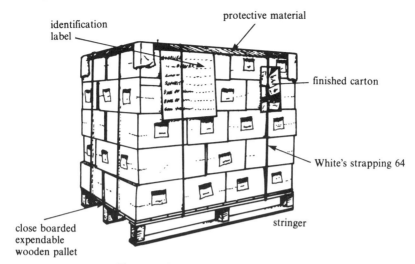

Figure 17.12   A purchasing specification.

has access to all the control results from the supplier. Conversely, he must allow the supplier to have access to all the information he requires since they must (and do) recognize that they are truly in partnership and that the supplier/user relationship changed immediately the contract was signed.

Such a quality assurance scheme will link the supplier of raw materials, the packaging supplier, and the user in two partnerships. In these, better ways of communicating the requirements are a primary objective of all parties leading to better quality, less waste and fuller understanding of each other's needs. The advantage of such a system is not only that it helps to maintain quality standards but also that it reduces the number of sorters needed to remove substandard material and provides a record of the quality of the work produced. The final inspection (quality audit) also makes it possible to obtain a truly representative sample of any batch that can be tested by the customer and saves him taking a further sample for quality assessment.

References

1. A. Cowan, *Quality Control for the Manager*, Pergamon (1964), Chap. 2.
2. A. Cowan, *Quality Control for the Manager*, Pergamon (1964), p. 30.
3. Based on a presentation by A. Wainwright, Pira course (1970).
4. A. Cowan, *Quality Control for the Manager*, Pergamon (1964), Chap. 6.
5. For example, Ministry of Defence specification DEF. 131A, *Sampling Procedures and Tables for Inspection by Attributes*, HMSO, London.

Further reading

S.M. Herschdorfer (ed), *Quality Control in the Food Industry*, Academic Press (1986), 4 vols.

18   Evaluation and testing of transport packages

Methods of evaluation

Evaluations may be carried out on complete packs (filled or empty), on pack components or materials, or on the product to be carried. Methods include:

(a) Visual examinations and estimates made from existing knowledge and experience. These are particularly relevant when the performance of similar packs or products is known, or when small numbers of items are to be dispatched.
(b) Non-reproducible tests, e.g. drop tests by hand in an office, tearing a closure tape between the fingers.
(c) Reproducible tests, usually (but not always) carried out in a laboratory.
(d) Field trials on larger numbers of packages over several routes.
(e) Observations of performance in actual distribution. The proof of any package lies in its performance, in the field for which it was designed, over a relatively long period of time; only in this way can we fully evaluate the package or the material from which it is made.

Classifying methods of evaluation in another way, there are basically four avenues of approach:

(1) Comparative testing—to compare the unknown pack, component, material or product with one whose performance is known. This is often the simplest approach and can determine not only whether the unknown is better or worse than the known but also provide a measure of the degree of difference.
(2) Assessment testing—to simulate the events likely to be experienced in service and deduce, from the results, what may happen.
(3) Investigational testing—to determine where the strengths or weaknesses of the packaged product lie.
(4) Observational testing—to observe and/or record the performance in field trials or actual distribution.

Journey hazards [1]

Distribution systems

Distribution systems exist in great variety and complexity, but however great the complexity, they may be considered to be combinations of a number of simpler elements. These simple elements are (a) the transport of packages from one point to another, with or without change of mode of transport (transport is considered to include the loading and unloading operations) and (b) storage. Points to be remembered are that the size, shape and weight of the outer container has a considerable effect on the hazards encountered during a journey, and that different methods of transport give different hazards.

Once it leaves the packaging line the package, and the product within it, will be subjected to a series of drops and impacts, crushing forces and vibrations, climatic and other environmental effects, before it reaches the point where the product is available for use by the customer.

The first thing to do in making an assessment of the hazards is to analyse the distribution system in terms of distances travelled, mode of transport, points of transfer or interchange, and time taken, including information about handling and other relevant details.

Figure 18.1 shows such analyses of some relatively simple distribution systems; one for consignment of anything between 48 and 144 fibreboard cases containing 24 cartons of a powdered foodstuff and another for 25 kg multiwall paper sacks travelling either palletized throughout or as individuals according to the customer's requirements. Neither is fully completed and the reader will find it instructive to study these and to attempt an analysis of a distribution with which he/she is familiar.

The likelihood of a hazard exceeding a certain level is also most important. For example, the chances of a package being dropped from above a given height must be considered when providing protection against drops. Protection against a rare drop from a height of 1.5 m or more may not (and usually does not) justify the increased cost of the package; a small percentage of damaged goods can often be tolerated, and may be preferable to increasing the cost of packaging.

In practice, the regular absence of damage in repeated consignments may be indicative of overpacking. Protection is required against the normal hazards and not the most severe that may be encountered. Exceptions to this are: very expensive goods, where the increased cost of packaging is low compared with the value of the product; those in which goodwill or the cost and difficulty of servicing are important; and some types of equipment for the military.

Obtaining data on journey hazards

Information on the journey hazards can often be obtained by observation of, for example, stacking heights in warehouses, or temperature and humidity records. Other information can be obtained by inspection of containers sent on journeys. Observation is less satisfactory for assessment of drops or other impacts, particularly where mishandling may be involved, unless precautions are taken. Instruments can be used for measuring and recording some of these.

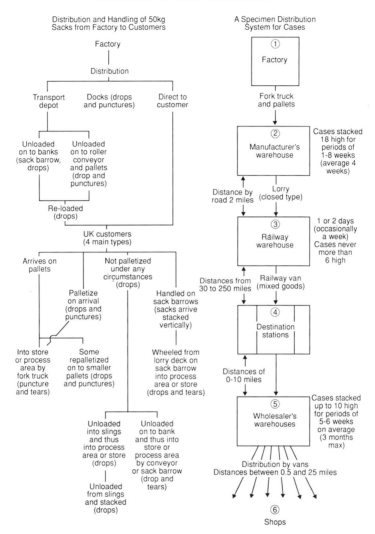

Figure 18.1

The effect of environment on packages

To design packages, and to develop test sequences, we need to know what effects various environmental factors have, and in particular the effect of these factors at different levels. The following notes discuss these with respect to drops, compression and vibration.

Impacts

Impacts may be from dropping or by shocks from other goods coming into dynamic contact during handling, etc. The drop hazard has been investigated more than most other hazards, and Table 18.1 gives an indication of the heights of drop which packages may experience in various handling systems.

The drop hazard is usually studied by drop recorders placed in packages, which are similar in all respects, including weight, to normal packs. The instrument inside the package is calibrated by dropping on to a standard floor before the package is despatched with a normal consignment of similar packages.

There are several types of instrument, from single drop counters (which only record that the package has been dropped above a certain height) to sophisticated equipment that records impacts in the three major axes of the package with a time base and provides a means of estimating the actual height and attitude of fall at impact.

Such data obtained from instruments or observations often have to be interpreted to make a realistic, and not too complicated, test sequence. For example, how do three 100-mm drops followed by one 900-mm drop compare with four 300-mm drops in terms of the damaging effect on a case and/or its contents? In both the total impact energy is the same, but will the effect be different?

For several types of content, and for sacks and fibreboard cases, experiments have been performed in which the package was dropped from constant height until a specified amount of damage was caused. Results of most such tests can be fitted by an expression of following type:

(No. of drops from height h_2 to cause specific damage)/(No. of drops from height h_1 to cause some damage)

i.e.
$$N_2/N_1 = (h_1/h_2)^a \qquad (1)$$

where the exponent a is usually between 2 and 3. If $a = 2$, and 36 drops from 150 mm produce failure, then $36 \times (150/300)^2$, i.e. 9, drops from 300 mm will produce the same failure. In other words, 9 drops from 300 mm produce the same effect on the package as 36 drops from 150 mm. Thus, if a package has

Table 18.1 Typical heights from which packages may be dropped

Weight range (kg)	Type of handling	Typical drop heights (m)
1–10	1 man throwing	1.0–1.2
10–50	1 man carrying	0.8–1.0
35–200	2 men handling	0.5–0.7
150–400	handled by light equipment	0.4–0.6
400 plus	handled by heavy equipment	0.3–0.5

a series of drops, the cumulative effect is proportional to

$$\left(\frac{h_1}{h_1}\right)^a + \left(\frac{h_2}{h_1}\right)^a + \left(\frac{h_3}{h_1}\right)^a + \cdots + \left(\frac{h_n}{h_1}\right)^a \qquad (2)$$

where h_1 is the height of the first drop of the package, h_2 the second, and so on.

An example. A package receives the following drops (all quoted in mm) in the order stated: 150, 300, 300, 150, 750, 450, 600. If $a = 2$, this is equivalent to

$$\left(\frac{150}{150}\right)^2 + \left(\frac{300}{150}\right)^2 + \left(\frac{300}{150}\right)^2 + \left(\frac{150}{150}\right)^2 + \left(\frac{750}{150}\right)^2 + \left(\frac{450}{150}\right)^2 + \left(\frac{600}{150}\right)^2$$

$$= 1^2 + 2^2 + 2^2 + 1^2 + 5^2 + 3^2 + 4^2 = 60 \text{ drops from } 150 \text{ mm}$$

or to

$$60 \times \left(\frac{150}{450}\right)^2 = 6.7$$

i.e. 7 drops from 450 mm.

If $a = 3$, we have the equivalent as

$$1^3 + 2^3 + 2^3 + 1^3 + 5^3 + 3^3 + 4^3 = 234 \text{ drops from } 150 \text{ mm}$$

or to

$$234 \times \left(\frac{150}{450}\right)^3 = 8.7$$

i.e. 9 drops from 450 mm.

Note that the higher the value of a, the greater the contribution of the higher drops to the total, and the less the relative effect of the lower drops. Also, the number of drops from 450 mm is not greatly altered by changing from $a = 2$ to $a = 3$, whereas the equivalent number from 150 mm changes by a factor of 4.

Several investigations on paper sacks have shown, for both butt and face drops, that equation (1) fits their data with the value of a lying between 2 and 3 for breaking the sack. On fibreboard cases containing dried peas, a figure of $a = 2.5$ was found for case failure. For cases containing powder, or metal cans, a value for a around 2 was found, for case closure tape failure and also for case failure itself. Values for a of 1, 2 and 2.8 were found for damage to cans, glass bottles and cartons respectively when packed in fibreboard cases, and of approximately 2 for damage to the cases containing the cans, bottles or cartons themselves.

In addition to a knowledge of the drop heights likely in transport and distribution, the value of the goods, the number being transported in each consignment and other factors will also have a bearing on the degree of safety that a manufacturer may wish to apply. If one is shipping cheap glass tumblers some breakage can be tolerated, whereas with items each worth £2500 if even 1

Table 18.2 Relation of drop height to package weight and assurance level

Package weight kg(lb)	Drop height at assurance level mm (in)		
	I	II	III
0–9.1 (0–20)	610 (24)	381 (15)	229 (9)
9.1–18.1 (20–40)	533 (21)	330 (13)	203 (8)
18.1–27.2 (40–60)	457 (18)	305 (12)	178 (7)
27.2–36.3 (60–80)	381 (15)	254 (10)	152 (6)
36.3–45.4 (80–100)	305 (12)	229 (9)	127 (5)
45.4–90.7 (100–200)	254 (10)	178 (7)	102 (4)

in 200 should suffer damage the financial loss could be considerable. This has been recognized by many test authorities, e.g. American Society for Testing of Materials (ASTM) in the USA [2] who define an 'assurance level' in the appropriate standard (Table 18.2). One thing is certain, if no damage complaints are ever received then the product is over-packaged and a cost reduction exercise will produce economies.

The assurance level is chosen on the basis of the value of the goods and/or the difficulty of replacement, etc. Level II is the norm, level I being selected where a larger safety factor is needed and level III for relatively less valuable or easily replaced items. Note the quite large differences between the figures in the Tables 18.1 and 18.2. This emphasizes the importance of investigating the actual distribution system concerned before decisions on the required design drop heights are made.

Horizontal impacts, such as those which result from shunting on trains and starting and stopping with road vehicles, can also be measured by appropriate instrumentation of a similar nature to that used for drops [3].

Crushing

Crushing is generally caused by stacking loads and these may occur in warehousing or in transit. In the latter instance the packaged product is also subject to vibration from the vehicle at the same time. As far as warehousing is concerned it is usual to assume that the load is evenly distributed in a stack of cases whether or not these are palletized. Under such conditions the maximum load will be experienced by the bottom layer of cases and each individual case in that layer will be supporting a load equivalent to the weight of one case multiplied by the number of layers

$$W = (H/h - 1)w$$

Where H = stack height, h = package height, w = package weight and W = the load to be supported. As might be expected, the effects of the time the stack will exist must be taken into account by the use of a safety factor.

The assessment is more complex when the stack is vibrated at the same time, as it is during transport, but in general the height of the stack is limited here by the height of the transport vehicle and this is rarely greater than 2.5 m in road, rail and air transport. Stack heights in ships, however, can be much greater where deep hold storage is concerned.

The load which a fibreboard case, for example, will support in a stack is not the same as its compression strength P measured under standard conditions. It is affected by several factors:

(a) The length of time the load is applied
(b) The moisture content of the board
(c) The way the load is applied to the case (i.e. pallet profile and stacking pattern)
(d) Previous handling of the package
(e) Support provided by fittings and contents.

The compression strength determined by the standard method and needed by a batch of cases will be given by $P = (n - 1)\ Wabcdef$, where a–e are factors corresponding to items (a)–(e) above, and f is a factor to allow for the variability in strength of a consignment of cases, so that a weak case from the batch placed at the bottom of a stack is unlikely to collapse. All factors except e are greater than 1; if the case is required to take all the load, then e will be 1, otherwise it will be less than 1. Thus the required compression strength of the case is normally greater than the compressive load by a factor which usually lies between 1.5 and 5. The factors for moisture content, time of loading and, to a lesser extent, stacking pattern, are known fairly accurately but those for previous handling and internal support depend much more on the individual package and distribution system. Hence, actual stacking tests are often included in package testing.

The vibration hazard

This can cause loosening of fastenings and components of items such as closures, abrasion of surfaces of items or print on packages, fatigue of cushioning materials, and, when combined with compression loads, more rapid failure of fibreboard cases at lower compression loads during transit. Generally, there is a lack of quantitative data on the effects of vibration on packages. An exception is the effect of compressive load and vibration on corrugated fibreboard cases.

Vibration damage may take two forms. First, failure due to resonance and second, failure due to fatigue. All structures will vibrate; each has its own natural or resonant frequency which will depend on its mass and stiffness. When the frequency to which the packaged item is subjected is small, compared with its natural frequency, the effect is slight. Also when the exciting frequency is large (three or more times the natural frequency), the item does not have time to react and again the effect is small. It is when the exciting frequency is close to the natural frequency that effects become serious because in this range the exciting force pushes the vibrating part of the item at exactly the right time and in the right direction. This is analogous to a child's swing, where if the swing is pushed at the right moment the child goes higher or similarly when the hand is passed through water in a bath, back and forth at the right rate the water will soon be over the ends of the bath and on to the floor. At lower or higher rates these effects do not happen.

In packaging we are only concerned with the vibrations that occur in transport systems and these are generally covered by testing [4] over the 1–100 Hz frequency range. Damage is most likely from vibrations in the frequency band 1–15 Hz because for many packages the resonant frequency of the packaged products falls into this range.

Package test equipment

Mechanical testing

Even when the climatic hazard is not regarded as relevant, package testing must be performed under standard conditions (see ISO 2206 or BS 4826 pt. 2). All packages should be tested in a controlled atmosphere, generally 23°C, 50% r.h.

Equipment to impact the package.   The variables here are the intensity of impact, which is dependent on distribution system, size of pack, etc., and the position of impact, which is selected from most likely or most vulnerable points. In *drop-test equipment,* impact intensity is varied by adjusting the drop height. The equipment is of three basic types: (a) drop arm (convenient but limited capacity), (b) sling and quick release (time-consuming) and (c) release-trap platform (erection time is long, difficult to operate). All these have a specified floor as dropping surface. They can all be used as part of a sequence, or as single tests, such as the Bruceton Staircase technique to establish the critical drop height. Details are found in ASTM D 775 and D 997, ISO 2248 and BS 4826 pt. 4. In the *inclined plane* technique, the impact is varied by the speed at the moment of impact. The equipment consists of a 'dolly' running down a plane at 10° to the horizontal to impact a buffer across the track. This is generally used by impacting the faces and edges only as part of a test sequence, but suffers because of poor reproducibility between different laboratories (ASTM D 880, ISO 2244 and BS 4826 pt. 5). In the *rolling test,* the intensity of impact is fixed by the dimensions of the pack. No equipment is needed except a standard floor. It is used for packs in the 36–90 kg (about 80–200 lb) range, and principally tests the efficacy of fittings (see ISO 2876 and BS 4826 pt. 11). In the *revolving drum* technique, impact and sequence depend entirely on pack dimensions, etc.

The equipment is a 2- or 4-m (7- or 14-ft) drum fitted with baffles, which is used for comparative testing of packs of the same weight and shape. This is not much favoured now, since it cannot be used for comparing packs of different sizes and weight and it has a poor reproducibility (see ASTM D 782; no ISO or BSI standard exists).

Equipment to vibrate the package.   The variables here are frequency, amplitude, time, overload, and whether or not the pack is fixed to the platform. Equipment is of two types: fixed-amplitude, small frequency variation machines with overload (LAB), or variable-frequency and amplitude electrohydraulic-type machines. LAB machines are used in testing fatigue effects on packaging, fittings and abrasion of the product in normal test sequences. Electrohydraulic-type machines are used in testing effects of resonance frequencies on packaging and contents (see ASTM D 999, ISO 2247 and BS 4826 pt. 6). The ISO and British Standards are both concerned with the fixed amplitude vibrators, but an International Standard for the more sophisticated vibration machines is under consideration.

Equipment to compress the package.   The variables here are the weight of the stack load, the method of supporting the load, and the time of stacking. The stacking test requires a platform, weights, and a deflection recorder, and is used as part of a test sequence. The *box compression test* controlling the rate of loading and measuring the load/deflection) is used to give quantitative results in short time (see ASTM D 642, ISO 2234, BS 4826 pt. 3, ISO 2872 and BS 4826 pt. 7).

Climatic testing

Cabinets and chambers.   These can be of two kinds: the *injection type* (see BS 3898) where relative humidity is maintained by a humidity-sensitive device controlling admission of moisture from an external source, and the *non-injection type* (BS 3718), where relative humidity is maintained by circulating air over a selected saturated salt solution.

The performance requirement limits for cabinets and chambers are as follows. *Temperature variation* (the difference in temperature between points within the working space at any one moment) $\pm 0.5°C$; *temperature fluctuation* (changes in temperature within the working space at one point during 30 min) $\pm 0.5°C$; *long-term stability* (difference in temperature at centre of working space over 72 h $\pm 1°C$). Injection-type cabinets have an unlimited moisture supply; non-injection cabinets will not maintain conditions if the salt is no longer in solution (i.e. if all the water is evaporated). Non-injection systems are not suitable for large chambers, as these have a large temperature variation (at

Table 18.3 Testing conditions in various standards (see ISO 2233 and BS 4826 pt.2)

	°C	%r.h.
Extreme cold	− 50	40
Very cold	− 18	40
Cold and dry	− 10	40
Hot and dry	+ 65	40
Normal temperature (UK)	+ 20	65
Normal temperature (USA)	+ 23	50
Wet temperate	+ 20	85
Warm and moist	+ 38	85
Wet tropical	+ 38	95

30°C, 90% r.h., a temperature variation of only 0.5°C causes a r.h. change of 2.7%).

It may be necessary to give the pack some preliminary mechanical handling testing before the climatic test, and it is important to examine seals, etc., to ensure the pack is properly formed and closed. In assessing overall performance of a package, the package is stored for a period (usually in a chamber) as part of a test sequence that includes simulation of other hazards. When testing fibreboard cases for compression strength under damp conditions, differences may arise through differences in r.h. within the chamber. These may be corrected by measuring the moisture content of the board at the time of test.

Shower booth, water spray test (ISO 2875, BS 4826 pt. 10).   This is used to simulate liquid water, rain, spray, etc., as part of a test sequence. The following are controlled: water pressure, angle of cone, consumption, drop size and position of pack.

Methods of using tests

Laboratory tests on transport packages aim to simulate or represent the distribution hazards. Appropriate application of these tests requires a knowledge of the stresses arising from the hazards, and the ability to reproduce these stresses by a particular test, or alternatively of producing damage identical to that observed in practice. The levels of intensity selected for the tests will thus depend on the above factors and also on the degree of assurance that the package should give (from the point of view of protection and containment of the contents or pollution of the environment) and the nature of the contents and the frequency and value of the consignments. Relevant test methods, and the factors requiring quantification before each test can be used, are given in Table 18.4.

Table 18.4 Methods of test and factors requiring quantification

Method of test	Relevant International Standard	Factors requiring quantification
Conditioning	ISO 2233	Temperature, relative humidity, time, pre-drying conditions (if any)
Stacking test	ISO 2234	Load, duration of time under load, attitude(s) of the package(s),[a] atmospheric temperature and relative humidity, number of replicate packages
Stack test using compression tester	ISO 2874	Load applied, duration of time under load, attitude(s) of the package(s),[a] atmospheric temperature and relative humidity, number of replicate packages.
Compression test	ISO 2872	Maximum load (where applicable), attitude(s) of the package(s),[a] atmospheric temperature and relative humidity, upper platen rigidly mounted or free to tilt, number of replicate packages
Vertical impact by dropping	ISO 2248	Drop height, attitude(s) of the package(s),[a] atmospheric temperature and relative humidity, number of replicate packages, number of impacts
Horizontal impact tests (inclined plane test)	ISO 2244	Horizontal velocity, attitude(s) of the package(s),[a] atmospheric temperature and relative humidity, profiles of impacting surfaces and use (if any) of an interposed hazard, number of replicate packages
Rolling test	ISO 2876	Atmospheric temperature and relative humidity, number of replicate packages
Vibration test	ISO 2247	Duration of test, attitude(s) of the package(s),[a] atmospheric temperature and relative humidity, load (if any) superimposed on the package(s), number of replicate packages
Low pressure test	ISO 2873	Pressure, duration of time at reduced pressure, temperature within test chamber, number of replicate packages
Water spray test	ISO 2875	Duration of time under spray, attitude(s) of the package(s),[a] number of replicate packages

[a]See ISO 2206

Performance test schedules [5, 6]

Performance test schedules are used for a number of purposes:

(a) Functional evaluations—will the package be adequate in performance?
(b) Investigation—what causes damage and how can it be corrected?
(c) Comparison—is package A better than package B?
(d) Determination of compliance with statutes, regulations, or an International Standard.

Multi test schedules are generally used for functional evaluations in the context of a complete distribution system. Single test schedules are generally used for functional evaluations in the context of a particular hazard or for investigations. Either type of schedule may be used for comparisons. Statutes, regulations, or International Standards may specify a test schedule. In compiling test schedules, in addition to the above factors, we must also take account of the *time* available for conditioning and testing; the *number* of packages available for test; *past experience* of the particular packages or of similar packages; and the *cost* of testing relative to other factors.

The international standard [5] was produced by the International Organization for Standardization and deals with the general principles and rules for compiling a transport test (Part 1) as well as listing the quantitative data (Part 2) to be used to decide the severity of the test procedure. This standard is designed to be used by national and trade organizations to devise suitable test sequences for particular products and/or distribution systems.

Well tried test procedures have been produced by the National Safe Transit Association [6] and the American Society for Testing and Materials [2]. These should be consulted for details but it must be realized that several factors have to be taken into account before deciding on any particular test procedure. These will relate to both the product and the distribution system to be employed.

Consideration of the distribution system in terms of the simple elements of which it is constituted will decide which tests to carry out in the test schedule. (If a particular hazard does not exist at a significant level, the test appropriate to this hazard should be left out.)

The six-step procedure is as follows:

(1) Identify the simple elements in the distribution system.
(2) Decide what hazards these elements involve.
(3) Decide which test will represent or simulate these hazards.
(4) Decide the basic values of the test intensities associated with the particular package and distribution system.
(5) Decide what test intensity modifying factors, if any, should be applied to the normal values.
(6) Place the tests thus identified into the order given in Table 18.5.

Table 18.5 Order of tests: recommended sequence

(a) Conditioning for testing (ISO 2233)
(b) Stacking (ISO 2234)
(c) Impacts (ISO 2248 and ISO 2244)
(d) Climatic treatment (ISO 2875)
(e) Vibration (ISO 2247)
(f) Stacking (ISO 2234)
(g) Impacts (ISO 2248 and ISO 2244).

A resonant frequency test should be carried out between test (b), stacking, and test (c), impacts, in order to discover whether resonant vibrations are a likely cause of damage when the complete, filled package is transported through the distribution system. However, this test may be omitted if previous experience indicates that damage from resonant vibrations is unlikely.
Other tests may be interposed in the test schedule as appropriate.

References

1. Based on methods and data presented by G.A. Gordon of Pira on numerous occasions and symposia; see also J.M. Montresor *et al.*, *Packaging Evaluation*, Newnes-Butterworth (1974).
2. ASTM D4169. *Standard Practice for Performance Testing Shipping Containers and Systems*, American Society for Testing Materials, Philadelphia, USA (1984).
3. J. Boulanger *et al.*, L'Emballage et toutes ces facettes, Laboratoire Général D'Essai, Paris (1988), pp. 370–371.
4. G.A. Gordon, Use of shock testing in developing transport packages, in *IAPRI Symposium*, Vienna (1984).
5. ISO 4180/1–1980, *Guide to Compilation of Performance Test Schedules—General Principles*; ISO 4180/2–1980, *Quantitative Data*, International Standards Organisation, Geneva (1980).
6. National Safe Transit Association, *Preshipment Test Procedures* (1982).

Appendix 1 European packaging legislation

Introduction

There is no separate branch of law that can be described as the 'law on packaging' in any of the EC member countries but statutes in all of them can affect packaging. Legislation concerned with the sale of goods, trade descriptions, transport, weights and measures, food and drugs as well as environmental issues all affect packaging practice. Many of these statutes deal with environmental issues, standardization and transport legislation. This chapter will only deal with these if they are relevant to food. The main thrust will be concerned with legislation related to food packages and their test requirements, marking, labelling and description.

The EC has had a considerable effect on the packaging user and the manufacturers and converters of packaging materials. As 1993 approaches, bringing with it the Single European Market of 350 million people, EC laws and regulations become increasingly important.

In 1955 EEC was created with the Treaty of Rome. The founders of the Common Market recognized that a closer union among the nations of Europe could remove many of the social and economic ills within the member states. Differences in the legislative provisions governing such details as the manufacturing conditions, handling, packaging and labelling of, for example, foods would hinder efforts to create such a union, hence the Treaty of Rome addressed this and requires that the community 'shall remove obstacles to the free movement of goods' by working towards the 'approximation of the laws of the member states' in other words, by *harmonization*.

Since its beginning, under Articles 43 and 100 of the treaty, the community has steadily built up a framework of common legislation. Article 43 employs direct legislation by means of a *regulation* and is concerned mainly with agricultural products and the Common Agricultural Policy. Such regulations are binding on member states from the day they come into force and overrule national laws. The measures adopted under Article 100 are *directives* and these have no force in any member state until incorporated into the national legislation. They are binding on member states, and the method of incorporation into national laws is left to the discretion of each national authority. A time limit always is specified to give the members a period to introduce the necessary legislation. Failure to comply would result in an action brought before the European Court of Justice.

The Court of Justice is one of the main bodies involved in the legislative action of the Community. The others are: the Council of Ministers, consisting of one minister from the government of each state; the Commission; the European Parliament; and the Economic and Social Committee (ECOSOC).

The Council of Ministers is the body that actually promulgates any legislation, agreement to some of which may require a unanimous vote. The Commission is responsible for initiating action and ensuring that it is correctly applied and the Commissioners are appointed by the member states for a four year renewable term. They act only in the interests of the community

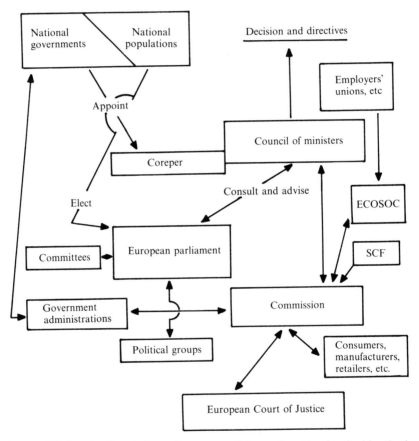

Notes   (i) Before the Commission makes a proposal it consults national authorities. (In the food area these are the Scientific Committee for Food (SCF), professional groups, trade associations, consumer organizations, etc.)
          (ii) After consultation with the Parliament, ECOSOC and Corper, the Commission decides when to submit a proposal to the Council.
          (iii) The Council takes the decisions and issues the directives.

Figure A.1   Legislative structure of the European Community with regard to the food industry.

and are not instructed by member governments. The European Parliament consists of members elected in and by the people of the member states. The ECOSOC is made up of representatives from employers' organizations, trades unions and a wide range of other trade interests. Both the European Parliament and the ECOSOC have an advisory function in the main, but the scope of the former has increased in recent times and is likely to continue to do so.

The way the system works and the relations between the main and various other bodies is illustrated in Figure A.1 using the food industry as an example. Harmonization in food is usually accomplished by either a *horizontal* or a *vertical* directive. Horizontal directives affect all food products across the board with subjects such as labelling, food additives, packaging methods and materials, and quantities packed by weight or volume. Vertical directives deal with particular products such as cocoa and chocolate products, honey, sugar, preserves, coffee and fruit juices. All directives may include temporary or permanent *derogations* to allow the continued sale of a particular national product or extra time for a specific country to adapt its laws. These derogations are generally negotiated and are subject to review.

Once any directive has been adopted by the Council of Ministers it is published in the Official Journal (OJ) of the European Communities and sent to each member state with the instruction to amend their laws within the prescribed period (usually about two years).

Food/packaging compatibility

Not many materials suitable for packaging are completely inert towards foods, and those that are inert often have to be used in conjuction with others that are not. Any substance which migrates from packaging into a food is of concern if it could be harmful to the consumer or has an adverse effect on taste. The care taken to produce wholesome and attractive foods must be matched by the care taken to see that their packaging is compatible. For many years food manufacturers and packaging suppliers have worked together to achieve this.

The basic UK laws and regulations in respect of the packaging materials for food have always been very simple. They require, essentially, that the packaging should not transfer to the food any constituents that would render the food unfit to eat. Until the late 1970s UK food packaging was controlled by food laws, which made it an offence to offer food for sale unless it was 'safe, fit to eat and of the substance and quality demanded.'

Since a large part of our food is obtained in packages, the extent of any interaction between the packaging material and the food must be known and kept very small. If the interactions are not small, the container is not efficient. For example, for obvious reasons wet foods are not packed in

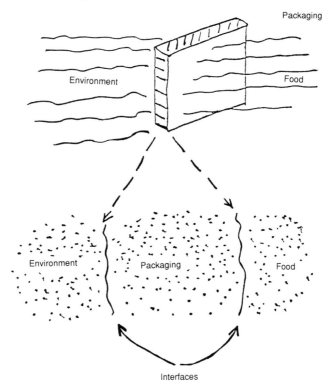

Figure A.2 Diagram showing the basic model for interaction between packaging and food.

unprotected paper or paperboard, nor are acidic foods, such as pickles, packed in uncoated metal cans. Small interactions are often easily detected by taste or smell and/or by sensitive analytical instrumental techniques. Such interactions can result from the migration of substances; movement can be either from packaging to food or vice versa (Figure A.2). Volatile substances can migrate without the material and the food coming into contact but where non-volatile substances are concerned there must be contact.

The basic equation for all toxic hazards is

$$\text{Toxic hazard} = \text{Quantity ingested} \times \text{Intrinsic toxicity}$$

For the hazard to be acceptable either the intrinsic toxicity of a contaminant must be negligible or the quantity ingested small enough to have no effect. Traditionally toxicology is the study of poisons and their effects; if a substance was poisonous it had an adverse effect in small quantities. Today, we are mainly concerned with the effects not of truly poisonous substances in the classic sense, but of minute amounts of materials of very low toxicity possibly ingested at irregular intervals over very long periods of time.

Food contact directives

In 1973, the EC initiated a programme for the harmonization of legislation on food contact materials and articles. Two major reasons for this action were identified:

1. to remove technical barriers to trade between member states
2. to protect consumers from any risks of components in packaging and other food contacting materials, such as china and earthenware, glass, metal and other household utensils, migrating into foods

In 1976 a framework directive (76/893, now replaced by Directive 89/109) on materials and articles in contact with food (Table A.1) was produced which prescribed the basic criteria of human health safety, i.e. that materials contacting food were inert, should not endanger health and cause no unacceptable organoleptic changes in the food. It also authorizes further directives of a specific nature and orders that there must be some indication when containers are intended for food use. This might be obvious from their nature (e.g. domestic hollow-ware) or by labelling (e.g. for food use) or by use of a symbol. If the packaging is not sold directly for retail use, this indication may be in the accompanying documentation.

Table A.1   The framework directive

This directive establishes two general principles:
  A. *The principle of the 'inert nature' of the materials and the 'purity' of the foodstuffs.* Materials and articles in contact must not transfer to food any of their components in quantities which could 'endanger human health', and/or 'bring about any unacceptable change in the composition of the foodstuff or a deterioration in the organoleptic characteristics thereof.'
  B. *The principle of positive labelling.* Materials and articles intended to come into contact with foods must be marked with the words 'for food use' (or with an appropriate symbol). Where there are restrictions on the use this should be indicated on the label.

Three specific directives on PVC followed. In 1978 Directive 78/142 controlled the limits of the monomer (VCM) in plastics to below 1 ppm and any migration into food to less than 0.01 mg/kg (10 ppb). Another two directives (80/766 and 80/432) prescribed the methods of analysis for determining VCM in plastics and food respectively. A further directive (80/590) described the symbol to be used on packages (Table A.2).

Regenerated cellulose film (RCF) is regulated in Directive 83/229 which was issued in April 1983. It includes two positive lists, one for coated film and one for uncoated. Most of the materials approved for use in manufacture are without specific limits, but there are some limitations on the softening agents (mono- and di-ethylene glycol) governed by the film and the coating thickness.

During the 1970s the Community worked on a directive to limit the overall

Table A.2 Main areas of active and probable EC legislation on packaging

Area	Subject	Directive no.
Food contact	1. Framework Directive on food contact materials,	76/893
	replacement for 76/893	89/109
	2. Limits of VCM	78/142
	3. Food contact symbol	80/590
	4. Analysis of VCM in plastics	80/766
	5. Analysis of VCM in food	81/432
	6. Migration test methods	82/711
	7. Regenerated cellulose film	83/229
	Amendment to 83/229	86/338
	8. Ceramics	84/500
	9. Food simulants for migration testing	85/572

Other materials and articles to be covered eventually by specific directives:

    Plastics including varnishes, etc.
    Elastomers and rubber
    Paper and board
    Glass   see Directive 86/C 124/07
    Metals and alloys
    Wood
    Textile products
    Paraffin or microcrystalline wax

Area	Subject	Directive no.
Prepacking units, etc.	Measuring instruments and control	71/316A
	Liquid packages	75/106A
	Bottles	75/107A
	Packages for solids	76/211A
	Labelling of food	79/112A
	SI units	80/101A
	Nominal quantities	80/232A
	Nominal quantities and capacities	86/96A
	Presentation and advertising	86/197A
Manufacture	VCM limits in working atmospheres	78/610
Consumer protection	Indicated prices for food	79/581A
	Product liability	85/374
Environmental	Beverage containers and liquid foods	85/339

or global migration of the constituents of plastics into food. A directive in 1982 laid down the rules for the testing of plastics for all purposes (Directive 82/711). This is the most controversial of all the EC food packaging directives. Originally (1978) it was planned to cover all relevant aspects; positive lists, specific migration and organolepsis. It soon became clear that this was a monumental task and the 'simplification' (sic) of an all-embracing global or overall migration limit was introduced, which was already in use in Italy and France. This was strongly opposed at first by the UK, Germany and Ireland and modified versions were proposed in 1980 and 1982. At one time it seemed likely that the concept might be dropped but the final draft of the Directive 89/109 which replaces 76/893 contains requirements for:

(i)   A positive list of monomers and other starting substances
(ii) Overall migration limits for the 4 specified food simulants
(iii) Specific migration

The positive list

The 600 or so substances on the positive list have been assessed by the Scientific Committee for Food (SCF) which is the expert advisory body to the EC on toxicological matters. The SCF graded each substance on a scale of 0–9 (see Table A.3) according to the information available and the Commission interpreted this information to produce two categories. The first contains all those for which a decision was possible and the second those for which the lack of toxicological, migration or other data made a decision impossible. The majority of substances in the first list have no restrictions on use but some have limitations on migration limits, etc.

The substances in the second category (50–60% of the total) must be re-examined within 3–5 years after publication of the list in the OJ. Within this time industry will have to produce the necessary data or there is a threat that the substance will be placed in category 5 of Table A.3 i.e. 'banned'.

Table A.3   Grading of substances on a 0–9 scale

0   Acceptable without limit (e.g. approved food ingredients)
1   Acceptable daily intake (ADI) or similar criterion established
2   TDI (tolerable or temporary DI) established
3   ADI or TDI not established but use accepted by other factors, e.g. organolepsis (taste or smell) would self-limit
4   ADI or TDI not established but acceptable 'if not detected by an agreed sensitive method'
5   Banned
6   'Suspicions about toxicity' exist; data insufficient, more tests needed. Two subdivisions are made here:
    (a) suspect carcinogens which 'should not be detectable in food by an appropriate sensitive method'
    (b) suspect toxins (other than carcinogens)
7   Some data exist but insufficient to establish ADI or TDI; further specified information required
8   Inadequate data available
9   Not evaluated because of an inadequate specification

Overall migration (OM)

This is the total migration of substances from the plastics into the appropriate food simulant. No attempt is made to identify the migrating substances which

Table A.4 Migration tests for plastics

Conditions during actual use	Test conditions
1. Duration of contact: more than 24 hours	
Temperature 5°C or less	10 days at 5°C
Temperature exceeding 5°C, up to, but not exceeding 20°C (labelling compulsory)	10 days at 20°C
Temperature exceeding 5°C, up to, but not exceeding 40°C	10 days at 40°C
2. Duration of contact: between 2 and 24 hours	
Temperature 5°C or less	24 hours 5°C
Temperature exceeding 5°C, up to, but not exceeding 40°C	24 hours at 40°C
Temperature exceeding 40°C	in accordance with national laws
3. Duration of contact: less than 2 hours	
Temperature 5°C or less	2 hours at 5°C
Temperature exceeding 5°C, up to, but not exceeding 40°C	2 hours at 40°C
Temperature exceeding 40°C, up to, but not exceeding 70°C	2 hours at 70°C
Temperature exceeding 70°C, up to, but not exceeding 100°C	1 hour at 100°C
Temperature exceeding 100°C, up to, but not exceeding 121°C	30 min at 121°C
Temperature exceeding 121°C	in accordance with national laws

may or may not be harmful to a consumer. It is claimed that if the OM is below $10\,mg/dm^2$ of packaging material in general, or $60\,mg/kg$ food for items of between 0.5 and 10 litre capacity at a temperature and for a time related to the actual conditions (Table A.4) of use, then that material will not constitute any hazard. Since no identification of the material migrating is made, the OM can at best be regarded only as a poor measure of the inertness of the material and has little to justify it on a scientific or toxicological basis.

Although the four food simulants have been specified (Table A.5) there is as yet no official procedure for carrying out the test—an unsatisfactory

Table A.5 The EC Food Simulants Directive 82/711

Materials: distilled water
             15% vol/vol ethanol
             3% acetic acid
             olive oil, sunflower oil or synthetic glyceride
Note: the selection of the simulant(s) to be used for any particular food is determined by reference to another directive (85/572).

Outline of procedure:
For the three aqueous simulants the procedure is relatively simple. After exposure to the material under test at the specified conditions for the prescribed time the OM is determined by weighing the residue after evaporating the food simulant to dryness. This is obviously not possible for the last simulant for fatty foods and a complicated procedure requiring considerable attention to detail is needed. This can and does give rise to unreliable results and much controversy exists.

situation for any analytical procedure let alone one as crude as the determination of OM. Ultimately, methods will have to be agreed or harmonization by 1992 will just be a political talking point. As has been pointed out specifying a limit without giving a method of measuring it would be like prescribing a speed limit on the roads of 30 mph, while the police wave on motorists doing 60 mph because they have no means of measuring the infringement.

Specific migration

There are about 20 substances in the current list for which a specific migration limit (SML) into the food or the appropriate food simulant is given. Where a monomer with an SML is used to produce a plastics material for food contact, a specific migration test for that monomer on the finished article must be carried out. Moreover the sum of all the specific migrations must not exceed the OM limit of $10 \, mg/dm^2$ or $60 \, mg/kg$ of food.

Thus there is still a considerable amount of work to be done before a satisfactory situation will be achieved. However, it is anticipated that the directive will have to be put into effect by member states by 1992.

We must now consider the implication of these directives on industry. There is no doubt that the increasing public concern over the safety and purity of food, even though it may to some extent be promoted by political considerations, will accelerate the introduction of any new legislation needed to implement EC directives. It is perhaps a little ironic that many of the initial problems in the food field were with virtually unpackaged foods—eggs, soft cheese, cooked meats, pies, etc.—and not with those foods that are preserved in their packaging. Nevertheless the legislation resulting from the food contact field will trigger changes.

Certificates of warranty from packaging suppliers will no longer be sufficient for a packer to demonstrate that he has taken due diligence. If we look at the supply chain (Figure A.3) we can see the complexity of the situation. The responsibilities may be allocated along the following lines. The monomers, additives and other ingredients used in polymer manufacture will obviously have to be on the appropriate positive list, and polymer manufacturers will be responsible for using only approved materials. If they want to use any new material not on the approved list they will have to get the necessary testing done and obtain approval (from the SCF?) before using it.

The packaging converter will need an assurance from his polymer supplier that the material is approved and complies with the regulations. However, he will still, probably, have to get the finished products tested for both OM and SM even though the polymer manufacturer will, almost certainly, be able to supply such data, because it will not have been obtained on the *finished product* and the new regulations insist on this.

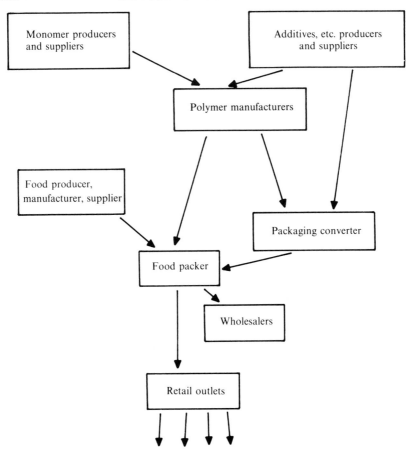

Figure A.3   The supply chain.

The food packer receiving film or containers will be in possession of the facts, from his suppliers, that:

(a)   the correct approved ingredients were used
(b)   the finished packaging has been tested and passed the tests for OM, SM, etc.

but this may still not be sufficient. It is the food that is being sold not the packaging and to prove diligence in a court the packer may be required to produce evidence that the food simulant is related to his food product and that his processing factors did not increase the OM or SM. The final link in the supply chain is the retailer and he may need to prove diligence by confirming the assurances of his suppliers with spot checks. After all the person who will be in court if the food is found to be illegally contaminated is the retailer and to defend himself he would probably need to prove that he took action to check the assurances from his suppliers.

Thus everyone in the chain is likely to be involved and will be required to have some system for checking their products either in-house or by an outside laboratory. It is likely that, in collaboration with Ministers in several countries Codes of Good Practice may be agreed with the regulating authorities in Europe as to who does what, how often and when. What is certain is that it could be a costly operation to achieve very little.

Outwith the field of plastics packaging harmonization has been achieved, as already mentioned, for both regenerated cellulose films (RCF) and ceramics, and the Commission is currently working on other materials (Table A.2). When all these have been adopted and are in force, harmonization of the laws on food contact materials could be considered complete.

As Table A.2 shows, other areas of importance to packaging are the nominal quantities requirements, the labelling of food (79/112A), claims about products in advertising (86/197A) and product liability. It is also probable that we may see European legislation on child-resistant and tamper-evident packaging but there has been no proposal to date.

Labelling

The directive on this subject was proposed in 1974 but was not adopted until the end of 1979. It then became incumbent upon member states to permit trade in goods labelled in accordance with its provisions after December 1981 and to prohibit those not so labelled after December 1983. In order to meet nationally established practices there are some 30 or so derogations in the directive from which member states can choose. This means that there are still differences among the 12 countries of the EC and harmonization is only partially achieved.

The directive is based on the principle that labelling must not mislead buyers as to the nature, identity, properties, composition, quantity, durability,

Table A.6   Labelling of food: Directive No 79/112

This directive prescribes that all labels shall include:

1. The name of the product
2. The list of ingredients and components
3. The net quantity (in the case of packaged foods)
4. The date of minimum durability
5. Any special storage conditions
6. Name and address of manufacturer, packer or distributor
7. Particulars of the place of origin (in any instance where failure to give such information might mislead the customer to a material degree, regarding the true origin or source of the foodstuff)
8. Instructions for use (when it is impossible to make proper use of the food without such information)

origin, method of manufacture or production of a food by attributing to it properties it does not possess or by suggesting that it has special characteristics when they are common to all similar foods. The main requirements are summarized in Table A.6. Most of these need little explanation but some comment is necessary on the *product name* and the *ingredients*.

First, if the product name is prescribed by Community or National Law then that is the one that must be used. When there is no prescribed name, a customary name may be used, that is to say the name which is customary in the state where it is to be sold. Some customary names are common to many states, e.g. spaghetti or frankfurters, while others are very localized, e.g. haggis or cornish pasties. If a customary name is not used the product must be described precisely so as to leave no confusion in distinguishing it from other products. Trade marks, brand names or fancy names must not be used alone without a fuller description. For example, although Coca Cola is an internationally known brand name, on its own it is not sufficient.

Ingredients must be listed in descending order of weight determined at the stage when the mixing takes place. Exceptions are volatile products and water which must be listed according to the amount present in the finished product. This allows for cooking losses. Less than 5% added water need not be declared. The names of the ingredients must be those used when they are sold separately. With some the generic name is permitted, e.g. fish, meat and cheese. Oils and fats must be distinguished as of vegetable or animal origin, and, later nutritional information may also be required. Figure A.4 gives examples of some ingredient lists.

Presentation and advertising

This is covered by Directive 86/197A and relates to claims made in the labelling of foods. In this context a 'claim' means any statement intended to promote the sale of the food by its characteristics, effects or properties. The following claims are prohibited:

1. Claims of measurable and objective characteristics which cannot be substantiated
2. Claims suggesting that adequate quantities of nutritious substances contained in the product cannot be obtained from a balanced everyday diet
3. Claims that a common food contains adequate quantities of all essential nutrients—unless authorized by the rules in force
4. Recommendations by medical professionals, authorities or other competent bodies in the public health field and testimonials connecting nutrition and health
5. Claims that exploit or arouse fear or anxiety or discredit other foods whether of a similar nature or not

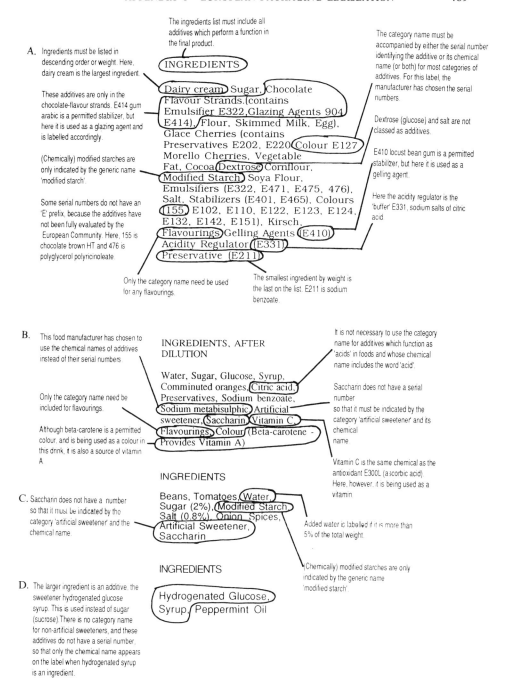

A. Ingredients must be listed in descending order or weight. Here, dairy cream is the largest ingredient.

These additives are only in the chocolate-flavour strands. E414 gum arabic is a permitted stabilizer, but here it is used as a glazing agent and is labelled accordingly.

(Chemically) modified starches are only indicated by the generic name 'modified starch'.

Some serial numbers do not have an 'E' prefix, because the additives have not been fully evaluated by the European Community. Here, 155 is chocolate brown HT and 476 is polyglycerol polyricinoleate.

The ingredients list must include all additives which perform a function in the final product.

INGREDIENTS

Dairy cream, Sugar, Chocolate Flavour Strands.(contains Emulsifier E322,Glazing Agents 904 (E414), Flour, Skimmed Milk, Egg), Glace Cherries (contains Preservatives E202, E220 Colour E127) Morello Cherries, Vegetable Fat, Cocoa Dextrose Cornflour, Modified Starch, Soya Flour, Emulsifiers (E322, E471, E475, 476), Salt, Stabilizers (E401, E465), Colours (155, E102, E110, E122, E123, E124, E132, E142, E151), Kirsch, Flavourings Gelling Agents (E410) Acidity Regulator (E331) Preservative (E211)

The category name must be accompanied by either the serial number identifying the additive or its chemical name (or both) for most categories of additives. For this label, the manufacturer has chosen the serial numbers.

Dextrose (glucose) and salt are not classed as additives.

E410 locust bean gum is a permitted stabilizer, but here it is used as a gelling agent.

Here the acidity regulator is the 'buffer' E331, sodium salts of citric acid.

Only the category name need be used for any flavourings.

The smallest ingredient by weight is the last on the list. E211 is sodium benzoate.

B. This food manufacturer has chosen to use the chemical names of additives instead of their serial numbers

Only the category name need be included for flavourings.

Although beta-carotene is a permitted colour, and is being used as a colour in this drink, it is also a source of vitamin A

INGREDIENTS, AFTER DILUTION

Water, Sugar, Glucose, Syrup, Comminuted oranges, Citric acid, Preservatives, Sodium benzoate, Sodium metabisulphic, Artificial sweetener, Saccharin, Vitamin C, Flavourings Colour (Beta-carotene - Provides Vitamin A)

It is not necessary to use the category name for additives which function as 'acids' in foods and whose chemical name includes the word 'acid'.

Saccharin does not have a serial number so that it must be indicated by the category 'artificial sweetener' and its chemical name.

Vitamin C is the same chemical as the antioxidant E300L (ascorbic acid). Here, however, it is being used as a vitamin.

C. Saccharin does not have a number so that it must be indicated by the category 'artificial sweetener' and the chemical name.

INGREDIENTS

Beans, Tomatoes, Water, Sugar (2%), Modified Starch, Salt (0.8%), Onion Spices, Artificial Sweetener, Saccharin

Added water is labelled if it is more than 5% of the total weight.

D. The larger ingredient is an additive: the sweetener hydrogenated glucose syrup. This is used instead of sugar (sucrose).There is no category name for non-artificial sweeteners, and these additives do not have a serial number, so that only the chemical name appears on the label when hydrogenated syrup is an ingredient.

INGREDIENTS

Hydrogenated Glucose, Syrup, Peppermint Oil

(Chemically) modified starches are only indicated by the generic name 'modified starch'.

Figure A.4 Ingredient lists. A: Black Forest gateau; B: Whole orange drink; C: Baked beans—reduced salt and sugar; D: Sugar-free mints

6. Claims that a food has become nutritious from the addition of substances added for technical and/or organoleptic reasons
7. References to doctors, pharmacists, medical instruments, the human body or organs—even if presented in a stylized form—intended to illustrate a physiological function

Product liability

The Consumer Protection Act of 1987 contains provisions on product liability which came into force in March 1988. The Act is the implementation in the UK of the EC Directive on product liability (85/374) which provides similar protection throughout the EC. The main emphasis is the change in responsibility for defective goods from the retailer to the manufacturer or producer. Also, in the past anyone injured had to prove negligence if they were to sue for damages successfully. Customers can already sue a supplier without proof of negligence under the Sale of Goods Act. Packaging suppliers are not affected in the case of a defective product unless the packaging has altered the essential characteristics of the product, but where the packaging itself was the product delivered to a 'place of work' and because it was defective (i.e., a product 'where the safety... was not such as persons generally are entitled to expect') injury occurred, then presumably an action for damages could be brought.

In conclusion, we must add that many of the directives which are produced by the EC stem from areas where there are in existence company, industry and/or national standards and these are frequently incorporated into the control processes on the directive and therefore into national regulations. It is therefore imperative that UK standards should be available for discussion at an early stage. BSI is attempting to coordinate the many trade practices which are currently used in the UK without official recognition and assist industry to produce rapidly, where it could be significant, a British Standard. Unless this is done we may find that the control criteria adopted stem from other EC countries that have existing national standards in the area concerned.

Appendix 2   USA legislation

The US Food and Drugs Administration (FDA) administers the Federal Food, Drug and Cosmetic Act (FFDCA) which is the controlling legislation for food, etc. This law is one of the most sophisticated pieces of legislation in the world in respect of controls on both direct and indirect additives in food. It has ramifications throughout the world and is used as a yardstick in many countries.

The main points can be summarized as follows. The present legislation was introduced in 1938 and was a revision of the original Food and Drugs Act of 1906. Under the 1906 Act, although it was unlawful to deliver adulterated or misbranded food in interstate commerce, the FDA were not authorized to make industry wide regulations or to set standards. The 1938 Act gave the FDA the authority not only to promulgate regulations but also to establish definitions and set standards for foods. It also extended the law to cosmetics and medical devices; clearance by the FDA was required before a new drug could be distributed and tolerance levels for unavoidable toxic substances were provided.

Later amendments in 1958 required, among other points, that components from packaging materials became subject to the same premarket approval as direct food additives. No statutory distinction is made between direct (i.e. deliberate) additives or indirect additives such as packaging components which may migrate into the food. Both are treated alike. In 1960 an amendment covering preclearance of added food colours was also made.

What is a food additive?

The FFDCA defines a food additive as 'any substance the intended use of which results...directly or indirectly in its becoming a component of, or otherwise affecting the characteristics of, any food, including any substance intended for use in...packaging'. In addition to packaging, substances migrating from any other sources, such as plant and machinery, processing techniques, transport, etc., are included.

There are categories of substances in packaging materials which may not be considered to be food additives and the most important are those which have been generally regarded as safe (GRAS) (Code of Federal Regulations, Chapter 21, parts 184 and 185). In another category are substances which have been previously approved by the FDA, so-called 'prior sanctioned' additives. The FDA publishes a list of GRAS substances and manufacturers can use recognized procedures to get substances placed on the GRAS list.

Once a packaging material has been cleared by the FDA, it must still comply with good manufacturing practice (GMP) for food contact materials as set out in Section 174.5 of the food regulations. There is, however, a lot of misunderstanding of the regulations on food packaging and consultation with an expert is always advisable.

The Delaney Amendment

This controversial provision is contained in Section 409 (c) (3) of the Act. It states that no additive may be deemed safe by the FDA if in a standardized toxicological test it produces cancer in animals.

If the FDA were to take a company to court under any part of the act they must prove:

    (i)   that the ingredient will migrate into the food
    (ii) the amount migrating involved is not generally recognized as safe

Expert testimony would be needed on both counts. Thus the burden of proof is with the FDA.

Finally it can be stated that the methods of examination of packaging materials for food safety and mitigation in the USA do not differ materially from those used in the UK and the EC and the general approach and philosophy are similar. In any actual case it is wise to consult an expert on the subject from the area of the world concerned.

Index

abiotic spoilage   195
acidity   188, 322, 336
acids   167, 315
acrylic multipolymer
       (XT polymer)   67, 69
acrylonitrile-butadiene styrene
       (ABS)   66, 69
active packaging   300
additives   491
adhesives   116, 159, 162–165
advertising   16, 33
aerobic bacteria   188
ales   342
aluminium   85, 287
antibiotics   316
antioxidants   316
Appert   1
apples   237, 238
apricots   238
ascorbic acid   336
aseptic processing   276–284, 334, 342
aseptic systems (table)   283
asparagus   238
auger fillers   107
avocado   239

bacon   319
bacteria   220, 241, 319
bag, filling and closing equipment   132
bagging   130
bag-in-box   133
baked foods   258, 273, 308
bananas   235, 239
Barex   67
barrier plastics   274, 390–402
beans   234, 239
beating (of pulps)   54
beef   253
beers   242
beetles   418
beetroot   238
berries   238
biodeterioration   187
birds   194
biscuit wrapping   152, 303, 412
bitumen   95
black plate   85
blanching   257, 269

blast freezers   249
blow moulding   386
board selection   376–383
boil-in-bag   252
bottles, returnable   225, 343
bottling   100, 339
bottling line   101
box compression failing load   172
bread   307
bread wraps   128, 412
breakfast cereal   303
brine   315
broccoli   238
butter   226

cabbage   238
cakes   412
campylobacter   216, 290
can closures   82
canning   117, 266–273, 278, 365–374
cans and tin boxes   85–94, 340
cans, general line   83
capping bottles and jars   114
    automatic   115
    roll-on   115
carboys   167
carrots   238
cartoning   133–141
cartoning systems   139, 374–383
cartons
    for liquids   134, 136, 341
    for solids   135, 137–139, 158, 251
case failing load   172
cases and crates   167
casing and sealing   116, 162–164
casks   343, 346
cauliflower   238
celery   239
cellulose acetate · 66, 69
cellulose acetate butyrate   69
cereals   306
cheese   322–325
cherries   238
chilling (fish)   218
Chinese leaves   238
chocolate   310
clarity (of glass)   79
classification (of food processes)   26–29

Clostridium botulinum 192, 266, 268, 285, 316
Clostridium perfringens 192
coatings 405
coconuts 238
coffee 305
Combibloc 279, 282, 325, 342
communication 5
compatibility 78, 421–425, 479
compression tests 172
containment 5
convenience 6
cook-chill products 287
cook-freeze foods 260, 287
corking 115
corrosion 269
cost 17–19, 81, 353, 356
cream 228
crisps 412
crown corks 115
cucumber 239
cup fillers 107
cured foods 317–320
curing 315
customer needs 33, 36–40

dairy products 226
dairy spreads 227
definitions 3
dehydration (frozen food) 250
dehydrated foods 299
Delaney amendment 492
design, of successful packagings 8, 10, 33
desserts 258, 273
deterioration indices 187–203, 219, 311
   table of 202
die stamping 50
diffusion 396
distribution hazards 8, 10–13, 22
distribution system (meat) 210
domestic freezers 250
dried foods 296–314
drop performance 170
drying methods
   freeze drying 298
   mechanical/spray/roller/fluid bed 297
   reduction of available water 297, 298
   sun or air 296
   vacuum drying 298

EC directives 481–490
economics (primary packs) 357–389
edge crush test (ECT) 175
eggs 227, 258, 303
electron beam 290
Elopak 224, 279
emballistics 411
engineering of packaging line 354

environment 263
enzymes 269, 336
ethylene (ripening) 235
ethylene vinyl acetate copolymer (EVA) 66

factory ships 217
fats and oils 331–334
FDA 491
fermented foods 321–330
fibreboard cases 170
Fick's Laws 392
figs 238
filling
   by weight 109
   cans 119
   gravity 105
   hot 80
   legal requirements 110
   liquids 104–106
   measured dose 105
   powders 106
   pressure 105
   vacuum 104, 107, 212, 221, 284
   volumetric 107
film bags 131
film wraps 252
films (properties) 391
filtration 337
fish 217, 255, 271, 310
   cured 320
   fermented 328
   smoked 321
fish farming 219
flask fillers 107
flavour loss 251
flexible materials 70–78, 278, 384–389
flexography 44
flexural rigidity 77
flour 347
food cans
   3-piece 86, 87
   2-piece 86–89
food preservation technologies 203
food simulants 424, 483
food spoilage and poisoning 192
form-fill-seal (FFS) machines 141–159
fragility 81
freezer burn 253
freezing 248, 249, 253, 256
fresh and chilled foods 205–230
friction, coefficient 75, 76
frozen foods 248–265
fruit processing 338
fruits 231–247, 257, 271, 309

gas transmission rate 413–416
glass containers 78–83
   economics of 359–364

gold blocking 50
grapefruit 239
grapes 238
graphics 33, 354
    check list 41
gravure printing 47
guidelines 228, 261, 284

hazards
    crushing 469
    impacts 467
    toxic 480
    vibration 470
heat penetration rate 267
heat processed foods 251–288
heat resistance 80
herbs and spices 245, 301
high pressure techniques 293
hot filling 80
humidity 240
hydrogen peroxide 281
hygiene 270

ice-cream 259
impact strength 77
inertness 78
ink-jet printing 49
injection moulding 388
insects 104, 197
instant beverages 298
internal pressure 80
ionising radiation 290
    detection of 292
ionomers 65
irradiated foods 288–293
    polymers for 292

jams and jellies 299, 310
journey hazards 464–470
juices (aseptic process) 282, 335–342

kraft pulp 54

labelling (bottles, jars) 116, 159–162,
        486, 487
    machinery 161
lamb 253
laminates 273, 274, 402
legislation 424, 478–490
    Europe 477–490
    USA 491–492
Letpak 273
letterpress printing 40
lettuce 238
liability 488
life cycles 347
lightweighting 362
line efficiency 22, 358

lithography 45
Lopac 67
low-acid foods 278

machineability 6
machinery 97–166, 357
management 23, 33
margarines 239, 332
marketing 3, 14–16, 160
marrows 239
materials utilization 20–22
mayonnaise 329
meat
    cured 317
    deterioration 190, 205, 206
    fermented 327
    frozen 252
    moisture relations 208
    prepackaging 211
    preparation 205
    smoked 321
    storage 201
    vacuum packages 212
    visible appearance 207
melamine formaldehyde 43, 69
melons 239
metal packaging 83–94
merchandising 34–36
mice 421
microbial growth 188, 265
microwave packaging 285
migration 313, 483–487
milk 222–226, 272, 280, 282, 302
modified atmosphere packaging (MAP)
    fish 221
    fruit and vegetables 242
    meat 214–216
    poultry 217
moisture sensitive foods 296–314, 405–411
moths 418
moulds 191, 309
mushrooms 238

nectarines 238
nuts 238, 246, 305
nylons 67, 69

oils and fats 333
olives 322
Omni can 276
onions 238
open-mesh bags 132
open top cans (ISO standard) 93
orange juice 335
oranges 239
oxidation 250
oxygen scavengers 299
oxygen transmission 401

packaging cycle 4
packaging definitions 3
packaging development 347–356
packaging line 67–69, 165
packaging materials, properties and
    forms 23–26
packaging needs of foods 6
palletizing 117, 179
paperboard 57, 60–61, 62
papers for packaging 56
parcels 127
parsnips 238
particulates 283
passion fruit 239
pasta 307
pasteurization 223, 268, 337, 339
paw-paws 239
peaches 238
pears 238
peas 234
permeability 395
pests 417–421
petfoods 225
pickles 322, 328
pillow packs 142, 149, 156
pineapples 239
plastics 62, 483
    acrylic multipolymer (XT polymer) 67,
        69
    acrylonitrile-butadiene-styrene (ABS)
        66, 69
    Barex 67
    cellulose acetate 66, 69
    ethylene vinyl acetate copolymer 66, 69
    Lopac 67
    melamine formaldehyde 63, 69
    nylons (polyamides) 67, 69
    phenol formaldehyde 63, 69
    polycarbonate 66, 69
    polyesters 68
    polyethylene
        LDPE 63, 69, 244
        LLDPE 64, 69
        HDPE 64, 69
    polyethylene terephthalate (PET) 68,
        69, 339
    polyethylene tetrafluoride (PTFE) 68
    polypropylene (PP) 64, 69, 159, 244
    polystyrene 66, 69, 157
    polyvinyl alcohol (PVA) 66, 69
    polyvinyl chloride (PVC) 65, 69, 156–
        157, 244, 481
    polyvinylidene chloride (PVDC) 65, 69,
        156
    polyvinylfluoride 68, 159
    Surlyn 65
    TPX 65
    urea formaldehyde 69

plate freezers 249
plums 238
pork 253
potatoes 239
poultry 205, 216, 154
preserves 337
pricing 16
printing processes 40–52
product assessment 8
property profile 99
protection and preservation 5
prunes 309
pulping 53
pulps (sulphate/sulphite) 54
pumpkin 239
Purepak 224, 279, 342

quality 427
quality assurance 458
quality checks, corrugated boxes 452
quality control 426–463
    cartons 438
    glass containers 447
quince 238

rancidity 332
ready meals 275
respiration 231, 237
retailing
    fish 219
    fruit and vegetables 242, 261
    meat 210
retorting 121, 273
retort pouches 273
returnable bottles 224
rhubarb 238
rigidity (of glass) 79
ripening 236
rodents 194, 421
round stick joinery 94

sachets 141, 149, 154, 156
safe stacking load 174
salad cream 329
salmonella 192, 216, 268, 290
sampling 435
sauces 328–330
sauerkraut 322
sausages 322
sealing (glass containers) 82, 146
seaming (tin boxes/cans) 88–93, 120
selection (machines and materials) 17
self-service 14
senescence 187
shellfish 217
shipping containers 167
shrink/stretch wraps 117, 182
silk screen 48

smoked foods 315, 320
snacks 275, 304
sorption isotherms 196–201
soups 302
sous vide 284
soya products (miso, tofu, tempeh, etc) 330
specifications 445–463
spinach 238
spirits 346
spoilage 187–203, 212, 220, 224, 249, 319, 329, 331, 337, 342
stacking performance 171, 177
staphylococcus 193
Stepcan 275
sterilization 80, 265, 276, 291
stiffness 175, 376–378
stoneware 95
stoppages 98
strapping methods 184
straw 94
strawberries 239
streptococcus 316
strip packs 143
sugar confectionery 310–314
susceptors (receptors) 287
swedes 238
sweet potatoes 239
sweets 313

tea 305
tea bags 155
tensile strength 55, 59, 73, 75
terminology (cans) 93
termites 420
testing
   flexibles 70–78
   paper 55, 59–60
   shipping containers 170–181
testing methods 471–476
TetraBrik 224, 279, 335, 342

textiles 95
thermal processing 121, 267
thermoforming 388
thermoform-fill-seal 156
thermoplastics 63
thermosets 62
thickness (films) 77
tinplate 83
TPX 65
trays 164, 252
tubs 333
twist-wrapping 126

ultrasonics 293
unit loads 182
unitizing 181–186
urea formaldehyde 63, 69
UV radiation 282, 293

vacuum filling 104, 107, 212, 221, 284
vacuum sealing 273
vegetables 245, 257, 272, 301
vertical form-fill-seal (f.f.s.) machines 141–148
vinegars 328
vitamin C 336
volumetric filling 107

water activity 188, 315
water in foods 195
water vapour transmission 400, 416
waxes 95
wheat 306
wicker baskets 96
wood pulp 53
wrap-around cases 162
wrapping 126–130

yeasts 191, 309, 337
yoghurt 325–327